Negotiation Analysis

Negotiation Analysis

The Science and Art of

Collaborative Decision Making

Howard Raiffa

with

John Richardson

David Metcalfe

THE BELKNAP PRESS OF HARVARD UNIVERSITY PRESS

Cambridge, Massachusetts, and London, England

First Harvard University Press paperback edition, 2007

Library of Congress Cataloging-in-Publication Data

Raiffa, Howard, 1924–
    Negotiation analysis : the science and art of collaborative decision making /
        Howard Raiffa with John Richardson, David Metcalfe.
        p.   cm.
    Includes bibliographical references and index.
    ISBN-13 978-0-674-00890-8 (alk. paper : cloth)
    ISBN-10 0-674-00890-1 (alk. paper : cloth)
    ISBN-13 978-0-674-02414-4 (pbk.)
    ISBN-10 0-674-02414-1 (pbk.)
    1. Negotiation in business. 2. Negotiation. 3. Decision making. 4. Game
Theory.
    I. Richardson, John, 1966– II. Metcalfe, David. III. Title.

    HD58.6.R342 2003
    658.4'052—dc21   2002074688

To my son, Mark Jay Raiffa,

who can read and assimilate whatever I write

at speeds that amaze me

H.R.

To my sons, Ben and Pete

J.R.

To my wonderful parents

D.M.

Contents

Preface

My purpose in this book is to try to synthesize four approaches to decision making, broadly conceived:

- Decision analysis (a prescriptive approach—how an analytically inclined individual *should* and *could* make wise decisions)
- Behavioral decision making (a descriptive approach—the psychology of how ordinary individuals *do* make decisions)
- Game theory (a normative approach—how groups of ultra-smart individuals *should* make separate, interactive decisions)
- Negotiation analysis (how groups of reasonably bright individuals *should* and *could* make joint, collaborative decisions)

Many researchers and practitioners concentrate on one or two of these approaches and ignore the others. But I believe their theories and stories are intertwined. The boundaries delineating them should be porous, but alas are not—and to attempt a synthesis is a daunting task. In my career as an analyst, however, I have worked in all four domains, and it occurred to me that I could draw on that experience in attempting a synthesis.

This book is primarily about analysis *for* negotiators and not about analysis *of* negotiations. It's primarily about how analysis can help a negotiator or intervenor do a better job. It's more prescriptively than descriptively oriented. It considerably deepens the analytical perspective of my 1982 book on negotiation.

What are my credentials for acting the role of self-appointed bridge spanner?

- I'm a card-carrying Ph.D. in mathematics, who thinks like a mathematician. This mainly means three things:

1. In trying to understand a complex phenomenon, I like to build up from the simplest case. If some aspect of, say, negotiation with several parties is in question, I want to think hard about the two-party counterpart to the problem. Does it have to entail negotiations? How about the analogue to individual decision making?

2. When I'm confused about what I should believe, I take seriously the idea of first specifying what I think is basic in the form of axioms or desiderata and exploring where this leads me.

3. I think in abstractions (Mr. AAA rather than Mr. Edwards).

- Despite my mathematical leanings, I believe I know how to communicate with nonmathematicians, having taught for over forty years in professional graduate programs in business and public policy.

- As a young academic, I taught mathematical statistics and converted myself—yes, by axiomatic probings—to the subjective, Bayesian point of view of probability. Like some converts, I remain a proselytizer for that view. It's the only act in town if you are serious about decisions under uncertainty.

- I wrote a book about game theory (with Duncan Luce) in 1957 that still sells, so I guess I'm also a bona fide game theorist. But for a long time I found the assumptions made in standard game theory too restrictive for it to have wide applicability. One trouble is that game theory shuns the use of subjective probabilities and invokes the restriction of common knowledge. Such limits are hard to swallow in seeking to put this elegant theory into practice.

- In trying to make Bayesian statistics operational, my colleague Robert Schlaifer and I investigated protocols for eliciting from experts (design engineers, market managers, and like) their probabilistic judgments about uncertainties and their deep-felt value preferences for consequences that are often initially incoherent and contradictory. Hence I've also worked in the domain that is now known as behavioral (or judgmental) decision theory.

- Nevertheless, I think of myself not as a game theorist, or statistician, or negotiation analyst but primarily as a decision analyst. I have thought deeply about decision making with multiple, conflicting objectives and, along with Ralph Keeney, wrote a book about this subject in 1976 that has had widespread applicability from engineering design to clinical medicine to political science and, nowadays, to negotiations.

- A rather remarkable (for me) interlude in my academic career from 1968 to 1975, and to a lesser extent to the present time, immersed me in real-world, practical negotiations and conflict resolution. From 1968 to 1972 I was involved in negotiations that led to the creation of the confidence-building, East-West think tank, the International Institute for Applied Systems Analysis (IIASA)—the "East" led by the Soviet Union and its satellites, and the "West" by the United States and its allies. From 1972 to 1975, as director of this newly created institute located just outside of Vienna, I had to mediate disputes between scholars from different cultures and disciplines working together on the same problem, be it water resources or global warming. Imagine this heady experience: I was coached by the chairman of the IIASA Council, Jerman Gvishiani (the son-in-law of then premier Alexei Kosygin) on how I should negotiate with the Austrian Foreign Ministry on the tax-exempt status of this unique institution.

- In 1975 I returned to my teaching duties at Harvard, determined to construct an intellectual base for the negotiating and mediating activities I found myself doing at IIASA. I taught large sections of MBA students on the art and science of negotiation and based the entire course on simulated exercises using eager student-subjects. The students learned what things worked, *and so did I.*

- Since the early 1980s, I, along with Roger Fisher—mostly Roger—helped launch and nurture (as a member of its Executive Committee) the remarkably innovative, interdisciplinary Program on Negotiation (PON), located at the Harvard Law School, representing a consortium of universities in the Boston/Cambridge area (MIT, Fletcher School of Diplomacy, Simmons, University of Massachusetts, Boston College, and others). PON not only features seminars on topics of negotiation (broadly interpreted) but disseminates teaching materials and simulation exercises through its clearing house, supports doctoral students writing dissertations on negotiating topics, publishes the *Negotiation Journal,* and supports with clerical assistance projects such as the writing of this book. Along with James Sebenius, for many years I codirected the Harvard Negotiation Roundtable, one of the many projects under PON's umbrella. The Roundtable concentrated its attention on selected topics in two- or three-year chunks (ranging from the manager as negotiator and mediator to global warming).

Thus my work in decision analysis, behavioral decision analysis, game theory, and negotiation analysis aids my efforts to develop and synthesize these four approaches to decision making. I think I got the story straight.

I acknowledge with pleasure the contributions of Max Bazerman, David Bell, Roger Fisher, Mark Gordon, John Hammond, Ralph Keeney, David Lax, Duncan Luce, Robert Mnookin, John Pratt, Tom Schelling, Robert Schlaifer, Jim Sebenius, Larry Susskind, Amos Tversky, Michael Watkins, Milton Weinstein, and Richard Zeckhauser. Some were my (pre- and post-) doctoral students (DB, JH, RK, DL, JS, MWa, MWe, RZ); some edited *Wise Choices*, the Festschrift dedicated to me (RZ with RK and JS); some were my coauthors on books or articles from which I draw heavily (DB, JH, RK, DL, JP, RS, AT, MWe); some interacted closely with me in developing pedagogical materials (DB, JH, RS, RZ); some used me as their consultant to them as they consulted on real problems (MB, MG, RF, RM, JS, LS, RZ); and all (but especially TS) contributed to my understanding of the field of decision sciences broadly defined.

For their continued interest and financial administrative support, I thank Marjorie Aarons and Sara Cobb, past Directors of the Program on Negotiation (PON), as well as Robert Mnookin, faculty chair of that program. I would also like to thank Kim Clark, the Dean of the Harvard Business School, for giving me partial financial support for my research efforts during my retirement years.

I'm retired and I like writing, but on my own, I would never have embarked on a project as demanding as this one without the able assistance of two remarkable contributors, John Richardson and David Metcalfe, who join me on the title page. John is a lecturer at the Law School, working with Roger Fisher, Bruce Patton, and Frank Sanders on their extraordinarily popular courses on negotiation and mediation. John is an exceptionally quick learner, and despite being a nonanalytical, lawyer type, he especially relishes those parts of this book that are mathematically challenging. He has consistently discouraged me from deleting some of the more mathematical passages. David spent eighteen months at Harvard on a PON Fellowship and was my mentee. During this period he finished his dissertation and was granted his doctorate from the other Cambridge, in England. He has an uncanny ability to structure messy problems and to discern structure in chaos. Together John and David helped me to better express my thinking in parts; they improved and expanded on my work in other parts; and they added new ideas to the book.

I take great comfort in knowing that someone has read this book with meticulous care and not only has straightened out the uses of "which" and "that" but has actually suggested major structural improvements. It's been a pleasure working with someone as capable, demanding, and insightful as Elizabeth Gilbert of Harvard University Press.

A *Negotiation Analysis* Supplement can be found at www.pon.harvard.edu/ raiffa/nas/.

Negotiation Analysis

I

Fundamentals

Part I, comprising Chapters 1 to 5, serves as an extended introduction to the book. Through it we begin to develop an analysis that facilitates negotiation by drawing extensively on:

- Decision analysis—individual prescriptive decision making
- Behavioral decision theory
- Game theory—interactive decision making

Sometimes we will think of negotiation analysis as the fourth approach to decision making.

Chapter 1 contrasts three perspectives of individual decision making:

- Normative analysis—how ideally decisions should be made (by super-rational individuals)
- Descriptive analysis—how real people actually behave (very often at variance with the normative abstraction)
- Prescriptive analysis—how real people could behave more advantageously with some systematic reflection

In group decision making, this threefold breakdown becomes more complicated. Much of the book gives partisan (prescriptive) advice to one decision maker in a group under the assumption that the other individuals are not privy to that advice and behave in a more descriptive fashion. In contrast to this orientation, the theory of games takes a jointly normative approach, assuming that both players' behavior is super-rational.

Chapter 2 is a tutorial that develops enough of decision analysis to be of use in the rest of the book, but it leaves additional topics for later development where needed.

Chapter 3 examines real behavior and indicates that many individuals at least some of the time do not act in conformity with rational behavior—especially when uncertainties loom large. We call such examples of "deviant" behavior *anomalies, biases, errors,* or *traps,* and we try to rationalize why and how they might occur.

Chapter 4 develops many of the concepts of the noncooperative theory of games in a tutorial, self-discovery, pedagogical style; but as in Chapter 2 it leaves special topics in the theory for just-in-time later development where needed.

Chapter 5 provides an outline of the rest of the book. It introduces the concept of idealized, joint behavior, in which all protagonists in a negotiation agree to negotiate in a truly collaborative manner, agreeing to tell the truth and all of the truth. We call this joint FOTE analysis—Full, Open, Truthful Exchange. We claim that in actual *deal* making, in contrast to *dispute* settling, this ideal is often approximated, and the FOTE examination sets up an ideal against which other approaches can be compared. In later discussion, we back away from FOTE analyses to consider POTE-like behavior, in which some protagonists tell the truth but not the whole truth—Partial, Open, Truthful Exchange—leaving their bottom lines hidden from the view of others.

There is a tension in negotiations between *creating* actions, designed to build a bigger pie, and *claiming* actions, designed to obtain a large share of the pie so created. The main concern of this book is the creating function, but the claiming function cannot be ignored. Part II (Chapters 6–10) is mostly about claiming behavior for two-party negotiations; Part III (Chapters 11–17) mostly concerns creating joint gains; Part IV (Chapters 18–20) studies external interventions—how facilitators, mediators, and arbitrators can help; and Part V (Chapters 21–27) complicates it all by introducing more parties into the fray.

1

Decision Perspectives

Our primary concern in this book is to suggest how people—perhaps you—might negotiate better. Sometimes we offer partisan advice to one of the negotiating parties, who may then profit at the expense of the others. Many times the advice we give to party A might also help B—helping yourself by helping others. We get more moral satisfaction from that part. But we won't limit ourselves to discussing only the feel-good part of negotiation. Sometimes we give advice on the process to all the parties about how they might choose to negotiate. Sometimes we give advice to external helpers (facilitators, mediators, arbitrators).

We hope our advice will help you, as negotiators:

- to understand the intricacies of the problems you face,
- to make decisions confidently in the face of complexity,
- to become able to justify your decisions convincingly to yourselves and to your constituents,
- and ultimately to conduct negotiations so that you will be satisfied with the consequences of your actions.

These benchmarks inform our purpose in crafting advice for you as negotiators, advice that deals primarily with how to help you and your constituents make better decisions in negotiations.

We are ourselves a bit uncomfortable writing as if we know better than you what you should do in your negotiations. When we say, "We advise . . . ," we usually mean, "Our analysis suggests that most people, including ourselves, would do better if they . . ." Our circumlocution is one of convenience, and an effort to avoid pedantry.

Our Approaches to Decision Making

We take a decision-making perspective; and, in so doing, we try to integrate four disciplinary approaches in our subject matter:

> Decision analysis: a prescriptive approach
> Behavioral decision making: a descriptive approach
> Game theory: an interactive prescriptive approach
> Negotiation analysis: mostly prescriptive

We shall try in this chapter to make fine distinctions among these four approaches. But despite heroic efforts to pull them apart, we ultimately come to the conclusion that the boundaries among them are so blurry that we need to integrate them in an overall approach. Such an integrated approach is a major goal of this book. Unfortunately the walls between these disciplines are not very permeable, and quite sophisticated practitioners in one domain know little about the other two related domains. Thus, for example, insights may be developed about some phenomenon in one domain that researchers think of as quite innovative without realizing that the "discoveries" are quite widely understood in the related fields. Fortunately, however, the task of integration is not overwhelmingly burdensome, because in this book we have only to draw on the more basic, less esoteric, aspects of each of these strands of inquiry. A little contribution from each discipline will collectively go a long way.

Other disciplines could be integrated with negotiation analysis as well: behavioral economics, social psychology, anthropology, and perhaps others. We don't mean to suggest that the steps toward synthesis that we take here are exhaustive. There is plenty more to be done!

Individual Decisions

The individual (unitary) perspective.   When we adopt an *individual* decision-making perspective, we view the world from the standpoint of a unitary decision entity. Note that the assumption of a unitary perspective does not preclude the existence of numerous individuals within the entity. For instance, we might, at one level of abstraction, advise a corporation in merger negotiations under the assumption that the corporation is a unitary decision entity. Descriptively, of course, this is not accurate. But we might prefer not to incorporate and encapsulate details of interdepartmental wrangling and boardroom spats. Adding too many details can complicate and hinder analysis. We want to tailor it to the primary needs of the decision maker. The unitary decision entity may be a family, or firm, or even a nation, depending on the problem. In summary, an individual decision-making perspective assumes the decision entity is monolithic—a blind eye is turned to internal conflict. When adopting an individual perspective, one employs the tools, techniques, and methods of decision analysis. A systematic discussion of this perspective is given in Chapter 2.

Prototypical decisions.   Consider the following cases from the perspectives of the individual decision makers:

Drill or not drill. Wilder Cather, an oil wildcatter, owns the rights to drill for oil at a given site. But there is a deadline to exercise these rights. The overriding uncertainty is the amount and ease of recoverability of oil deposits at the site. He could, at a considerable expense, take seismic soundings that could inform him about the chances of finding oil.

Surgery or radiation. Sue Vivor is getting conflicting advice about what to do about her newly discovered breast cancer. Should she follow the advice of her surgeon and undergo a radical mastectomy, or try a less invasive regimen of radiation and chemo-therapy? Probabilities matter, but there are so many more concerns that she has to weigh in her difficult decision.

Invest or not. Mr. Little must decide whether to buy the store property up for sale and open a pharmacy. He has to act soon, if he acts at all. His main concern is whether Wal-Mart will open a competing enterprise in the mall not very far away. Mr. Little must worry about what Wal-Mart will do, but Wal-Mart has no concern at all about what Mr. Little will do.

In the wildcatting and medical examples, the uncertainties are not in control of someone else, some sentient being who is thinking what the decision maker is thinking and might do. It's more complicated in the investment problem. Mr. Little must think about what Wal-Mart might do, but his analysis and subsequent action will be of little import to Wal-Mart. The decision problems of Little and Wal-Mart are only loosely coupled. Little does not have to think about what Wal-Mart is thinking about what Little is thinking.

Group Decisions

Viewing the world from a group decision-making perspective emphasizes the interactions and dynamics between multiple unitary decision entities (see Figure 1.1). The unitary entities do not need to consist of individual decision makers, but in the group setting there are, by definition, two or more unitary entities that interact to some degree. Adopting a group decision-making perspective involves more complexity than the individual perspective. Broadly defined, two approaches have developed in examining group decision making: the theory of games and the theory of negotiation.

Interactive decision making: Theory of games.   Imagine a market with a few firms competing to outdo one another by competitive pricing, advertising, pro-

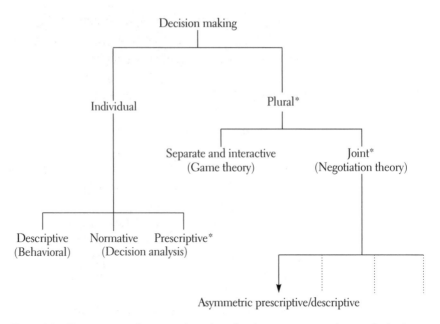

Figure 1.1.   Perspectives on decision making. Asterisk indicates primary emphasis in this book.

motion, and the introduction of innovative products. We're now in the domain of *interactive decision making,* or the theory of noncooperative games. The essence of the theory involves a set of individual decision makers (players, in the vernacular of game theory), each constrained to adopt a choice (strategy) from a specified set of choices, and the payoff to each player depends on the totality of choices made by all players. Each player must choose, sometimes not knowing the choice of the others. Each must think about what the others might do and realize that the others are, in turn, thinking about what the rest are thinking. The essence of this perspective is that although the individual decision entities make their choices separately of each other, the payoffs they receive are a function of all the players' choices. The players are caught in a web of strategic interaction in which they have only partial control over the payoffs they receive. Individual, or separate, decisions interact to determine joint payoffs.

Joint decision making: Negotiation theory.   Consider the case of a community dispute in which the parties voluntarily come together to see if they can jointly agree to do something as a group. Imagine them meeting around a table discussing possible compromises. We're now in the realm of *joint decision making,* or the domain of negotiations.

　　Roughly, and we are purposely rough here, both negotiation and game situations involve group decision making. Games involve multiple individuals making separate decisions that interact. The payoffs for each individual are dependent not only on their own decision but also on the decisions of the other individuals, and vice versa. Negotiations, in contrast, involve multiple individuals cooperating to arrive at a joint decision. The joint decision entails joint consequences, or payoffs, for each individual.

That's the difference, roughly, between games (separate interactive decisions) and negotiations (joint decisions). But the differences blur in reality.

Fuzzy boundaries.   Imagine that Abe and Betty have had a severe disagreement and are now negotiating a settlement to their dispute. Emotions erupt and Abe threatens to break off negotiations and to act in a way that will be hurtful to Betty. But if negotiations break down, Betty has alternatives too. If both Abe and Betty take this route, their negotiation will be transformed into a competitive game. In most negotiations you don't have to settle. Before you enter into a negotiation, it is prudent to think hard about what you might expect if there is no accord. What is your best alternative to a negotiated agreement? But this leads you into thinking what the parties might do separately, and you are now back in the domain of game theory.

Underlying every negotiation structure is a game-like component. And the more one studies negotiations, the more one realizes the artificiality of the borders between the two perspectives. Unfortunately, most negotiators are not experts in the theory of games and few game theorists are experts in negotiation theory. We believe that using both tools produces useful insights for crafting negotiation advice.

To complicate matters even further, in a negotiating context decisions arise that can best be analyzed from the perspective of the individual decision maker. Here are some:

- The decision to start negotiations
- Given a final (really final) offer by the other side to settle a dispute, the decision to reject it and take your chances in court
- The decision to bring in a mediator
- The decision to bargain separately with your potential customers or to involve them in an auction mechanism[1]

Descriptive, Normative, and Prescriptive Orientations

The Individual Case

In explicating the differences between the descriptive, normative, and prescriptive orientations of decision making, it will be helpful to start with the case of

---

1. There is another aspect of game theory, called by its founder, John von Neumann, the *cooperative theory*, that features more than two parties who maneuver to form coalitions. This theory will also be reviewed in the last part of this book, dealing with the dynamics of coalition formation. When we refer to game theory we usually have in mind its noncooperative portion.

the single, individual decision maker, saving for the next section complications arising because of group interactions.

Descriptive decision making: How decisions are made. The study of descriptive decision making is concerned with how and why individuals think and act the way they do. It is the domain of the behavioral decision theorists—mostly psychologists. Good descriptive analyses lead to good predictions of actual behavior. Descriptive work is a highly empirical and clinical activity that falls squarely in the province of the social sciences concerned with individual behavior. Scholars can study this domain without any concern whatsoever for trying to modify behavior, influence behavior, or moralize about behavior.

Descriptive analysis entails the following sorts of questions:

- How do people actually think and behave?
- How do they perceive uncertainties, accumulate evidence, learn and update their perceptions?
- How do they learn and adapt their behavior?
- What are their hang-ups, biases, internal conflicts? How do they talk about their perceptions and choices?
- Do they really act as they say they do?
- Can they articulate reasons for their actions?
- How do they resolve their internal conflicts or avoid resolving them?
- Do they decompose complex problems, think separately about component parts, and then recompose or integrate these separate analyses? Or do they think more holistically and intuitively?
- What are the differences in thought patterns for people of different cultures, genders, and experiences?
- What is the role of tradition, imitation, and superstition in decision making?

Normative decision making: How decisions should be made. Normative theory suggests how idealized, rational, super-intelligent people *should* make decisions. Such analyses abstract away known cognitive concerns of real people, such as their inner turmoil, their shifting values, their anxieties and lingering postdecision regrets, their dislike of or zest for ambiguity, their inability to do intricate calculations, and their limited attention span.

The hallmarks of such normative analyses are coherence and rationality, which are usually captured in terms of precisely specified desiderata, or axioms. Applied mathematicians and mathematical economists dominate the normative decision-making scene. Its advocates don't believe that people actually act that way, but feel that perhaps they should. Few steps are taken to suggest how mere mortals might try to act in accordance with these rational axioms in their everyday life. The theories are very elegant.

The vast majority of economics and game theory has a normative complexion. To add in descriptive realities makes life too complicated. Normative theories are sometimes used as first-order approximations of real-world, empirical behavior. But empirical veracity is sacrificed on the altar of theoretical parsimony.

Prescriptive decision making: How decisions could be made better. Prescriptive analyses ask: What can a real person actually do to make better decisions? What modes of thought, novel perspectives, decision aids, conceptual schemes, analytical devices, words of advice, are practically useful? In seeking to craft useful theory, prescriptive analyses are not concerned with conceptual ideas and techniques that are useful for idealized, mythical, super-rational, de-psychologized automata. Instead, prescriptive proposals must be useful for real people—warts and all.

Because real people are different, good advice has to be tuned to the differential needs, capabilities, psyches, foibles, fallibilities, and emotional makeups of the individuals for whom the prescriptive advice is intended. Prescriptive advice should be evaluated by its pragmatic value—by its ability to help people make better decisions. To reiterate, advice should promote an understanding of problems, confidence in decisions, justification for decisions, and, one hopes, satisfaction with consequences. Prescriptive analysis should be informed by descriptive and normative theories. We can think of the prescribers as playing the role of engineers, whereas normative theorizers are the pure scientists.

The case of transitivity. Let's use transitivity as a vehicle to illustrate the differences between the descriptive, normative, and prescriptive perspectives.

Normative theory posits that if alternative A is preferred to alternative B, and B is preferred to C, then A *should* be preferred to C. Such is the principle of transitivity. Without transitivity, modern economics would barely exist. But from the perspective of empirically observed real behavior, the normative assumption of transitivity is highly contentious.

A descriptive theorist might indignantly point out: "Transitivity is not always observed in human decision making. True, some of the people, some of the time, might exhibit transitivity, preferring A over B, B over C, and A over C. But there are numerous instances of real people violating transitivity. Some people make intransitive choices, preferring A over B, B over C, and C over A." This circularity really gets the goat of our normative theorist, who replies, "Descriptively this may be the case, but your naive, misguided, ill-advised decision maker really ought not to make intransitive choices. It's just not rational!" Why? "Well," says the normative idealist, "a person who insists on defending intransitivities can be turned into a money pump. Starting with A, he would pay a small sum to move to C—after all, he says C is preferred to A—then another small sum to go to B (since B is preferred to C) and another to go back to A (since A is preferred to B) and on and on." The describers understand this argument, but respond, "So why aren't you rich, then?" They would rather spend their efforts to help explain why such intransitivities arise in practice, and what one can do about it.

Let's examine how the prescriptively inclined think about the transitivity principle. Assume our decision maker, Art, is having trouble deciding whether to choose alternative A or alternative C. The problem is a tough one, because there are many features to compare, and there are uncertainties attached to the consequences of selecting either alternative. What's more, in making his choice, Art needs to compromise among a fistful of conflicting objectives. This poses a real headache for Art, especially because A and C will entail very different consequences. To assist Art, a prescriptive theorist might creatively concoct another alternative, say B, that has some features in common with A, and some with C. And let's suppose that Art may find it easy to say that, all things considered, for him, A is better than B, and B is better than C. Art concludes, by way of concocted alternative B, that the principle of transitivity has revealed that he should prefer A over C. It doesn't matter whether B is a real or hypothetical alternative. A behavioral theorist might identify pitfalls in using this kind of procedure, but on the whole, it provides helpful advice for Art.

Take another example. Betty has trouble comparing A and B. But she can successively make small modifications to A, changing A to equally desirable A', and then to A", and changing B to equally desirable B'. Betty may now have no trouble in deciding that A" is better than B', and conclude that she should therefore prefer A to B, after all. Analysis applied as prescriptive advice has assisted her to make up her mind. If she were a hyper-rational automaton she wouldn't need this advice, because she would have known all along that A is better than B and that she prefers the former to the latter.

Group Orientations

The situation gets more complicated when researchers study how individuals behave in a group setting. Analysis might entail mixtures of descriptive, normative, and prescriptive orientations.

Symmetrically descriptive. A researcher with a symmetrically descriptive orientation might be interested solely in describing the behavior of all the negotiators or all the players in a game context, without having any interest whatsoever in prescribing how they should behave. Such researchers can be very analytical about their subject matter; they can use esoteric descriptive and interactive models of behavior, involving simulations or mathematical models. Some of these researchers may be interested in important cases of negotiation from a historical perspective. For example: How do real people, with all their idiosyncrasies and bounded rationalities, actually behave? How do they learn? How is trust created? How is it destroyed? This descriptive perspective is the primary interest of story-tellers, historians, psychologists, sociologists, anthropologists, political scientists, and positive economists.

Symmetrically normative.   In the group context as in the individual, game the-orists—mostly applied mathematicians and mathematical economists—examine what ultra-smart, impeccably rational, super-people *should* do in competitive, interactive situations. With a symmetrically normative orientation, they are in-terested not in the way erring folks like you and me actually behave, but in how we should behave if we were smarter, thought harder, were more consistent, were all-knowing. Advice is given symmetrically to all parties about how to play certain intriguing games. This advice is built on an edifice of behavioral as-sumptions that together constitute a decision-making persona of "selfish ratio-nality." No attempt is made to adapt the advice to complex behavioral realities. Instead game theorists investigate illuminating oversimplifications, to find out what insights can be gained from that pristine imaginary situation that might be useful in the real world.

   In game theory, each party has to think about what the other party is think-ing about what the first party is thinking about—and so on, ad infinitum. The advice given to all parties must give rise to an *equilibrium* situation: if the theory says that party A should choose strategy 1, and party B should choose strategy 2, then 1 must be a good retort against 2, and 2 must be a good reply against 1; oth-erwise, the advice would be self-destructive and hence counterproductive. More about this later.

Asymmetrically prescriptive/descriptive.   The researcher working from the asymmetrically prescriptive/descriptive vantage point is concerned with studying and understanding the behavior of real people in real negotiations, so that he can better advise one party about how it should behave in order to achieve its best outcome. This type of analysis takes a prescriptive perspective for the bene-fit of one party, and views the competing parties in descriptive terms from this fo-cal party's vantage point. Drawing on normative principles and a description of the other parties' behavior, decisions, and attributes, advice is given to the rele-vant side. The advice can range from what to wear and how to present oneself to intricate analysis of what complex calculations to make. Of course, if all parties are getting such advice, the advice given to one party will have to reflect the fact that the advice is also being given to the other parties.

   I, Howard Raiffa, started my career as a game theorist doing research of the symmetrically normative variety, but later became increasingly involved in giving partisan advice to one party about how it could and should behave, given its descriptive probabilistic predictions about how other parties might behave.

Descriptive, normative, and prescriptive orientations for external helpers. In externally descriptive studies, one might investigate how external helpers actually behave in negotiations. What are the similarities and differences in the descriptive behaviors of these people? Our concern in this book is mainly with determining how external helpers (facilitators, mediators, and arbitrators) *should* or *could* behave in order to help the negotiating parties in some impartial, balanced way. These can be thought of respectively as externally normative or prescriptive orientations. Distinctions between mediators, arbitrators, and facilitators will be discussed extensively in Part IV.

An effective external helper should understand the negotiation process from the vantage points (symmetrically descriptive, symmetrically normative, and asymmetrically prescriptive/descriptive) we have outlined. Since external helpers have aspirations, ideals, values, judgments, and constraints of their own, they can be thought of as another negotiator in the game—albeit a special type of negotiator—trying to maximize their own payoffs. The trick for the other negotiators is to choose an external helper whose motivations and incentives are compatible with their own.

Core Concepts

A more descriptive title for this book would be *Analysis for Negotiation,* featuring the use of decision analysis, behavioral decision theory, and game theory. It's not *Analysis of* but *Analysis for.*

These three facilitating disciplines each have a rich literature of their own and layers upon layers of intricate techniques and observations. Unfortunately these domains are often not in communication with each other. Our task is to examine the basics in each of these fields and show how they can help in doing analysis for negotiations. We like to think that these disciplines, including negotiations, are part of an emerging field: the science of decision making.

We have offered a taxonomy of decision making by first splitting the domain into individual versus group problems and then splitting groups into game theory (where several individuals act separately, but each is affected by the totality of individual choices) and negotiations (where several individuals must act jointly by selecting a common negotiation contract).

In the individual domain, we have distinguished among a triad of perspectives: normative (what should be done), descriptive (what is done), and prescriptive (what realistically could be done to improve matters). Our primary, but not exclusive, concern is with prescription (advice giving). The group domain splits between games and negotiation. Game theory (mostly) gives normative advice to all parties, whereas in negotiation analysis we concentrate on giving prescriptive advice to one of the negotiators after reflecting on the descriptive behavior of the other negotiators. The question is how you might want to behave given that the others will behave as usual and without coaching advice. Sometimes we offer prescriptive advice for an external intervenor (facilitator, mediator, or arbitrator).

2

Decision Analysis

This chapter offers prescriptive advice to the unitary decision maker. As we have seen, the unitary decision maker can be a couple, or a single company, or a single country—as long as it poses as an undivided entity speaking in a monolithic voice. It is a party who can decide among the available choices without anyone else's permission. In common usage, much of what we say in this chapter falls under the heading of decision analysis.[1]

Generic Decisions in Negotiation Contexts

Actors in a negotiation setting have to make vexing individual choices, which include the following occasions or situations.

The decision to negotiate. AAA must decide whether or not to start negotiations with BBB to resolve a festering dispute. Starting negotiations carries costs in money, time, and pride, and it isn't clear what will happen if the two parties do negotiate.

What to do if negotiations fail. AAA and BBB are thinking about merging their companies. AAA should keep in mind what might happen if no agreement is reached with BBB—what is the best alternative? Going with CCC or DDD or going it alone? Uncertainties accompany each of these alternatives and it's not clear what AAA's objectives should be.

The decision to go to court or accept a settlement. Plaintiff AAA must decide to accept Defendant BBB's "last and final offer" or go to court. It's not

---

1. In this beginning tutorial we concentrate on decision *statics* as opposed to decision *dynamics*. The latter involves linked decisions over time and the sequencing of interrelated decisions that are often elaborated with decision trees. We don't entirely ignore dynamics in this book, but offer just-in-time discussions of special topics whenever needed.

clear whether AAA will win a trial or what the jury will award her if she does win.

With whom to negotiate? Should AAA approach BBB or CCC first? If she contacts BBB, and if XYZ happens, then what? Should AAA act now or gather further evidence, at a cost?

The decision to use a third-party helper. AAA is disappointed in how the negotiations seem to be going. Should AAA suggest that they bring in outside help? If so, what role should AAA suggest the intervenor play?

Choosing a style of negotiation. Negotiator AAA is not sure how principled negotiator BBB will be in the crunch. There are many possible negotiating styles that AAA can employ. Which will yield the best results with the particular party across the table? Should AAA be open and honest or act strategically, with some dissembling?

The decision to . . . (Add your own.)

We advise negotiators to learn the theory and practice of unitary, prescriptive decision making.

The PrOACT Way of Thought

Let's introduce Ms. AAA, who will be our decision maker. A problem is brewing and she wants to think deeply about which choice or strategy she should adopt. How should she proceed?

In this book we advocate a way of thinking about choice problems.[2] It involves five elemental steps:

1. Identify the *Pr*oblem
2. Clarify the *O*bjectives
3. Generate creative *A*lternatives
4. Evaluate the *C*onsequences
5. Make *T*radeoffs

The italicized letters in this list form the acronym PrOACT, used to refer to this way of thought for analyzing individual decision problems. The acronym is also short for *proactive*.

In a negotiation context, our protagonist, AAA, must be reminded that sometimes she must be proactive in making decisions about negotiations. She will be confronted by many decisions without clear or easy answers. In such situations, most negotiators put off making decisions. But as the saying goes, doing

2. We draw on Hammond, Keeney, and Raiffa (1998), from which the PrOACT system is taken.

nothing is doing something. And is it the right thing? "Reactives," in contrast to "proactives," tend to wait for more and more information, or allow the other party to set the agenda. Others jump to a decision without stopping to think it through. But smart choices just don't happen; they usually have to be nourished. PrOACT is a way to make this happen.

Identifying the Problem

Statisticians talk about errors of the first kind (rejecting a true statement) and of the second kind (accepting a false one). It is just as important to avoid the "error of the third kind": solving the wrong problem. Thus the first task for a decision maker is to challenge her understanding of the choice facing her. We suggest a few simple steps to help her, and you, define the problem.

- Examine the "trigger" that caused you to realize you had a decision to make.
- Don't restrict yourself to solving only the immediate problem you face.
- Question the constraints in your first statement of the problem. Look for ways to open it up to a wider range of possibilities.
- Ask friends or experts for their thoughts. They are unlikely to share exactly the same limitations on their thinking that you have.

Imagine a successful author of mystery novels. We'll call her Audrey. She is almost done with another book, and she has decided to leave her current publisher. At first blush it seems that her problem is choosing which company she would like to publish her next book. This may well be the right problem, but she ought to look at other possible specifications of it. She might step back and ask, "What is the best way to communicate this story to my audience?" This question may prompt her to consider publishing her work over the Internet, turning the novel into a screenplay and selling it to a movie studio, or serializing it in a magazine, all in addition to thinking of the many different publishers she might consider.

Ultimately the definition of your problem will depend on the objectives, interests, desires, and wants that you wish to satisfy. Taking a census of those desires is the next step in making a smart choice.

Clarifying Objectives

Objectives are the criteria by which you will judge the decision. They are the hopes, needs, desires, and fears that motivate you. They are the things that have value for you.

"Objectives" = "interests."   Unfortunately, negotiation theorists and decision theorists use different terminology for the same idea. In their justly popular negotiation book, *Getting to Yes*, Roger Fisher, William Ury, and Bruce Patton emphasize the role of *interests*. Larry Susskind, another top-notch mediator, talks of *"interest*-based negotiations for joint gains." Decision analysts talk about *objectives* instead of *interests*. But they mean the same thing. For now we'll use the decision theory lingo. In later chapters, however, we'll revert to the standard terminology in negotiations and also use *interests*.

The trick is to think broadly about your objectives. It is easy to focus on a few, especially when they are easily quantified, like money, and forget about others. Returning to the case of Audrey the novelist: what are her objectives? Money, sure, so she will want a publisher with deep pockets. But she will want other things as well. She wants a publisher with a strong distribution network, who can get the book into good display areas in a lot of stores. She wants good editors and other staff (she'll be spending a lot of time with them). But those are means and not ends. She wants fame and recognition. She wants to feel like a successful author. She may want a prestigious imprint to raise her status in literary circles.

Why? and what?   At this stage you want to concentrate on listing as many objectives as you can think of. For each objective you record, we suggest that you ask the *Why* question: why am I interested in this? This helps probe your deeper, more basic objectives and sort out means and ends. You also might want to ask the *What* question: what do I mean by that? What do you mean, for example, by saying you want to live in a nice neighborhood?

In negotiation contexts there are additional objectives that don't come up in the ordinary case. These may include your feelings about your ongoing relationships with the other parties, the question of how AAA's perceptions of BBB's payoffs might affect AAA's payoffs—reflecting empathy or vindictiveness—and moral concerns about using threats, exaggerations, and strategic misrepresentations. Finally, you might want to identify an operational set of evaluative objectives for further analysis. This process will be elucidated shortly when we discuss how such objectives are used in some simple examples.

Generating Creative Alternatives

Your alternatives are the many ways you might try to meet your objectives. They are actions you could take to satisfy your desires. In negotiation parlance, a distinction is made between the alternatives each party might pursue individually (external to the negotiation) if negotiations break down and the (internal) alternatives that might be jointly negotiated and jointly pursued. The term "alternatives" is reserved for choices external to negotiations and the term "options" is used for collective choices internal to negotiations. The generation of alternatives is usually a solo act, whereas the generation of options may involve joint deliberations.

How?   The big message about alternatives is simple. All too often we focus too narrowly on the first one that comes to mind. Force yourself to be more creative by asking, "Are there other things I might do?" Prod yourself to think imaginatively and creatively about generating alternatives by systematically reviewing your objectives and asking, "*How* might this be achieved?"

The more the better.   It's important to break away from perceived constraints and to think more imaginatively. For example, in some contexts, it might be helpful to ask the question: "What would you do if you had all the money in the world?" It is surprising how often that question triggers responses that don't require much money. Occasionally being deliciously irresponsible can pay big dividends. In a negotiation, the generation of alternatives or options might also be done collectively, and a lot depends on the atmospherics of preplay negotiations. We'll have a lot to say about this later on.

To simplify our presentation, suppose our novelist, Audrey, decides that she would like to publish with a traditional publishing house and confines herself to publishers of mass market fiction. Of these, she decides that three merit serious consideration: PPP, QQQ, and RRR.

Evaluating the Consequences

Having generated a good list of alternatives, our decision maker must then evaluate the merits of various possibilities. In most complex situations there can be several competing evaluative objectives.

Conditional analysis.   We recommend undertaking conditional analyses: how good is alternative X with regard to evaluative objective Y? Imagine constructing a matrix with the columns corresponding to alternatives and the rows to evaluative objectives, where each cell of the matrix records how well that alternative fares with regard to that objective. The entries in the cells could be quantities or verbal descriptions or a scaled value like high, medium, or low.

In our example, Audrey knows that publishers offer very standard royalty rates. When they want a book, they compete by offering generous advances against future royalties. She guesses (or makes preliminary contacts with each publisher) about how much they might offer her for her next book. She then thinks about how she would rate each of the publishers on the evaluative objectives of prestige, editing help, distribution, and advance, and she comes up with a chart (see Table 2.1).

Table 2.1   Conditional evaluation of alternatives on evaluative objectives

| | Audrey's alternatives (publishers) | | |
Evaluative objectives	PPP	QQQ	RRR
Prestige	Good	OK	Great
Editing help	Fair	Fair	Great
Distribution	Good	Fair	Good
Advance	$600K	$550K	$400K

Dominance.   At this level of abstraction it may not be clear which alternative is "best," but some alternatives may be clearly eliminated as noncontenders. We can see that PPP is better than QQQ on three objectives, and at least as good on the third. So much for QQQ. In the vernacular we say that QQQ is *dominated* by PPP.

Making Tradeoffs

Audrey still has to consider PPP and RRR. Each is better on some objectives. What's her preference overall, all things considered? It should depend on the strengths of her two remaining possible preferences vis-à-vis her objectives, and on how important the various objectives are.

Audrey could examine the two options and "go with her gut." And intuitive decisions are frequently right. But they are also frequently wrong. She can be more confident of making a good decision if she goes further with the analysis. In a context of collective decision making, as is the case in negotiations, it may be important to articulate the thought processes of the evaluators in order to be creative about suggesting other alternatives as compromises. We consider two ways to do this.

Costing out.   One is to look for real-world ways of converting one objective into the same currency as another. For example, RRR offers better editing help; PPP offers more money. Both are important. Thinking about it, Audrey remembers that she knows a freelance editor who is just as good as the staff editor at RRR. A quick call, and the freelancer offers to work on Audrey's book for $20,000. Audrey could accept the higher monetary offer from PPP and hire the editor out of her own pocket. That transaction would make the two publishers effectively equal on that objective, so Audrey can simply cross off editing from her objectives and subtract $20,000 from PPP's advance.

Equal swaps.   The technique of costing out independent sources of the same good won't work for every objective. RRR also has more prestige than PPP. Audrey can't single-handedly affect the reputation of the whole publishing house. How can she judge between money and prestige? She might try asking, "For each objective, how much money would it be worth to me for a wizard to wave his magic wand and improve that alternative?" Audrey is being asked to make a swap: money for prestige. The process reduces a two-dimensional comparison, money and anxiety, to a one-dimensional comparison, money only. Hard, granted; but thinkable.

Partial analysis or full?   Most problems don't require the full treatment: problem definition, objectives, alternatives, consequences, and tradeoffs. The conclusion might be obvious after a partial analysis. This is often the case when the PrOACT approach is applied to unitary decisions in negotiation contexts.

Uncertainty and Risk

We now introduce uncertainties. For some alternatives you may not know what the consequences will be. You may be presented with a clear-cut choice that is shrouded in uncertainties. In order to make a wise decision, you may choose to formalize your thought process by assessing your judgments about uncertainties and examining your attitudes toward risk.

Defining Risk Profiles

To each noncertain alternative, we attach a risk profile that exhibits:

- possible outcomes
- the likelihoods of these outcomes (ideally expressed in probabilistic assessments)
- the resulting consequences associated with these outcomes (for the chosen alternative)

Consider a schematic presentation of a risk profile for an alternative X (see Table 2.2). The possible outcomes, 1 to n, are listed in column 1. In a concrete case there might be names for each outcome—such as "trial lost" or "trial won with a large jury award." Column 2 is reserved for preliminary remarks about the likelihoods of these outcomes—perhaps a ranking of their plausibilities. The remaining columns are designed to record a description of the consequence associated with each outcome for alternative X. In the table we illustrate the case where there are three evaluative objectives that help describe the outcome. The cell entries might feature a few key words or values on some scales.

Table 2.2 Schematic risk profile for alternative X

1	2	3	4	5
		Evaluative objectives		
Outcome	Likelihood	OBJ-1	OBJ-2	OBJ-3
1	L_1	C_{11}	C_{12}	C_{13}
$\vdots$	$\vdots$	$\vdots$	$\vdots$	$\vdots$
i	L_i	C_{i1}	C_{i2}	C_{i3}
$\vdots$	$\vdots$	$\vdots$	$\vdots$	$\vdots$
n	L_n	C_{n1}	C_{n2}	C_{n3}

Table 2.3 Audrey's risk profile for alternative PPP

Outcome	Likelihood	Evaluative objectives			
		Prestige	Editing	Distribution	Monetary
Status quo	Most likely	Good	Fair	Good	Excellent
Minor change	Middle	Very good	Good	Excellent	Very good
Major change	Least likely	Poor	Fair	Fair	Poor

Back to Audrey. Let's assume that there is a complication confronting Audrey in her dealings with publisher PPP. An internal reorganization is looming and it is not clear which of three outcomes will prevail: the status quo will be preserved; a minor shakeup of management will occur; a major overhaul of the company will take place in the next six months. Audrey must commit herself before she knows what will happen. Her risk profile for the alternative PPP is shown in Table 2.3.

The separation principle. Let's imagine you are comparing alternatives X and Y, and you have the risk profile for each displayed. With this presentation in front of you, it might now be clear what your choice is. But suppose not. You might want to think more systematically about your choice. You might wish to separate judgments about *uncertainties* from judgments about *values*. In a more dramatic example, professional judgments about the uncertainty of an accident with a nuclear reactor may be tainted by the expert's attitudes about nuclear power. For your own job of comparison, separating uncertainties from values may make it possible for you to enlist the opinions of scientific experts who can contribute information and insights about the uncertainty domain. You may, however, be reluctant to use their expertise on the value domain. Perhaps these experts have too narrow an orientation to be of help with the value side of the ledger.

Table 2.4   Analytic reductions for the schematic representation of the risk profile for
             alternative X

1	2	3	4	5	6
			Composite evaluations of consequences		
Outcome	Probability	Ordinal values	Monetary values	Desirability values	Utility values (BRLTs)
1	p_1		m_1	d_1	u_1
$\vdots$	$\vdots$		$\vdots$	$\vdots$	$\vdots$
i	p_i		m_i	d_i	u_i
$\vdots$	$\vdots$		$\vdots$	$\vdots$	$\vdots$
n	p_n		m_n	d_n	u_n
			EMV	EDV	EUV

Expected *monetary* value   = EMV = $p_1m_1 + \ldots + p_im_i + \ldots + p_nm_n$.
Expected *desirability* value = EDV = $p_1d_1 + \ldots + p_id_i + \ldots + p_nd_n$.
Expected *utility* value   = EUV = $p_1u_1 + \ldots + p_iu_i + \ldots + p_nu_n$.

Uncertainty analysis.   Our analysis proceeds to Table 2.4. To make the risk
profile more amenable for precise comparison, in the second column, you, acting alone or with expert advisers, might assign subjective numerical probabilities
to the likelihoods of the outcomes, reflecting your best judgments. With this
added formalization, it might now be clear how to compare the X and Y profiles.

Value analysis.   You might want to record more precisely how you feel about
the consequences in the last three columns of Table 2.4. Somehow you have to
combine the conditional evaluations of the outcomes with respect to each of the
evaluative objectives into a composite preference judgment. We suggest four different value scales: ordinal, money, desirability, and utility, which we now discuss in turn.

Ordinal Ranking

Imagine listing the descriptions of the consequences for profiles X and Y and
now recording, all things considered, your preferences: what you like best, next
best, next best, and so on. Your rank ordering of *all* the consequences associated
with the possible outcomes for alternative X is depicted schematically in column
3 of Table 2.4. To do this bit of preferential analysis, you must take into account
not only how well each consequence does with respect to each of the evaluative
objectives but how important those objectives are to you. You are using your
mind's intuition to do this synthesis.

Now imagine that for each alternative not only have you recorded (as shown
schematically in Table 2.4) numerical probabilities in column 2 but in column
3 you have listed your ordinal rankings of all the consequences. With this infor-

mation in front of you, it may be clear which alternative you prefer. Remember that you also have to keep in mind your probabilistic judgments about outcomes and your attitudes toward risk. Not an easy task, but doable in some circumstances.

Expected Monetary Value

Let's back up once again to Table 2.3 and indicate how column 4 of Table 2.4 gets into the act. In many problems money is important. Perhaps the most important evaluative objective in Table 2.3 is to maximize a monetary return. Life would be simpler if that were the only objective of concern, but other objectives, like anxiety or guilt or time pressures, may be part of your overall agenda. One way of incorporating those ancillary but important issues is to impute a monetary value to them—incremental to your present assets—and assess an adjusted (incremental) monetary equivalent to each outcome (that is, to each row of the table). An example might help: one possible outcome of Robert's going to court is that he will win his case and get a jury award of $1,000,000. But how about his anxiety along the way and the disclosure of some embarrassing facts about himself? How much money would Robert be willing to forgo to have those ancillary concerns eliminated? Let's say $100,000. So Robert's equivalent monetary return, which incorporates anxiety and disclosure, is adjusted to $900,000. We'll further develop and use this technique in Chapter 8.

A risk profile with probabilities in column 2 and (equivalent) monetary values in column 4 can be thought of as a *lottery with money payoffs*. A choice between risk profiles X and Y (both described with money values) is tantamount to a choice between two lotteries, and we will adopt this terminology.

Choice between lotteries with monetary payoffs. Let's reconsider a risk profile for alternative X with probabilities in column 2 and incremental monetary payoffs in column 4—with columns 3, 5, and 6 deleted. Suppose alternative Y is similarly described with different numbers and you have to make a choice between them. The standard choice criterion is to compare the expected monetary values (EMVs) of the two lotteries. The EMV is just a weighted average of the monetary payoffs in column 4 using the probabilistic weights in column 2. The EMV is the average return you would receive if the lottery were conducted not once but independently repeatedly an infinite number of times. The EMV is also referred to as the *mean* of the distribution of monetary payoffs. A numerical example is provided in Table 2.5.

Why compare EMVs? The EMV is responsive to all the data, it is easy to compute, it combines probabilities and monetary values in an intuitively appealing way. Weighted averages (that's what EMVs are) are used extensively, and if we did not suggest the use of EMVs probably you might suggest their use without any prompting from us. Furthermore, we are shortly going to deepen our scaling of consequences to utility values and in that case we'll be able to justify the use of *expected values* as a result (or theorem, if you want to be a bit fancier) of the theory of probability.

There are many contexts in which the EMV criterion of choice is appropriate. If the incremental monetary amounts of the lottery are small in comparison with the decision maker's asset position, then the EMV is a defensible criterion of choice. A millionaire could justifiably use EMVs for lotteries with incremental payoffs in the hundreds of dollars. But there are limitations to the use of EMV in choosing between lotteries. For example, consider the lottery with just two equally likely outcomes: $0 and $1 million. Its EMV is: (.5 × $0 + .5 × $1 million) or $500,000. But a decision maker might much prefer $400,000 outright to that lottery. Or the decision maker might prefer the status quo to taking a fifty-fifty gamble in which he or she could lose $1 million or gain $3 million.

Limitations of EMVs.   EMVs implicitly assume that incremental monetary jumps have uniform intrinsic value. But that isn't always true. Indeed, the difference in happiness between making $10,000 a year and $20,000 may be greater than the difference between making $100,000 and $200,000. Returning to our mystery writer, perhaps in Audrey's case getting $600,000 will allow her to retire, buy a nice house in New Hampshire, and live a comfortable life. If she gets an extra million dollars beyond that, she might take more lavish vacations, but essentially her life will be unchanged. On the other hand, if she were to receive only $120,000 for her book, she would have to get another job or, even worse, write another book—disaster! In terms of her expected happiness, Audrey would actually be better off taking the security of the standard publisher's offer than going with no advance and a higher royalty rate—despite the lower EMV. EMV also abstracts away the decision maker's attitudes toward risk. A husband and wife, having identical wealth positions, may react quite differently to some gamble.

The conclusion: EMV is not a universally appropriate indicator to maximize. It is especially suspect when the range of possibilities is very broad (high upside and low downside); when losses as well as gains are possible; and when the risk profile includes very small chances at big gains.

Expected Desirability Value

One trouble with monetary values is that it does not record your strengths of preference for different amounts of money. For example, you might prefer consequence A to B to C but it may be of concern in your synthesis that A is much better than B which is only slightly better than C. This will be captured when we record your strengths of preference.

Intensity of feeling.   We now consider column 5, labeled *desirability values*, in Table 2.4, which records not only your ordinal rankings but the intensities of your feelings. If the consequence associated with outcome A is somehow given a desirability value of 65, say—don't worry for the time being how such scores are assessed—and B a desirability value of 40 and C a desirability value of 38, then this informs us that for you, there is a much bigger gap between A and B than between B and C. If somehow you can fill in meaningful numbers in column 5 of Table 2.4 for each alternative risk profile, then it may now be clear whether you prefer X to Y. We remind you once again that this synthetic judgment must somehow take into account your probabilistic assessments and your attitude toward risk (still to come). It will interrupt our development too much if we dwell now on how such desirability values can be derived, but roughly speaking there are three ways:

1. It can be done holistically without much ado, which is not very different from what a teacher does in assigning grades to exam questions and then totaling these.
2. Or you can go back to Table 2.3. By assigning conditional values within each cell of the table, you effectively score how conditionally desirable each consequence is in terms of that evaluative objective; then you weight the objectives and take a weighted average. We'll have a lot more to say about this so-called additive value scheme later on.
3. You can also assess for each consequence (that is, each row of Table 2.3 or 2.4) an adjusted monetary equivalent and then convert the monetary scale into a desirability scale that reflects the decision maker's strengths of preference for money.

We illustrate the third option here and the others in later discussions. Let's call upon Audrey once again. She now has the choice between (a) the safe option of going with a standard publisher with the security of an advance, or (b) going with a new, aggressive publisher that is willing to give her higher royalty rates with no down-side protection of an advance, or (c) joining some friends in launching an Internet publishing company and using her book as an initial enticement (see Table 2.5).

Audrey—perhaps with guidance from an analytical consultant—makes a first-cut analysis of her problem by examining, for each alternative, its associated risk profile and assigning numerical probabilities to likelihoods and adjusted monetary values to the consequences. She then examines the EMVs of the three alternatives and determines that, from a financial perspective, the Internet option is best, the higher royalty/no advance option is next, and the standard is third. But she realizes that she is leaving out of the analysis her intensity of feelings for incremental monetary amounts and her attitudes toward risk. "After all," she muses, "that first $100,000 will be much, much sweeter than the twentieth

Table 2.5   Audrey's choices

	Standard publisher			Aggressive publisher			Internet publisher	
Prob.	Monetary value ($000s)	Desirability value	Prob.	Monetary value ($000s)	Desirability value	Prob.	Monetary value ($000s)	Desirability value
0.2	300	35	0.2	100	15	0.5	0	0
0.6	500	50	0.5	500	50	0.2	500	50
0.2	700	62.5	0.3	1,000	75	0.3	2,000	100
1.0			1.0			1.0		
EMV	500			570			700	
EDV		49.5			50.5			40

Note: EDV = expected desirability value; EMV = expected monetary value.

such increment." So again with the possible help of an analytical consultant, she converts dollars into desirability units using Table 2.6. The range of monetary values she encounters in her three risk profiles goes from $0 to $2,000,000, and she norms her desirability scale by assigning a desirability of 0 to $0 and a desirability of 100 to $2,000,000. She feels that going from $0 to $500,000 is equally desirable to her as going from $500,000 to $2,000,000 and accordingly she scales the monetary consequence of $500,000 at a desirability level of 50. Stated somewhat differently: her judgmental mid-desirability point between $0 and $2,000,000 is $500,000. She also determines that her judgmental mid-desirability point between $0 and $500,000 is $200,000 (and therefore she scales the desirability of $200,000 at 25). In a similar manner her mid-desirability point between $500,000 and $2,000,000 is $1,000,000 (and therefore she scales $1,000,000 at desirability level of 75). The plot of desirability level as a function of incremental monetary gain is shown in Figure 2.1.

Now let's compare three risk profiles with numerical probabilities and desirability values shown in Table 2.5. Just by perusing the data, it may now be clear which is the subjective winner. But suppose not. Suppose there is a need for another step of analysis. What comes next on the agenda of analytical interventions is to examine the expected desirability value (EDV) of each profile and to choose the larger.

The EDV for the standard publisher, for example is:

$$.2 \times 35 + .6 \times 50 + .2 \times 62.5 = 49.5.$$

For the three alternatives, the EDVs are respectively 49.5, 50.5, and 40.0. So in terms of EDVs, it seems that going with the aggressive publisher is best for Audrey and going with the Internet is worst. This captures the decision maker's subjective strengths of preference for the consequences but it still leaves out her attitude toward risk. That's next on our agenda.

Table 2.6 Audrey's desirability for money

Money ($000s)	Desirability
0	0
200	25
500	50
1,000	75
2,000	100

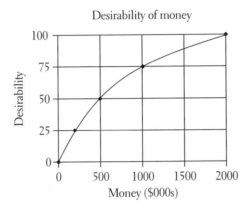

Figure 2.1. Desirability of money.

The limitation of EDVs. Why compare EDVs? They have the same redeeming features as EMVs. A husband and wife with identical financial assets may agree on the conversion of money to a desirability scale but differ in their responses to the following question: would you prefer receiving $500,000 outright or taking a fifty-fifty gamble with payoffs $0 and $2,000,000? Even though $500,000 was the mid-desirability point between $0 and $2,000,000 for both, it may not be the gambling midpoint. She might say that despite her recorded desirability levels, she would much prefer the sure thing of $500,000 to the two-valued lottery. He might think otherwise. EDVs still abstract away a key ingredient in an analytical resolution of risky choices. We now move on to discuss this last ingredient under the title of utility theory.

Expected Utility Values

Risk profiles with BRLT consequences. The technique we are about to introduce may seem very strange. Be patient. The punch line is worth it. We have seen that evaluating possible outcomes in terms of expected monetary values or even expected desirability values leaves us unable to compare outcomes with different payoffs and different probabilities. We need a new scale that will justify using expected values for risky choices.

We shall show how we can assign a utility value to each outcome of a risk profile and justify the use of the expected utility value (EUV) as a suitable summary measure for choice among risk profiles. Imagine a risk profile whose consequences are special lottery tickets, called Basic Reference Lottery Tickets (BRLTs, pronounced "brilts"). On the front of the ticket is a number between 0.0 and 1.0. On the back of the ticket is this message:

> This ticket entitles the bearer to a prescribed probability at the prize W and a complementary[3] probability at L. The probability at W is shown on the reverse side of this ticket. W = $2,000,000 and L = $0.

If you are in possession of a .35 BRLT you would have a .35 chance at W and a complementary chance $(1 - .35)$ or .65 at L. To cash it in, the executor puts 35 balls marked W and 65 marked L into an urn and a random drawing is made: a drawing of W nets you $2 million; and a drawing of L nets you nothing ($0). For the length of this discussion we will keep L and W fixed. We will think about a lot of BRLT's with different chances of winning, but always with the same prizes.

Monotonicity, continuity, and substitutability.   Let's be a bit concrete in an abstract sort of way. Imagine that you, the decision maker, confront a risk profile of three outcomes whose uncertainties have already been processed in terms of probabilities. The consequence associated with A is described by aaaaaaa in the third column of Table 2.7. Similarly with outcomes B and C. You have to choose between this risk profile and another one not shown. How might you proceed? We'll start you off. Let's choose two reference consequences, W and L, such that you prefer W to L; and assume that L and W are chosen so that any consequence in the profiles you have to compare falls between L and W in your preference rating. Now introduce the family of Basic Reference Lottery Tickets with W and L as reference outcomes. A BRLT that features on one side the number .38, say, will be called a .38-BRLT, and in this make-believe world of ours, this .38-BRLT can be resolved by a random device that yields W with (objective) probability of .38 and complementary probability of .62 for L.

Now for *monotonicity.* Consider the family of *u*-lotteries, where *u* is some number between 0 and 1. The 0-BRLT is sure to produce L and the 1-BRLT is sure to produce W. As *u* increases from 0 toward 1, the *u*-BRLT becomes monotonically more desirable. We rule out the guy who prefers a .7-BRLT to a

---

3. "Complementary" in the sense that $P(L) = 1 - P(W)$.

Table 2.7 Illustrative risk profile reducible to a lottery with BRLT prizes

Outcome	Probability	Consequences	BRLT equivalent
			$(W = \ldots, L = \ldots)$
A	.5	aaaaaaa	.4
B	.3	bbbbbbb	.6
C	.2	ccccccc	.9

.8-BRLT because 7 is his lucky number. Thank heavens you are not so mystically inclined.

Turning toward *continuity*, let's look at consequence B described by bbbbbbb. B is preferred over *u*-BRLTS with low *u*-values and is less preferred to *u*-BRLTs with high values. As *u* increases from 0 to 1, "dispreference" for the BRLT is turned into preference by passing through indifference. That's what we mean by continuity. Not only in our imaginary world, but in your real world, we assume there is some *u*-number—suppose it's .60—so that you are indifferent between receiving consequence B outright or getting a .60-BRLT. It would not help one bit to say that you are indifferent between B and any *u*-BRLT from .55 to .65, because by monotonicity you already have asserted that you strictly prefer higher *u* values. You might legitimately assert that you are not sure about your break-even *u*-value, but you are confident it is somewhere in the interval between .55 and .65.

Substitutability says that if you are indifferent between consequence B and a .60-BRLT then the desirability of the profile will not be altered by substituting the .60-BRLT for consequence B. And if you are indifferent between A (described by aaaaaaa) and a .4-BRLT, as well as C and a .9-BRLT, then we can substitute these BRLTs for the consequences without changing the desirability of the profile.

So imagine you are considering the lottery in Table 2.7 that yields a .5 chance at a .4-BRLT, a .3 chance at a .6-BRLT, and a .2 chance at a .9-BRLT. By the laws of probability this reduced lottery, if executed, will yield W with probability

$$(.5 \times .4) + (.3 \times .6) + (.2 \times .9) = .56,$$

so that we are entitled to say the lottery is equivalent to a .56-BRLT. Notice that the .56 value is the expected BRLT value, that is, an average of the BRLT values weighted by the probabilities of getting those BRLTs. There is nothing ad hoc in using expected values of BRLT payoffs. Indeed the beauty of the analysis is that the BRLT commodity was chosen to justify the use of expected values. This was definitely not the case for monetary or desirability units.

Let's put some of these ideas to use in the simplest of monetary lotteries. We'll consider the lottery that has two equally likely monetary payoffs, a and b, which we write as $<a, b>$. Suppose we ask our author Audrey how she feels about the lottery $<\$0, \$2$ million$>$. Earlier Audrey indicated that going from $0 to $.5m was equally desirable as going from $.5m to $2m. Suppose on further probing, and deep soul-searching, Audrey would rather get $.5m outright than take the lottery $<\$0, \$2m>$. She feels that her break-even point for the lottery is $.4m. She is quite risk averse. We can also say that Audrey's subjective mid-risk point for the interval $0 to $ 2m is $.4m. Remember the mid-desirability point for the interval $0 to $2 million was $500k. Now on to the task of constructing Audrey's utility curve for money (see Figure 2.2). We successively ascertain that

- her mid-risk point for the interval $0 to $2,000k is $400k;
- her mid-risk point for the interval $0 to $400k is $150k;
- her mid-risk point for the interval $ 400k to $2,000k is $800k;

and from these subjective assessments we complete Table 2.8.

Now return to Audrey's choice among three alternatives (calculations shown in Table 2.9). With risk aversion factored in, over and above desirability concerns, Audrey is better advised to keep with the standard publisher (an expected utility of .582) rather than switching to the aggressive publisher (expected utility .547). A close call, but clearly the Internet publisher with expected utility of .38 is far worse than the other two.

Change of scales.   There is something arbitrary about the utility analysis insofar as L and W were arbitrarily chosen. What would happen if we chose as refer-

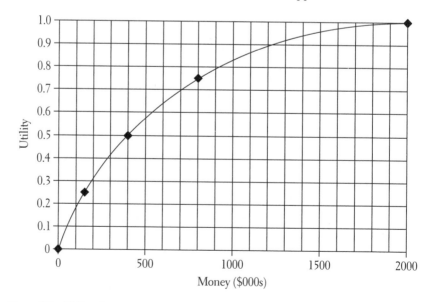

Figure 2.2.   Utility of money.

Table 2.8   Audrey's utility for money

Money ($000s)	Utility
0	0.00
150	0.25
400	0.50
800	0.75
2,000	1.00

Table 2.9   Audrey's choices using utilities

Standard publisher			Aggressive publisher			Internet publisher		
Prob.	Monetary value ($000s)	Utility value	Prob.	Monetary value ($000s)	Utility value	Prob.	Monetary value ($000s)	Utility value
0.2	300	0.4	0.2	100	0.18	0.5	0	0
0.6	500	0.6	0.5	500	0.6	0.3	500	0.6
0.2	700	0.71	0.3	1,000	0.8	0.2	2,000	1
1.0			1.0			1.0		
EMV	500			570			550	
EUV		0.582			0.547			0.38

Note: EMV = expected monetary value; EUV = expected utility value.

ence prizes W′ and L′, where L′ < L < W < W′? The u-BRLT equivalents would change and the expected BRLT values of the lotteries to be compared would change, but not their comparisons. If lottery X is preferred to lottery Y when done in {L, W} accounting, the same will prevail when done in {L′, W′} accounting. Going from {L, W} accounting to {L′, W′} accounting is like changing from a centigrade to a Fahrenheit scale. If you change the prizes from {L, W} to {L′, W′}, then all the u-BRLT equivalences will undergo a positive linear transformation, going from u to $a + bu$ for some constant a and positive constant b. This is akin to the transformation from Fahrenheit to centigrade using the formula

$$C = \frac{5}{9}(F - 32) = a + bF,$$

where $a = -160/9$ and $b = 5/9$.

Notice that in Table 2.8 if the entries are transformed by multiplying each entry by a positive constant b and adding a constant a, then the expected value would be similarly transformed and the comparisons with alternate lotteries undergoing the identical transformations would remain fixed. It is in this sense that we say that the utility numbers are meaningful up to a positive linear transformation.

In the literature the numbers in column 6 of Table 2.4 are referred to as "utilities" or "von Neumann utilities," in honor of the great mathematician John von Neumann who introduced utilities in his seminal work with Oskar Morgenstern on the theory of games. Actually, the great English philosopher, mathematician, and logician Frank P. Ramsey, who died at the tender age of twenty-eight, got the story straight a couple of decades before von Neumann. These utility numbers capture not only one's strengths of preference but one's attitudes toward risk. What we earlier called "desirability values" are sometimes called "Marshallian utility values." In this book utility assessments will be of the von Neumann kind that reflect risk attitudes.

Core Concepts

This chapter has offered a set of prescriptions to the unitary decision maker. Individuals negotiate, and often these individuals have critical decisions to make, such as: Should I negotiate? With whom? Should the person I am negotiating with and I terminate negotiations? Should we bring in a mediator? And so on.

Our prescriptions fall into two classes: decisions under certainty (or where uncertainties can be conveniently side-stepped) and decisions under uncertainty. For the certainty class we make use of the PrOACT way of thought, which argues the merits of being *proactive*. We use the term PrOACT as a mnemonic to recall the five basic ingredients of good decision making: specifying the *problem* or opportunity; examining *objectives* (interests); formulating creative *alternatives*; conditionally evaluating *consequences* on some evaluative objectives; and grappling with *tradeoffs*.

In decisions under uncertainties, an alternative X has an associated risk profile that specifies the possible uncertain outcomes and their likelihoods and consequences. Analysis can often proceed by separating uncertainty analysis and value analysis. On the value side of the ledger we contrast the use of expected monetary value (EMV), expected desirability value (EDV), and expected utility value (EUV) to choose among competing risk profiles. EMV abstracts away both strengths of preferences (for incremental monetary amounts) and attitudes toward risk. EDV addresses the former limitation but ignores risks; finally EUV rises to the challenge by taking into account both strengths and risks.

3

Behavioral Decision Theory

How do real people make decisions? Do they structure the problem, define their objectives, create alternatives, specify consequences, and quantify their trade-offs? Do people follow the prescriptive model set out in the previous chapter? Not too amazingly, they do not. Most people, most of the time, don't stick to the sage advice of prescriptive decision theorists. Why not? People simply may not know about decision theory, may not have the time to model their decision problems, or may not have the skills or inclination to do so. For a plethora of reasons, people rely on their behavioral "intuition" to make decisions rather than following rational prescriptions.

We need to take these behavioral predilections into account when crafting advice for negotiators. To do so we compile a list of the deviations from rationality that frequently occur in negotiators' decision making. There are two reasons for studying these deviations: (1) they may help you to change your behavior in negotiations; (2) they can shed light on how you can exploit others' mistakes; and (3) they may help you understand behavior of your own as well as of others.

At this point in the book we limit ourselves to a discussion of behavioral errors, biases, anomalies, and heuristics affecting individual decisions. We deliberately exclude departures from rationality born of dyadic interactions or social situations. This initial discussion is followed up in later chapters with a list of behavioral anomalies relevant specifically to negotiation. We often give prescriptive advice to one party under the assumption that the other negotiators are behaving naturally and somewhat irrationally without the benefit of this advice.

Decision Traps

Although the words describing these nonrational behaviors tend to be pejorative, the reality of human decision making is not at all bad news. Most of the time, most people do pretty well without any coaching. After all, the human mind is a wonderful synthesizer of information and we're all experienced decision makers. We've all learned how to correct egregious errors. We're genetically programmed to behave in ways that maintain our species. We also develop rules of thumb, or heuristics, that guide our actions without a lot of conscious thought. Much of our decision-making behavior is driven by a desire to economize on how much time we actively spend making choices. Yet despite the fact that we can get along relatively well without any advice, there is still significant room for improvement. For example, because we're a trifle lazy, we often take a heuristic learned appropriately in one context, and apply it inappropriately in another. It makes life easier to do this, but we don't obtain the best result that way.

In reading through the list of behaviors in Box 3.1, when we say "We do this or that bad thing," the *we* refers to *most people, some of the time.* Some items on this list will be important in discussions about negotiations; many are intriguing in their own right.

Most of the content of this chapter reports on the work of behavioral psychologists, some of whom might take exception to our assertion that some behaviors are errors or biases or anomalies. They would argue that it is not the behavior that is wrong and needs modification, but rather it's the prescriptive theories that need to be modified. We'll underline these points of disagreement as we go along.

The Anchoring Trap: First Impressions

A seller knows the buyer is vague about what price he is willing to pay. She suggests something in the high range, trying to anchor his thinking in the high registers. She'll work down from there, but he'll still think he's getting a bargain when they finally agree.

Here's a simple example. *Is the population of Turkey above or below 30 million?* The subject doesn't know but on reflection says, "I'm not sure but I guess above." The questioner continues, "Well what's your best guess?" And the hapless subject replies, "I dunno, somewhere around 40 million." Now if we can replay this scenario with an initial figure of 80 million substituted for 30 million. This shifts the *anchor* for the respondent from 30 million to 80 million, and the best guess may well come out being something like 65 million. A big difference.

The Status Quo Trap: Sticking with the Past

John has a portfolio of stocks that he knows is not balanced. But he has it. He can change it without paying financial commissions but . . . not today. Maybe

---

Box 3.1   Decision Traps

We hate deciding:
  We don't want to give anything up.
  We don't want to make a mistake.
  We procrastinate.
  We look for quick and partial solutions (we satisfice).

We have no structure for thinking about deciding.
We are reactive, not proactive:
  We wait for external triggers.
  We don't internally generate opportunities.

We often solve the wrong problem: we are unduly influenced by how questions are
  posed or framed.
We don't generate creative alternatives: we allow ourselves to be anchored by the
  first possibility we think of.
We don't clarify our objectives, fears, or interests:
  We focus on one objective and fail to consider tradeoffs among objectives.

We include only what we can handle formally:
  We let the hard drive out the soft.
  Quantitative data is privileged over qualitative.

We see, hear, and recall what we want to: we selectively seek confirmatory evi-
  dence.
We erroneously apply heuristics developed in one context for other contexts.
We get anchored by first impressions.
We pursue sunk costs.
We are erratic about discounting the future: we're more worried about the near fu-
  ture than the far future and it matters little when something will occur at a far
  horizon.

---

later. Starting from a clean slate, John would certainly not choose his current
holdings. Does he stick with them out of laziness? Yes, but there is also a reluctance to change.

When I (HR) was a young faculty member at Columbia, a colleague of
mine had an offer to go to the University of Chicago. He was favorably inclined,
but before he told the dean, another offer came in from the University of California, Berkeley. I know he preferred Chicago to Columbia, and Berkeley to Columbia, but he couldn't make up his mind between Chicago and Berkeley.
Guess what. You're right, he stayed at Columbia.

The Sunk-Cost Trap: Trying to Recoup Losses

Ann has made, in retrospect, a bad choice in buying stock XYZ. It's down. Should she take a loss and admit her error to herself or should she stick with it until it recoups?

Banker B has lost a bundle on investment Q. Intolerable. B decides to double up his investment. Not because he would do this de novo if he were not already hooked, but because he will not have to take responsibility now for his earlier mistake. And his luck might turn. Incidentally, if banker B will lose his job if he acknowledges his losses with investment Q, it may be rational, from his perspective (not his company's), to play the double-up game.

The Confirming Evidence Trap: Seeing What You Want to See

In a highly controversial arena, we are already partially committed to a position. Inundated with new information, we selectively choose to pay attention to those sources that tend to confirm our position. We might even be present when the other side speaks, but we choose not to hear them. Don't let the facts get in the way of what we hold dear! We act as if our minds are made up when we have no real basis for doing so.

The Framing Trap: Solving the Wrong Problem

It's old hat that the way you pose a question influences the responses you get. It makes a big difference if you phrase the question in terms of lives saved or deaths caused.

Consequence A can be framed as a "gain" when the reference position is in terms of net assets; it also can be framed as a "loss" when the reference is shifted to an incremental change from the status quo. The different frames can significantly affect the psyche of the responder.

The way you pose the question may make certain evaluative objectives more salient and thereby influence the response. "Can I pray while smoking?" is different than "Can I smoke while praying?"

Prescriptive Advice, Given Behavioral Errors

If this is the way people behave, then what do we prescribers suggest to remedy the situation?

1. Be knowledgeable about biases and traps.
2. Be aware when you make decisions.
3. Rethink your first response.

It's easier said than done. Most errors or biases we make are made without any systematic thought. We do what we always have done or what comes naturally.

Here we preach doing what's not natural, to be far more conscious of the decisions we make and to reflect on whether they make any sense.

Uncertainty Anomalies

We now turn our attention to judgment in decision making under conditions of uncertainty—when we have only a rough idea of the likelihood of different sets of consequences resulting from our choice (see Box 3.2). We continue to examine errors, biases, and anomalies.

On Thinking Probabilistically

Most people simply don't bother to think probabilistically. It confuses them to ponder uncertainties and the probability of different outcomes resulting from their decision. Probabilities are not operational for them. More likely than not they will use ambiguous terms, such as fair possibility, pretty good chance, not impossible, or most likely. Using this vocabulary carries some danger, because further noise is being introduced into an already noisy system. You don't have to go all the way to communicate less ambiguously. You might say:

- This event is more likely than that event.
- This event is more likely than not—the probability of this event occurring is greater than .5.
- This event is more likely than drawing a red ball from an urn containing three red balls and one black.

With a little practice one can learn to be more precise about the state of one's imprecision.

Box 3.2   Decision Traps Involving Uncertainty

We ignore uncertainties:

> We narrow our concern to a single outcome (most likely, or the best, or the worst).
>
> We don't think probabilistically.
>
> We take refuge in ambiguity and use imprecise language (for example, "this event is not very likely").
>
> We use the complexity of uncertainty as an excuse for not thinking or not deciding.

We think mystically:

> Some people or numbers are lucky.
>
> The stars are lined up right.
>
> Our team lost because I didn't wear my baseball cap when watching the game on TV.
>
> Of course I pay extra to use my favorite numbers when playing the lottery.

We think we can outguess random outcomes (like the throw of dice):

> Gambler's fallacy no. 1: Five reds have appeared in a row and it's time for a change.
>
> Gambler's fallacy no. 2: Let's keep playing with the hot hand; let's not change, the dice are running hot.
>
> We are unduly surprised by coincidences (like someone winning a lottery twice in a lifetime—this ain't chance!).

We ignore the dependencies among uncertainties.

We mix up conditionalities: the probability of A given B, and B given A.

We assign a higher probability to the joint event (A and B) than to A alone.

We ignore base rates in assessing probabilities based on evidence.

We undervalue sample evidence.

We believe the size of a random sample should depend on the size of the population, believing you can't tell anything from a random sample of 1,000 taken from a population of 250 million.

We sometimes ignore small probabilities, and other times we magnify the importance of small probabilities (p is not equal to 0; therefore p is possible; therefore the event is possible; therefore we should treat the event as probable).

We behave as if all small probabilities are the same.

We calibrate poorly in assessing probabilities. We are surprised surprisingly often.

We want to slant our probability assessments to account for vagueness.

We distort our assessments of probabilities to account for the severity of the consequences:

> Prudence above honesty.

We are not sure how to think about how risk averse we should be.

We worry about the regret we might feel if, after the fact, we think we acted too greedily. We worry about the disappointment we might feel if, after the fact, we think we acted too cautiously.

Conditional Ambiguities

The conditional probability of A given B—written as $P(A|B)$—is often confused with the conditional probability of B given A—written as $P(B|A)$.

Here's an example. A medical specialist is queried about the conditional probability of a newborn's having a certain eye defect (ED) if the mother had German measles (GM) in her first trimester of pregnancy. The doctor thinks of all the babies he or she has seen with this eye defect and observes that most of the mothers had had German measles in a critical period. So the answer is "very, very high," and if you push some doctors, they might say "a probability higher than .9." Of course the specialist is assessing the probability of German measles given the eye defect—$P(GM|ED)$—rather than the probability of the eye defect given German measles—$P(ED|GM)$.

The Monty Hall conundra. Consider:

The Monty Hall Conundrum #1

Monty Hall, a TV host, tells a successful contestant that she has won the rights to possibly winning a car. The car has been randomly placed behind one of three doors, A, B, or C. Monty asks the contestant to choose the door that she wants to open—hoping the car is behind it. Let's say she picks door A. Before she opens the door, Monty intervenes and says, "One of the other doors certainly has no car behind it. I'm now going to open one of those other doors with no car." He then opens door B, showing there is no car there. "Do you want to remain with your initial choice of door A, or would you rather switch to C?" asks the tricky showman.

Would you switch?

The argument in the contestant's mind runs like this: all the doors are equally likely to be concealing the car, so my choice of A has a $\frac{1}{3}$ probability of winning. After Monty showed no car behind door B, then doors A and C are equally likely; so now door A has a $\frac{1}{2}$ chance of being lucky. But C also has $\frac{1}{2}$ chance. So no, there is no motivation for me to switch my choice. And whether I switch or not, my chances of winning have now improved from $\frac{1}{3}$ to $\frac{1}{2}$.

There's something fundamentally wrong with this line of reasoning. Even if Monty didn't provide the evidence that the car was not behind one door, the contestant could have imagined this to be the case. So simply thinking about this altered state of affairs would raise the probability of winning the car from $\frac{1}{3}$ to $\frac{1}{2}$. This simply cannot be the case! There is something drastically wrong.

Here is the proper way to think about the problem. The evidence of the empty door B has no inferential significance whatsoever concerning the chance that door A will be lucky. The success of door A remains $\frac{1}{3}$ before and after Monty's new information is given. Hence the conditional probability of "not-A" (or equivalently success with the composite event "B or C") remains ⅔. But we know that the answer isn't B, so the conditional probability of success with door C is ⅔. We conclude that the contestant should indeed switch.

Are you convinced? Try the following variation told to me by my son, Mark.

The Monty Hall Conundrum #2

There are three potentially lucky contestants who line up in front of doors A, B, and C. Monty tells them that the prize of a car awaits one of them. The randomized choice is performed secretly. To maintain the suspense, Monty tells his contestants that he will open at random one of the two doors not containing the car. He opens door B. He now asks the contestant in front of door A if he wants to swap places with the contestant in front of door C.

What would you do?

Contestant A thinks like this: this variation is the same as the original problem. I'm now convinced I should switch to improve my odds of winning.

But this time the original answer is *not* the right answer. The contestants in front of doors A and C are in symmetrical positions. If it makes sense for one of them to switch, it makes sense for both of them to switch. This is ridiculous. There is a subtle difference between the two versions. Contestant A should think through the problem like this: when Monty selected B, he could have selected A if indeed A was empty, but he didn't select A, and therefore there is some inferential evidence relevant to the information Monty revealed. My chances of being lucky with A should now rise. Since C is in exactly the same position, my chance of winning is the same as his, that is, ½.

The Overconfidence Trap

Most nonspecialists, when they use probabilities, don't calibrate very well. In early experiments that I ran, subjects were asked to consider a set of almanac-type questions, such as How many doctors and physicians are registered in the Yellow Pages of the Boston phone book? For each question, subjects were asked to supply a numerical interval large enough so that 95 percent of the true answers would fall in that interval. When, for any particular question, the subject supplies a lower number L, and an upper number U, the subject is saying that for him or her the judgmental probability that the true answer is between L and U is .95. The subject should be choosing L's and U's to be surprised about 5 percent of the time. But a startling thing happens.

Most subjects—around 80 percent—use uncertainty intervals that are much too tight. Averaging over all subjects and all questions, the number of surprises is not 5 percent but more like 40 percent! Indeed, for the number-of-physicians question, half the respondents were surprised—the true value fell outside their L to U range—and of these surprises half were surprised at the low end and half at the upper end. People are by and large too overconfident. With practice subjects improve, but slowly. With most physical judgments (heights, weights, and distances) subjects get frequent feedback. Not so with probability assessment, and so it should not be surprising that subjects calibrate poorly.

The Conjunction Fallacy

It's not common, but there are cases where people assess a higher probability for a joint event, A and B, than for the event A itself. In formal terms, they are saying $P(A \text{ and } B) > P(A)$. How come? Here's an example.

The probability that a severe economic depression and a nuclear war will occur may be deemed higher than the probability of a nuclear war (through any possible cause). It's easy to see how this might happen: the respondent might not be thinking about the possible triggering mechanism of a severe depression when thinking about nuclear war. In cases like these most thoughtful people will say that this is an error of oversight and duly change their responses. In many other cases of logical fallacies, we'll see a good deal of resistance to changing one's mind.

A bit of probability theory. Let A and B represent two states of the world. They are mutually distinct but not necessarily exhaustive. With the knowledge we have at hand, we assess their probabilities of states A and B as $P(A)$ and $P(B)$ respectively and say, "the *prior* odds ratio for A versus B is $P(A)$ divided by $P(B)$." We now observe new information (NI), and our prior probability judgments of A and B become posterior judgments $P(A|NI)$ and $P(B|NI)$; the *posterior* odds of A versus B is the ratio of these two quantities.

Another important definition: the "likelihood ratio" for the new information is the ratio of the [conditional probability of the new information given that state A is true] over the [conditional probability of the new information given that state B is true]. In formal terms, this is represented as $P(NI|A)$ divided by $P(NI|B)$. Probability theory tells us that:

$$[\text{posterior odds}] = [\text{prior odds}] \times [\text{likelihood ratio}].$$

This result is proved using Bayes's theorem going from prior to posterior probabilities.

The Base-Rate Fallacy

Consider this question: *Donald Jones is either a librarian (L) or a salesman (S). His personality can best be described as retiring. What are the odds he is a librarian?*

Most subjects conclude that the odds are greatly in favor of librarian (L) over salesman (S). But before we learned that Donald had a retiring (R) personality, the prior odds for librarian over salesman is about one in a hundred for the U.S. population. That's called the base rate. In answering this sort of question many people forget about the base rate and use the likelihood ratio as the posterior odds. The likelihood ratio is P(R|L) divided by P(R|S), and this ratio might be a value of ten or so. So the posterior odds should be about one to ten and not ten to one.

Forgetting about base rates can be particularly pernicious in medical trials. Patients can have HIV or not. The prior odds ratio of a subject drawn at random from the male population having HIV—that is, [P(HIV) divided by P(not HIV)]—is very small. A positive test for HIV may have a likelihood ratio of a hundred to one—that is, [P(+ test|HIV) divided by P(+test|not HIV)] = 100. But the posterior odds of HIV for a subject drawn at random and who scores positive on the test may still be small.

Underestimating the Value of Sample Evidence

Professor Ward Edwards, a psychologist at the University of Michigan, has investigated the intuitive reactions of many subjects to experimental, probabilistic evidence. In one of his experiments he poses the following problem.

Small Sample Experiment

I have two identical canvas book bags filled with poker chips. The first bag contains seventy green chips and thirty white chips, and I shall refer to this as the predominantly green bag. The second bag contains seventy white chips and thirty green chips, and I shall refer to this as the predominantly white bag. The chips are identical except for their color. I now mix up the two bags so that you don't know which is which, and put one of them aside. I shall be concerned with your judgments about whether the remaining bag is predominantly green or not. Now suppose that you choose twelve chips sequentially at random, replacing each withdrawn chip before selecting the next from this remaining bag, and it turns out that you draw eight green chips and four white chips, in some particular order. What do you think the odds are that the bag you have sampled from is predominantly green?

At a cocktail party a few years ago I asked a group of lawyers, who were discussing the interpretation of probabilistic evidence, how they would respond to Edwards's experiment. First of all, they wanted to know whether there was any purposeful malice in the actions of the experimenter. I assured them of his neutrality and told them that it would be appropriate to assign a .5 chance to "predominantly green" before any sampling took place.

"In that case," one lawyer exclaimed after thinking a while, "I would bet the unknown bag is predominantly white."

"No, you don't understand," one of his colleagues retorted, "you have drawn *eight* greens and *four* whites from this bag. Not the other way around."

"Yes, I understand, but in my experience at the bar, life is just plain perverse, and I would still bet on predominantly white! But I am not really a gambling man."

The other lawyers all agreed that this was not a very rational thing to do—that the evidence was in favor of the bag's being predominantly green.

"But by how much?" I persisted. After a while a consensus emerged: "the evidence is meager, the odds might rise from 50–50 to 55–45, but as lawyers we are trained to be skeptical, so we would slant our best judgments downward and act as if the odds were still roughly 50–50."

The answer to the question "By how much?" can be computed in a straightforward fashion, and there is no controversy about the answer. The analysis goes like this. Let us denote the predominantly green and white bags as GB and WB, respectively. We then have $P(GB) = .5$ and $P(WB) = .5$. Let A stand for the event "eight greens and four whites, in the particular order $\{g, g, w, g, w, g, g, w, w, g, g, g\}$." The order is unimportant and we include it just for the sake of concreteness. Then keeping in mind that there are seventy greens in GB and thirty greens in WB, we have:

$$P(A|GB) = .7 \times .7 \times .3 \times \ldots .7 = (.7)^8 (.3)^4$$

$$P(A|WB) = .3 \times .3 \times .7 \times \ldots .3 = (.7)^4 (.3)^8$$

And the likelihood ratio is $[P(A|GB)$ divided by $P(A|WB)] = (.7)^4/(.3)^4 = 29.6$.

Since the prior odds of predominantly green over predominantly white is $P(GB)/P(WB) = 1.0$, the posterior odds is just the likelihood ratio, giving $P(GB|A)/P(WB|A) = 29.6$ (to one) and from this we calculate the probability the bag is predominantly green to be $[29.6/(29.6 + 1)]$ or .967. That's right, .967! The bag is predominantly green, "beyond a reasonable doubt," whatever that may mean. This story points out the fact that most subjects vastly underestimate the predictive power of a small sample. The lawyers described above had an extreme reaction to the problem, but even my statistics students clustered their guesses around a .70 probability of a predominantly green bag.

Getting Mystical about Coincidences

Art is a skeptic about probabilities. "Seventeen years ago I was supposed to go to Paris and because of a storm I ended up in Copenhagen. I had to stay an extra day because of a horrendous travel mix-up and I got lost when I wanted to visit the Tivoli. I took a trolley car instead of a cab, because, stupid me, I left my wallet in my room. The trolley was full and when it lurched to a stop I fell smack into the lap of a lady who became my future wife; she in turn had never taken that trolley ride before. Now, that couldn't have happened by chance." From an a priori standpoint, the chance of a concatenation of those rare events that led Art onto the lap of his future wife is staggeringly low. But the number of potentially surprising outcomes that could occur is also astonishingly large. It's a little like observing a particular random sequence of fifty heads and tails and then observing that a really rare event (this particular sequence) has just occurred.

Some people are so surprised at the occurrence of some rare event (for example, Ruth wins two separate national lotteries) that they believe in divine intervention and shun any consideration of probabilities. There are legions of people who mystically think that they can influence the outcomes of physical forces by following some ritual, like wearing a sweater backward to bring on the rains. Why? It happened that way once before. Oh well.

Johnny once had his sweater on backward while watching television when slugger Magee hit a homer. Now if Johnny listens without his sweater on backward and slugger Magee fouls out, Johnny feels guilty. There is simply no physical causative connection between Johnny's living-room behavior and the realities in the stadium. But Johnny wears his sweater on backward and notices that it works better if he also has one shoe on. Oh well.

Choice under Uncertainty

The two most serious attacks against the theory of subjective expected utility (SEU), described at the end of the previous chapter, are the paradoxes of Nobelist Maurice Allais and Daniel Ellsberg. They show in some key problems that most subjects make choices that violate the theory. After these incompatibilities are pointed out to them, even sophisticated subjects often refuse to switch their behavior. Something must be wrong: either the people are making a mistake or subjective expected utility theory is not a good guide to making choices. We now describe these paradoxes and argue that it is the people who should change their initial choices.

The Allais Paradox

Problem 1. You have the choice between two alternatives, A1 and A2.

- A1 yields a certain payoff of $1 million
- A2 yields payoffs of $5 million, $ 1 million, and $0 with probabilities of .10, .89, and .01, respectively.

Problem 2. You have the choice between two alternatives, A3 and A4.

- A3 yields payoffs of $5 million and $0 million with probabilities of .10 and .90, respectively.
- A4 yields payoffs of $1 million and $0 with probabilities .11 and .89, respectively.

For both questions, how would you choose? (See Figure 3.1.)

The attack on expected utility value. In practice most subjects choose A1 over A2, arguing that A1 is a certainty, so why gamble? Most subjects also choose A3 over A4, arguing that in both cases one is forced to gamble and that $5 million is much more than $1 million; moreover, they believe the chances of .10 and .11 are practically the same.

Allais pointed out that the simultaneous choices of A1 and A3 violate the principles of expected utility theory. Furthermore, when this is pointed out to knowledgeable subjects, they still don't want to change their choices. So much the worse for the theory, these people say. We'll try to defend it.

First of all, let's show why the theory suggests that the choices of A1 and A3 are incompatible.

There are three monetary consequences involved in the two choice problems: $5m, $1m, and $0m. Now let's introduce BRLTs with winning prize W = $5m and losing prize L = $0m. Let $1m be equivalent to a z-BRLT, where z is some number to be determined by the decision maker. In utility vernacular: $5m is given utility of 1.0; $0 is given utility of 0.0; and $1m is assessed at utility z. The expected utility values of the four Allais alternatives are:

EUV of A1 z
EUV of A2 $.1 + .89z$

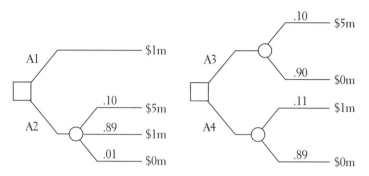

Figure 3.1.   Allais paradox decision tree.

EUV of A3   .1
EUV of A4   .11z

The comparison of A1 versus A2 depends on:

$$[z \text{ versus } .1 + .89z] \text{ or on } [.11z \text{ versus } .1] \text{ or on } [z \text{ versus } \tfrac{1}{11}].$$

The comparison of A3 versus A4 depends on:

$$[.1 \text{ versus } .11z] \text{ or on } [\tfrac{1}{11} \text{ versus } z].$$

So we see that if z is such that it favors A1 over A2 in Problem 1 (that is, if $z > \tfrac{1}{11}$), then to be consistent A4 must also be chosen in Problem 2.

But most subjects, argues Allais, choose A1 and A3 and don't want to change even after they are informed of the EUV results.

Defense on normative grounds.   Imagine an urn containing a large number of identically shaped white balls, and suppose each ball has a pair of payoffs such as ($5 million, $1 million). We'll say the left payoff is $5 million, and the right payoff is $1 million. The balls are identical except for their payoffs. Now imagine that you must first choose Left or Right, and then draw a ball at random from the urn. Your payoff depends on your choice of Left or Right and on the ball you pull out. Before you choose Left or Right you have the privilege of studying all the balls in the urn and recording anything you want and making any calculations you want before you decide. Let's say that in a specific case, after you study the pairs of numbers on the balls, you decide on Left.

Now we introduce a new ball, an orange-colored one of identical shape and texture to the white ones, but with *identical left and right payoffs,* and we put this orange ball in the urn. The desirability of saying Left or Right and drawing a ball from this newly constituted urn might change, but would your preferred choice of Left change? We think it should not: if you draw the orange ball, it doesn't make any difference whether you said Left or Right; if you draw a white ball, you're back to the old problem in which you preferred Left to Right.

This same argument would hold if we added more than one orange ball, each with identical left and right payoffs. And since your preferences of Left or Right should not be affected by the addition of identically marked balls, then your choice should not be affected by the deletion of balls with identical payoffs. Think about it for a moment. This is not simple stuff. In fact it's at the heart of entire schools of philosophical thought.

Adding or deleting balls with the same left and right payoffs should not change your preferred choice of Left or Right.

The Sure-Thing Principle. Now let's return to the Allais paradox. Put 100 balls in an urn, each with two payoffs:

Number of balls	Left payoff	Right payoff
10	$1m	$5m
89	$1m	$1m
1	$1m	$0m

Notice that the Left payoff is $1 million on all, and the Right payoffs are: $5 million on 10 balls, $1 million on 89 balls; and $0 on 1 ball. Referring back to Problem 1, Left is tantamount to choosing A1 and Right to choosing A2. Check that you agree with this. The Sure-Thing Principle says that if we have to choose Left or Right our choice should not depend on eliminating those 89 identically labeled balls. This leaves us with 11 balls:

Number of balls	Left payoff	Right payoff
10	$1m	$5m
1	$1m	$0m

That's the essence of the choice.

Let's take a different tack. Start with the stripped-down urn containing 11 balls shown above, and now suppose we add to this urn 89 identically marked balls each with Left and Right labels of $0. We then arrive at the choice:

Number of balls	Left payoff	Right payoff
10	$1m	$5m
89	$0m	$0m
1	$1m	$0m

Notice that Left depicts A4 and Right depicts A3. If you pick A1 in Problem 1, to be consistent with the Sure-Thing Principle, you should pick A4 in Problem 2.

"Not so fast," remarks the skeptic. "I find your normative arguments instructive but not completely compelling. I would feel just awful if I were to take A2 in Problem 1 and end up with $0. Not only would I suffer pangs of regret but my wife and mother would never forgive me for being so greedy."

To this exclamation we retort, "But if you took A3 in Problem 2 and you ended up with $0 wouldn't you have the same pangs of regret? You could instead have selected A4, which would have given you an increased chance of a positive return."

"Are you kidding? My kibbitzers would attribute my action not to greed but to bad luck. There's a world of difference between $0 consequence resulting from A2 (because of the presence of A1) and the $0 consequence of A3. One zero is not like the other zero."

Should we throw out the theory? Some certainly would. We would prefer to complicate the description of the payoffs in order to capture the cognitive psychological nuances. The $0 payoff of A2 becomes $0 with deep psychological stress from postdecision regret. The $0 payoff of A3 and A4 doesn't carry that psychological baggage. We believe that a good many anomalies or inconsistencies in expected utility theory come from inadequate descriptions of the consequences. The behavioralist, who is trying to understand and predict behavior, may not like adding such cognitive concerns into the descriptions of consequences, because you can then explain almost anything and you sacrifice the strong predictive validity of the theory. But as a prescriber, if we're giving advice to someone plagued with anticipated ex ante worries about possible ex post regret, then we would like to reflect that concern in our advice—even though when acting on our own, we prefer to purge our thoughts of such regret feelings.

The Ellsberg Paradox

Problem 1. An urn contains 50 red and 50 white poker chips. You must announce a color, red or white, and then draw a chip at random. If you match, you get $100, otherwise nothing. How much is this option worth to you?

Problem 2. An urn contains an unknown number of red and white chips. You don't have the foggiest idea of the number of chips or the relative numbers of red or white chips. You must announce a color—red or white—and then draw a chip at random. If you match, you get $100, otherwise nothing. How much is this option worth to you?

These two problems are the same except that in the second problem you are vague about the contents of the urn. Would you rather draw from the known or the unknown urn? How much would you pay to go from Problem 2 to Problem 1?

The attack on subjective probabilities.   In practice most subjects prefer drawing from the known urn. A modal value for the worth of Problem 1 is $40 and only $15 for Problem 2. In the first problem you have an objective, crisp .5 chance at $100. In the case of the second urn you have a vague, subjective, squishy .5 chance at $100. But if you feel, in the second problem, that drawing a red is just as likely as drawing a white, then the *subjective probability* of a match is .5—like the first problem. The counterargument is that in the first urn the problem is crisp, while the second urn offers a vague problem, so allowances need to be made. What do you think now?

A normative defense of subjective probabilities.   We would like to argue that the worth of participating in Problem 2 is at least as high as taking part in Problem 1.

In Problem 2 it makes no difference if you decide to grab a chip in your hand without looking at it and then announce your color. Do you agree? Now, with your other hand, toss a coin: heads you announce red, tails you announce white. What's the objective chance of your announcement matching the color of the chip? Well, it must be .5, mustn't it? So if you agree to toss a coin to choose your color, the value of Problem 2 should be the same as Problem 1. But you may have some information that would lead you not to toss the coin, and in that case the value of the second problem would be worth more. Are you convinced? Most subjects feel uncomfortable with this argument.

The Ellsberg paradox is important for the foundations of decision theory. Consider Figure 3.2, which depicts in decision-tree format the Ellsberg choices.

In Problem 1 the probability assessment for Red at chance nodes 2 and 3 is an objective .5. No problem. But for Problem 2 some might want to assign a probability of .4 to Red at node 2 and a probability of .6 to Red at node 3. Why? They want to register the feeling that no matter what they call, Red or White, in their mind there is less than an objective .5 chance of winning. The subjectivist, on the other hand, feels that at nodes 2 and 3 Red and White are subjectively equally likely and therefore each must be assigned a probability of .5 and .5. Some otherwise astute folk want to assign .4 for Red and .4 for White at nodes 2 and 3, thus violating the requirement that probabilities for mutually exclusive and collectively exhaustive events should sum to unity.

We often negotiate with vague uncertainties. It behooves us to ponder this Ellsberg paradox in order to get us to think straight about our own behavior and to make allowances for the behavior of others.

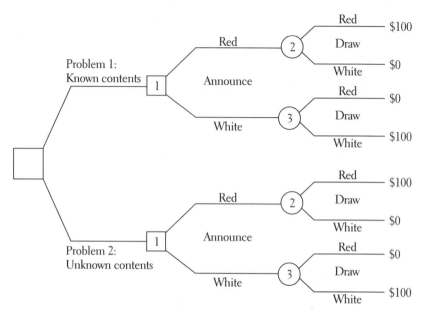

Figure 3.2.   Ellsberg paradox decision tree.

Descriptive theorists happily demonstrate that people are incoherent in their probabilistic assessments; normative analysts generally remain aloof and do not get involved in empirical measurements; and it is the prescriptive analysts who must learn how to make sense out of seemingly incoherent measurement responses. On the probabilistic side we tend to view incoherent responses as errors in perception that should be corrected. Matters are more complicated on the utility side of the ledger. There is much more reluctance to change.

Core Concepts

This chapter has compiled a list of deviations from rationality that frequently occur in negotiations. There are two reasons for studying these deviations: (1) they might help you to change your negotiating behavior; (2) they can shed light on how you can exploit the faulty behavior of others. We refer to these deviations as behavioral errors, biases, anomalies, or traps.

Although our labels for these nonrational behaviors tend to be negative, when humans make decisions the results are by no means all bad. Without the luxury of expert advice, the majority of people still manage. They use their minds, they learn from experience, and they correct themselves as they go along. Our genetic programming encourages behavior that helps our species continue to exist. We have also developed rules, or heuristics, to guide us so that we don't have to think about what we do so much. We may misapply these regulations, however, using the rule for one domain in another. This may seem to facilitate matters, but the results are problematic.

The chapter first examined anomalies when uncertainty is not the main issue. These include:

- Being unduly influenced by first impressions (the anchoring trap)
- Unduly sticking with the past (the status quo trap and the sunk-cost trap)
- Seeing what you want to see or hearing what you want to hear (the confirming evidence trap)
- Being unduly influenced by the way the question is posed (the framing trap)

To police your actions and try to behave differently, it is important to be knowledgeable about such biases and traps, to be aware of them when you make decisions, and to rethink your first response to avoid these traps.

When uncertainties are present, behavior becomes more anomalous, because feedback itself is so uncertain. It's hard to learn. Most individuals simply ignore uncertainties, and the precise language of probabilities does not enter into the everyday analysis of alternatives.

The chapter has also listed a set of common errors, anomalies, and traps:

- Confusions about conditional probabilities: mix-ups between $P(A|B)$ and $P(B|A)$.
- The Monty Hall problem: in considering $P(A|B)$, the conditioning event B is so subtle that it is ignored.
- Ignoring base rates.
- Underestimating the power of a small sample.

The very issue of what is rational is challenged by sophisticated critics. Allais shows that in some carefully crafted choices involving monetary gambles with known, crisp, objective probabilities, not only do people not follow the dictates of expected utility theory but they insist on behaving in this "irrational" way despite being told of their egregiously aberrant behavior. We defend the expected utility theory, and show that (1) some people do change their behavior when their illogical reasoning is made transparent; or (2) their behavior can be shown to be consistent with expected utility theory provided that the consequences are more fully described (a zero outcome with a lot of associated guilt or regret must be treated as different from other zeroes without these psychological encumbrances).

Another serious condemnation of subjective utility theory comes from Ellsberg, who shows that uncomfortable feelings about vagueness cannot be captured by the probability calculus. We try to show that such aberrant behavior should be modified.

The Allais and Ellsberg paradoxes are the most compelling attacks against subjective expected utility (SEU) theory, but we remain stalwart in our judgment that the theory can be adequately defended.

4

Game Theory

Game theory studies how rational actors ought to behave when their separate choices interact to produce payoffs to each player. We refer to game theory as interactive, separate decision making because by definition the players make their decisions independently of each other (there is no collusion), but these separate choices interact to determine payoffs for each side. We present a preliminary, basic discussion of game theory in this chapter and resume the theoretical development as we need it in the rest of the book.

Game Theory and Negotiation Analysis

The analytical approach taken in orthodox game theory is primarily symmetric normative: it specifies what each of the rational players should do given a set of tightly defined assumptions. This approach leads to some elegant mathematical solutions. These conceptualizations certainly have their merit, but our primary concern in this book is with prescriptive relevance rather than with normative elegance. Although game theory is beautiful in its own right, we will look into it for what it can tell us about negotiation. As we go along we will relax the strict assumptions of classical game theory and adopt some novel perspectives to permit more practical understandings of specific interactions.

In departing from orthodox game theory, we are not throwing the baby out with the bath water. Game-theoretic thinking offers powerful insights into the problem of negotiation. It focuses the mind wonderfully on the thoughts and likely actions of the other side. It helps us consider how the other negotiator might respond to our suggestions and what unintended effects our actions might provoke.

The Strategic Essence of Interactive Thinking

Let's start by defining the assumptions that game theory makes in structuring situations for analysis:

- You have to act. (Doing nothing is an act.)
- Your payoff depends both on what you do and on what other designated players do.
- You do not know what they *will* do—but you know what they *could* do.
- They do not know what you will do.

That's what game theory is all about.

To help you think about what your strategy should be as a player, we will present a number of games. From this presentation, you can draw out general principles that will help guide you in more realistic situations—in negotiations, in competitive marketing, in strategic industrial competition, in competitive pricing, or even in military strategy.

Why is game theory useful? We will use it as a subsidiary area of inquiry that can be applied to negotiation. Game theorists, of course, do not believe that their work is merely a part of negotiation analysis—and indeed, it is applicable to many other fields. Game-theoretic insights have made valuable contributions to economics, psychology, and even to evolutionary biology.

The Rules of the Game

Let's start simply. In the simplest case there are just two players, each having two alternatives. Nonetheless, all sorts of interesting lessons can be gleaned from this simple situation. The two-by-two matrix games are alike in that they all have:

- *Fixed strategies.* You have to choose one of two prespecified strategies. There is no innovation, no creation of alternatives.
- *Two alternatives.* You are concerned about the choice to be made by just one other player. Simplest case: he or she also has just two alternatives.
- *Perfect information.* For each choice of alternatives (one chosen independently by you and one by the other player) there will be a joint consequence. You and the other player have accurate knowledge of all possible consequences and of each other's preferences.
- *Common knowledge.* You know the other player's possible choices; he or she knows yours; you both know the other knows, and vice versa. The choice sets for each are common knowledge.

- *Simultaneous choices.* Each of you must choose simultaneously; or equivalently, the second chooser does not know the choice of the first chooser.
- *No cheap talk.* There is to be no preplay discussion, known as cheap talk, between the players.

So that's the setup of the simplest, nontrivial class of games. An amazing variety of games fall under this very restrictive set of assumptions. Our approach will be to start with a series of simple two-by-two games and let you discover some guiding principles about how to play them.

Consequences and the Matrix Display

We'll follow the customary display of two-by-two games by exhibiting them in a matrix format. We imagine a row player with two choices, Up and Down; and a column player with two choices, Left and Right. The four possible outcomes, or consequences, can then be displayed in a two-by-two matrix as shown in Figure 4.1.

In particular examples, the consequences may include such factors as: immediate cash payments to each; deferred cash flows to each; shares of the market; prestige factors; goodwill concerns; obligations; and more besides. We now assume that each player assigns a utility value to each of the four consequences, and that these values include any feelings of malevolence or benevolence toward the other player. A benevolent player would prefer a nice outcome for the other; the malevolent one wishes ill to the other. These utility values reflect not only the players' preferences for the four consequences but their strengths of preference as well. More about this shortly. It is further assumed that these pref-

	COLUMN PLAYER	
	C1 (LEFT)	C2 (RIGHT)
R1 (UP)	Payoffs associated with (UP, LEFT)	Payoffs associated with (UP, RIGHT)
R2 (DOWN)	Payoffs associated with (DOWN, LEFT)	Payoffs associated with (DOWN, RIGHT)

(ROW PLAYER)

Figure 4.1. The 2×2 matrix display.

erences are *common knowledge*. And that, dear readers, is one hell of an assumption.

Let's give the players names: Rowena for the row player and Colin for the column player. These names should help your memory and let the personal pronouns aid us in differentiating the actors.

Game 1: Indeterminacy

Now see Game 1. Payoffs follow the format (R, C). Thus in Game 1, if Rowena were to choose Down as her choice and Colin Left, then Rowena would have a payoff of 12 and Colin a payoff of 7. By payoff we mean the player's utility value associated with the resulting consequence. Each pair of strategy choices results in a pair of associated utility payoffs. When we talk about shared or joint payoffs we certainly don't mean the two players "share" one payoff value or receive the same value. Payoffs always come in pairs, one payoff for Rowena and one for Colin. They may be the same, but more likely, they will be different. The aim is not to get more than the other side but to get as much for your side as possible. Remember, all concerns of malevolence or benevolence are already embedded in these numbers and all are common knowledge.

If you were Rowena, which choice would you take—Up or Down? If you were Colin, which choice would you take—Left or Right?

Think through what you would do as each player. Then think: what lessons are there to be learned in this exercise? Once you have done that, you can check your thoughts against those of two previous players and a game theory expositor who talks them through the analysis. Because game situations involve recursive predictions about what the other side will do, it is easier to talk about them by using the convention of a dialogue between the parties.

EXPOSITOR: O.K. Rowena, what did you do?

ROWENA: I chose Down. I thought Colin would choose Left so I would end up with a score of 12.

EXP.: O.K. Colin, what did you do?

COLIN: I chose Right.

ROW.: Why did you do that? Darn it. I end up with 5 rather than 12. I felt sure he was going to take Left. I would have, if I were he. Are you just trying to be mean to me?

COL.: Not at all. I just figured that you would think I would take Left and therefore you would take Down. And if you take Down I'm better off with Right than with Left. I'm sorry you got 5 and not 12, but 9 is better than 7 for me.

COLUMN
PLAYER

	LEFT	RIGHT
UP	(4, 5)	(10, −6)
DOWN	(12, 7)	(5, 9)

ROW
PLAYER

Game 1

ROW.: Well, I could have thought the way you did and figured you'd choose Right. If I took Up, you would have gotten a nasty −6. I should have done that!

COL.: But you didn't. Why stop there? If I could have reasoned that you would take Up, I would have chosen Left. This is complicated. We could go round and round. I'm thinking what you're thinking about what I'm thinking, on and on. Where does it stop?

There are no simple prescriptions for some games. You must think what the other player might do and he or she is thinking likewise. You must think about what the other person is thinking about what you are thinking, and so on. In some games, there is no end to that cycle of strategically interdependent thinking.

Remember: in Game 1,

Down (for Rowena) is best against Left (for Colin) (12 > 4).
Up is best against Right (10 > 5).
Left is best against Up (5 > −6).
Right is best against Down (9 > 7).

This game resembles "rock, paper, scissors." One can outguess one's opponent in any given round, but there is no way to improve one's chances of outguessing the opponent over time. No simple fixed strategy for either player is clearly better than any other. And if the game were repeated, the players would constantly randomize their choices, causing the outcome to move from box to box.

We now will introduce a sequence of games, starting simply and getting more complex as we go along.

Game 2: Dominance

Consider Game 2. Remember all information is common knowledge. The game is to be played once, with no preplay communication.

If you were Rowena, which choice would you take—Up or Down? If you were Colin, which choice would you take—Left or Right?

    EXP.: Well, Rowena, what choice did you make?

    ROW.: I chose Down.

    EXP.: And Colin, what column did you choose?

    COL.: I chose Left.

    ROW.: Well, at least this time you did what I expected.

    EXP.: Would it have made any difference, Rowena, if Colin had not chosen Left? Suppose you didn't know the column payoffs?

    ROW.: For me, in this case, Down is best against Left and Down is also best against Right. So no matter what Colin does, I'm better off with Down. What he does can make a slight difference in my payoff, but it does not affect my choice.

    EXP.: In this case we'll say that Down *dominates* Up or Up is dominated by Down.

    COL.: A similar story holds for me. Left is best against Up or Down. So Left dominates Right.

Let us summarize what we've learned. "Dominance" refers to a situation in which one strategy has a higher payoff for the player choosing it, regardless of the other player's strategy. In other words, strategy X dominates strategy Y if you are always better off with X. In this case:

Rowena's Report
    Down dominates Up; Down is better than Up against all Column choices:
        a. Down is better than Up against Left (12 > 4).
        b. Down is better than Up against Right (5 > 3).

Colin's Report
    Left dominates Right:
        a. Left is better than Right against Up (3 > 0).
        b. Left is better than Right against Down (8 > 4).

If players could collude, they would choose the pair {Down, Left} yielding the joint payoff (12, 8). It is the best for each of them. And it is also where they will end up without collusion. Thus the outcome of this game is predictable—

COLUMN
PLAYER

	LEFT	RIGHT
UP	(4, 3)	(3, 0)
DOWN	(12, 8)	(5, 4)

ROW
PLAYER

Game 2

perfectly predictable, if both players conform to our definition of rational behavior.

Game 3: Iterative Dominance

Now see Game 3, where the same rules apply as before. Both players enjoy common knowledge, it is a one-shot, or single-play, game, and there is no opportunity for preplay communication.

If you were Rowena, which choice would you take –Up or Down? If you were Colin, which choice would you take—Left or Right?

Now the players' reports become more complicated. Rowena's has two parts.

Rowena's Report I

Up does not dominate Down, because Down is better than Up against Left (10 > 0).

Down does not dominate Up. Up is better than Down against Right (5 > 3).

Not clear what to do; it depends on what Column player does.

Colin's Report I

Right dominates Left (4 > 2 and 8 > 3)

or

Left is dominated by Right. Left is a noncontender.

Rowena's Continuation

Column player should choose Right since it dominates Left.

If Column player chooses Right, then Up is better than Down. (5 > 3)

COLUMN
PLAYER

	LEFT	RIGHT
UP	(0, 2)	(5, 4)
DOWN	(10, 3)	(3, 8)

ROW
PLAYER

Game 3

The pair {Up, Right} is in *equilibrium*: no motivation for either player to change if the other holds fixed.

The point of this game is relatively simple, so we can dispense with the dialogue momentarily. Since Right dominates Left, Rowena "expects" that Colin will take Right. Rowena's choice of Up depends on Colin's being smart enough to take Right. If Rowena has doubts about Colin's sagacity, then she might have second thoughts about Colin's ability to do what's best for him. Standard game theory assumes the players are ultra-rational, so Rowena has nothing to worry about if Colin is the "standard" player. Colin can now think what Rowena might be thinking, but this should not upset the wisdom of taking Right.

Thus the outcome of Game 3 between two rational players can be predicted, as long as Colin doesn't have any reason to change his move once he thinks that Rowena knows what it will be. All you need is one player with a dominant move.

Game 4: Equilibrium

Now examine Game 4, and notice that neither player has a dominating strategy.

If you were Rowena, which choice would you take—Up or Down? If you were Colin, which choice would you take—Left or Right?

In Game 4 neither player has a dominant strategy. Does that mean that there is no good advice on how to play? Not really. First we should notice something. If Rowena plays Up and Colin plays Left, then both players have an incentive to change if the other holds fixed. {Up, Right} is better for Colin and {Down, Left} is better for Rowena. We can expect that at least one of them will change. So if they play multiple rounds, we don't expect that they will stay in {Up, Left}. The same holds true of {Down, Right}.

COLUMN
PLAYER

	LEFT	RIGHT
UP	(4, 3)	(10, 6)*
DOWN	(12, 8)*	(5, 4)

ROW
PLAYER

Game 4

However, if they land in either {Up, Right} or {Down, Left}, we would expect that they would stay put. Neither player is better off changing as long as the other stays put. In either of those boxes we would say that the players are in *equilibrium* (indicated in the matrix by asterisks in the appropriate boxes). This is a very important concept. Much of game theory is devoted to finding strategy pairs in equilibrium. They are important because many stable economic situations in the real world are thought to result from equilibrium dynamics. Situations only remain stable when no critical player wants to change its choice—provided, once again, the other holds fixed. When you find an equilibrium situation, you need to be very careful about getting into it. Once you are there, there is a tendency to stay put provided the other decision makers do not change their strategies.

Choice of equilibria. In this particular game there are two equilibrium pairs: {Up, Right} and {Down, Left}. The payoff with {Down, Left} is (12, 8), which is better for each than the joint payoff of (10, 6) for {Up, Right}. If the players could talk it over in preplay discussion, then they would commit themselves to {Down, Left}, yielding 12 to Rowena and 8 to Colin.

Smart players will also think ahead and do the same without benefit of preplay discussion. The Row player acting independently would be well advised to take Down, and the Column player similarly to take Left. Empirically, most business school student-subjects come to the {Down, Left} solution on the first try.

Game 5. Asymmetric Prescriptive/Descriptive Analysis

On to Game 5, which is quite like Game 4, but oh so very different.
If you were Rowena, which choice would you take—Up or Down? If you were Colin, which choice would you take—Left or Right?

COLUMN
PLAYER

	LEFT	RIGHT
UP	(4, −100)	(10, 6)*
DOWN	(12, 8)*	(5, 4)

ROW
PLAYER

Game 5

Game 5 is almost the same as Game 4. There are two equilibrium pairs, and {Down, Left} is better for both players than {Up, Right}. So should the advice to both players in Game 5 be the same as in Game 4? Not quite.

Colin might think: I'm afraid of that −100. Should I take Right to protect myself? Rowena might think: Colin might be afraid to take Left. Perhaps I should take Up in anticipation of his taking Right. Colin might think: I can see why she might take Up in anticipation of my taking Right. I think I might go Right.

How can Colin resolve this conundrum? One way is to apply the methods for making a decision under uncertainty that we discussed in Chapter 2. In that chapter we assumed that the uncertainties were not the result of another's conscious decision, but of more impersonal forces. The market will decide how a book does, the weather will determine whether the crops get enough rain. But the same logic can apply to the actions of another conscious decision maker.

Colin thinks: My alternatives are Left or Right. With Left there are two possible outcomes: Rowena could choose Up or Down, with consequences to me in utility values of −100 and 8, respectively. If I assign a probability of p to the outcome Up and $1 - p$ to Down my expected utility values are:

$$-100p + 8(1 - p) \text{ or } 8 - 108p \quad \text{for Left}$$

and

$$6p + 4(1 - p) \text{ or } 4 + 2p \quad \text{for Right.}$$

To find the break-even probability, p, he sets

$$8 - 108p = 4 + 2p \quad \text{or} \quad 4 = 110p \quad \text{or} \quad p = 4/110 = .036.$$

Thus Colin should take Right if he thinks that the chance of Rowena's choosing Up is larger than .036. He thinks: Out of 1,000 Rowena-type players, would there be more than 36 who would choose Up in Game 4?

For fun, you might try working out the corresponding analysis for Rowena. If Colin picks Left she wants to pick Down, and if he picks Right she should choose Up. How many Colins do you think would choose Right and how many Left? On the basis of your estimate, which should Rowena choose?

Empirical results. In a class of 110 MBA students at Harvard, 48 percent of Rowena players chose Up in Game 5, and 39 percent of Colin players chose Left, to their chagrin.

Game 6: Commitment and the Advantage in Going First

Up until this point we've analyzed the games under the assumption that both players move simultaneously. Let's relax that rule. Consider Game 6.

As either Rowena or Colin, would you want to go first or second? And what would your move be?

There are two equilibria: {Down, Left} and {Up, Right}. But unlike Games 4 and 5, each player prefers one of the equilibria. Rowena likes {Down, Left} and Colin likes {Up, Right}. But each prefers the equilibrium that favors the other over the other two boxes. The player who moves second has no choice but to accede to the first player's choice of equilibrium. Any player who can choose first has an advantage.

This game has been dubbed "the battle of the sexes," to describe a situation in which the husband wants to watch a baseball game, and his wife wants to attend a ballet performance. Whatever they do, they must do it together, because

COLUMN
PLAYER

		LEFT	RIGHT
	UP	(0, 0)	(1, 2)*
ROW PLAYER			
	DOWN	(2, 1)*	(0, 0)

Game 6

going anywhere alone yields a zero payoff. The first one to decide leaves no choice to his or her spouse.

Game 7: Threats and Communication

For Game 7, let's restore the rule about simultaneous moves, but change another one. Would it be to your advantage to communicate with the other party before play starts? In some of the previous games we have seen that it would help to talk. The players can agree to the better equilibrium in Game 5 if they have a chance to communicate.

As Rowena, would you want to talk first before choosing? What might you say? As Colin, would you want to talk first?

Sometimes you may be better off not negotiating. Up dominates Down and Left dominates Right, yielding a clear result that has a payoff of 5 units to Colin. If Colin comes to the negotiating table, he may be forced by Rowena to choose Right under the threat that Rowena would otherwise take her dominated alternative, Down. The threat hurts Rowena, but it hurts Colin more, and he is vulnerable.

Game 8: Doomed by Rationality

Consider Game 8 under standard rules of common knowledge, simultaneous choices, and no preplay communication. Think hard about this one.

As the row player, what alternative would you choose? Up or Down? As column player, what alternative would you choose? Left or Right?

Let's return to the dialogue format.

COL.: Oh, this is a tough one. It seems as if . . . well, Right dominates Left and Down dominates Up, so it seems that . . .

EXP.: Excuse me, let's see, you're saying that Right dominates Left because . . . ?

COL.: Because 10 is better than 5 against Up and −2 is better than −5 against Down. So you'd think that I would want to go Right and that Rowena would want to go Down, because 10 is better than 5 and −2 is better than −5 for her; but that puts us with a payoff of (−2, −2). At the same time, there is the upper-left corner, which gives us 5 and 5. It seems crazy that we should end up with −2 and −2. I am going to go Left and hope that Rowena sees what I see, which is that there is another box there that's better for both of us.

COLUMN
PLAYER

	LEFT	RIGHT
UP	(1, 5)	(4, 1)
DOWN	(0, −10)	(1, −20)

ROW
PLAYER

Game 7

COLUMN
PLAYER

	LEFT	RIGHT
UP	(5, 5)	(−5, 10)
DOWN	(10, −5)	(−2, −2)

ROW
PLAYER

Game 8

EXP.: O.K., so you chose Left. O.K., Rowena, what did you do?

ROW.: Sorry, Colin, but I chose Down.

COL.: But, Rowena! How could you?

ROW.: Listen, I'm here to maximize my payoff, and knowing that you choose Left, I have only one choice. I could get 5 or 10, and my responsibility to my constituency is to maximize my return, all things considered. I'm acting as an agent for a boss.

EXP.: Let me interrupt here. In this game Down dominates Up. And similarly Right dominates Left. The {Down, Right} pair of strategies is in equilibrium. It's the only equilibrium. O.K., Colin, how do you react to all of this?

COL.: I feel as if I tried to do the smart thing and Rowena took advantage of that.

ROW.: That's nonsense! I didn't take advantage, I simply looked at what my payoffs were. You could see them just as well as I could. There is no reason whatsoever that I should choose 5 rather than 10. Down dominates Up.

COL.: You're saying that no matter what I do, you still want to choose Down; but if I reasoned the same way, then you wouldn't be getting your 10, you'd be getting -2.

ROW.: Yeah, but if you took Right, then I would have a choice between -5 and -2 and I prefer -2.

COL.: Yeah, but if you took Down, then I would have a choice between -5 and -2 and I prefer -2.

EXP.: Colin, before choosing yourself would you want to know what Rowena chose? Would it help you to know?

COL.: Sure it would.

EXP.: Well, suppose you found out that she was going to choose Up.

COL.: If we were playing one game, I guess I would choose Right.

EXP.: So, if you had espionage and you found out she was going to pick Up, what would you do?

COL.: I would choose Right, as I just said.

EXP.: And if your espionage service told you that she chose Down, what would you do?

COL.: I'd still take Right.

EXP.: But that's what we said before.

ROW.: This is a trap; we're trapped in $(-2, -2)$ but yet there's another box $(5, 5)$ that's better. How do we get out of that trap?

COL.: Well, we could try to make a side agreement.

EXP.: O.K., but the point is, it is a trap. It's called a *social trap*. There is something disturbing here. There should be a rule against such games! The anomaly lies in the game itself. *Rational* behavior calls for players to use their dominating strategy; *rational* behavior calls for the row player to choose Down and the column player to choose Right.

COL.: You're saying it's rational to make choices that lead to inferior outcomes?

EXP.: Yes, it turns out in this situation that two rational players do worse than two irrational players.

ROW.: What's rational about doing poorly?

EXP.: Well, let me repeat, if you were the Column player and you were the agent for a principal and the principal says, I want you to get as much as possible, what would you, Colin, do if you knew that Rowena took Up? You would take Right? What would you do if she took Down? You would take Right. Right?

ROW.: One of my economics professors used to say that individual rationality can sometimes lead to an inferior collective outcome. Is
this a case in point?

EXP.: Well, we could say that individual rationality can lead to group
irrationality or poor outcomes. That's the dilemma; that's why they
call it a social dilemma, or a social trap. We're going to spend a lot
of time looking at this situation.

COL.: This is probably a game where we would want to talk to each
other first.

EXP.: Absolutely. If you could talk to each other, what would you do?

ROW.: Try to come to some kind of an agreement. We would want
preplay communication with binding agreements. With binding
agreements you could lock into the upper left.

The Prisoner's Dilemma

The two-person dilemma game is by far the most celebrated of all allegorical
games. At the mid-century mark, in 1950, a lot of the folks working in game theory knew about this dilemma game—it was, as they say, folk knowledge. But it
wasn't called the prisoner's dilemma at that time. The interpretation of the game
as a dilemma of two prisoners was introduced by Al Tucker around 1953, and literally thousands of papers, experiments, and doctoral theses have been based on
it since that time. It is important because its message is so clear: uncoordinated,
rational, self-interested behavior can result in awful outcomes. It's the essence of
a social pathology, and evidence of its structure can be found extensively in our
society. But before jumping ahead, let me first describe Tucker's interpretation
of the dilemma.

A district attorney knows that two prisoners are indeed guilty of a crime, but
he doesn't have acceptable evidence to convince a jury of this fact. The alleged
criminals know this. The district attorney presents the following choice-problem
to each of the prisoners separately. The prisoners are kept separated. Each is
given the choice of not confessing or confessing to the crime they have committed. If neither confesses, they will be held for one year on a minor charge: possession of an illegal weapon. If they both confess, each will get a three-year sentence, less than the maximum sentence for their crime. If one confesses and the
other doesn't, then the confessor will get off scot-free and the nonconfessor will
be given five years. These payoffs are as shown in Figure 4.2.

To give the game the structure we want to discuss, it is critically important
that each prisoner prefers going free himself and having his buddy in crime get
five years over both of them getting one year. They are each worrying about
themselves. No honor among thieves.

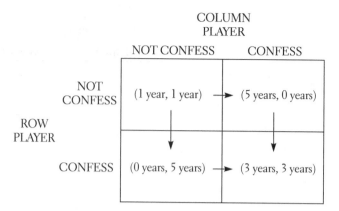

Figure 4.2. The prisoner's dilemma.

Remember that they are held incommunicado and that each has little loyalty to the other. Naturally the game is to be played once. It's like Game 8. Confess dominates Not Confess, since Confess is better than Not Confess no matter what the other prisoner chooses to do. So they both confess and each gets three years. That's the social trap.

Game 9: Chicken and the Temptation of Mutual Annihilation

Let's turn our attention to Game 9, the game of "chicken" or "macho-chicken."

EXP.: This game is often confused with Game 8, the dilemma game, but structurally it is quite different. In this game the Macho choice (Down and Right) is best against the Chicken choice (Down and Right) and the Chicken choice is best against the Macho choice. The Chicken-Chicken outcome, cell I, is far better than the Macho-Macho outcome, cell IV. There are no dominating strategies for either side. What about equilibrium pairs?

COL.: The problem, as I see it, is that the joint strategy {Up, Left}—or Chicken-Chicken—is not in equilibrium. There's an incentive for each side to switch to Macho if the other side sticks to Chicken.

ROW.: But it looks like cell II is in equilibrium and so is cell III. So we have two equilibrium pairs, and I as the Row player prefer cell II whereas Colin prefers cell III. It's a little like an earlier game. But I think it's clear that both players should be Chicken.

COL.: But if I'm certain you'll be Chicken, and I'm acting for my boss, I should be Macho.

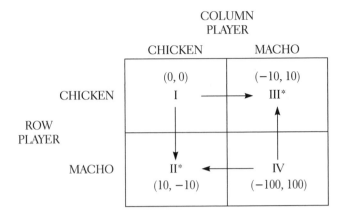

COLUMN
PLAYER

	CHICKEN	MACHO
CHICKEN	(0, 0) I	(−10, 10) III*
MACHO	II* (10, −10)	IV (−100, 100)

ROW
PLAYER

Game 9. Chicken and the danger of mutual annihilation.

EXP.: Can you think of interpretations?

COL.: It's like Game 6 but with the added complication that cell I is much better for each of us than cell IV.

The terms Macho and Chicken suggest the interpretation of the two teenagers who are driving speedily toward each other. If they both choose the Macho strategy they crash and are killed. If you keep steadfast with the Macho strategy and the other guy flinches and is a chicken, you are a hero and the other guy loses his rep. If you both flinch, it's O.K., but not ecstatically impressive for either.

ROW.: And if I think the same way?

COL.: Then poof, up we go to heaven.

ROW.: Or hell. So all things considered, we should each take the Chicken alternative.

COL.: But here we go around again. There's no way out of the trap. The more convinced you become to be Chicken, the more temptation I have to be Macho.

Game 10: The Iterated Dilemma Game and Tacit Collusion

Let's return to the dilemma game (Game 8), but this time imagine that you are to play the game repeatedly. No talking. At each round you simultaneously submit the choice of C for "Cooperate" or D for "Defect." Imagine that to begin with, you will play the game a large number of times. Since the game is symmetric (in the sense that each player is confronting an identical problem), we can concentrate on the behavior of the Row player. Think of yourself as Rowena. Imagine that you are an agent for a principal who tells you to maximize the sum

COLUMN
PLAYER

	Cooperate (C)	Defect (D)
Cooperate (C)	(5, 5)	(−5, 10)
Defect (D)	(10, −5)	(−2, −2)

ROW
PLAYER

Game 10. Same as Game 8, but repeated for multiple rounds.

of your returns. Do not empathize with the other player. But neither should you, on the other hand, be purposively malevolent. You will be judged by your boss by your accumulated scores.

What would your strategy be? In particular, how would you start?

Suppose on the first trial you choose C and he chooses D. What would you do on the next trial?

Suppose you each play C on the first three trials and on the fourth he switches to D. What then?

Suppose that you have established a pattern of Cs against Cs and the experimenter says that there are just two more trials remaining. What do you do?

Let's consider Game 10 via some dialogue.

ROW.: It wouldn't make any sense at all to get −2 at each trial.

COL.: If we could have preplay communication and could make binding agreements, we could take (C, C) at each trial and each get a stream of 5's.

ROW.: Yes, but can't we do this without talking?

COL.: How do I know you won't double cross me at a given trial?

ROW.: Because if you double cross me, I'll punish you in succeeding trials. We should be able to get a tacit collusion by playing (C, C) over time. Certainly, that's what I would try to do.

EXP.: O.K. Rowena, suppose that you and Colin start out with (C, C) at trials 1 to 3. But on trial 4 you choose C and he chooses D. He gets 10 instead of 5 at that trial and pushes you from 5 to −5. What would you do?

ROW.: I would be as mad as hell. I would punish him severely.

EXP.: Well what would you do on trial number 5?

ROW.: I would use D.

EXP.: On trial 6?

ROW.: I would use D. I would continue with D for quite some time. He started this fight.

COL.: "Hell hath no fury like a woman . . ."

ROW.: (Heatedly) That's quite an inappropriate, sexist remark. What would you do if the roles were reversed?

COL.: I would use D on trial 5, maybe also on trial 6, but on trial 7 I would switch back.

ROW.: But I want to teach you a lesson not to start taking advantage of me.

COL.: That may be the message you're wanting to convey. But that's not the message I may be receiving. On trial 4, I might have taken D because I might not have analyzed the problem clearly enough. Then when you play D on trials 5, 6, 7, 8, 9, and 10 when I am try-ing to signal you with a few Cs, what am I supposed to think? This lady's vindictive. So now it's my turn to get mad.

ROW.: I'm getting mad just listening to you. Why the hell did you take D at trial 4? It was your sexist interpretation of my response that's bothering me.

EXP.: Enough of that! Suppose the pattern of play is as shown below, where "C" means cooperate (do not confess), and "D" means de-fect.

	1	2	3	4	5	6	7	8	9	10	
Row	C	C	C	C	D	D	C	C	D	(?)	...
Column	C	C	C	D	C	C	D	D	C		...

EXP.: All right. Suppose we have the pattern shown here through trial 9. What would you do, Rowena, on trial 10?

ROW.: I don't see why I should go back to C. He started the defection and he has taken advantage of me on trials 4, 7, and 8. I want to punish him once more on trial 10 to tell him that his behavior was irresponsible.

COL.: I would take C on trial 10 but if she used D on 10, I would use D on 11.

EXP.: What's the message here?

ROW.: Don't start messing around.

COL.: It's damn hard to get back into a steady collusion after someone breaks the pattern.

ROW.: And therefore don't break the tacit collusion.

EXP.: *Trust is fragile.* Once broken it is hard to restore. But at the same time, even though the players may not trust each other, each might be willing to maintain a (C, C) pattern. They do this not because they empathize with each other, but because they trust that the other side will be wise enough to act in its own selfish interest by not rocking the boat. This is an important observation: *empathetic* trust is important but not always indispensable in developing and maintaining a *working* trust.

What about end play? What do you do when you establish a (C, C) pattern, say, and you learn that the repeated plays are soon to end?

ROW.: I would ignore that information and continue to play C — if we had established that pattern.

COL.: What would you do if I switched to D when there were two trials left?

ROW.: I would be mad as hell again and switch to D also for the last trial.

COL.: But you know I will take D on the last trial. My boss said I have to maximize my return. I thought you surely would take D on the last trial, so I thought I would have nothing to lose by taking D with two trials to go.

ROW.: But that's silly. You could use the same logic if you knew there were just three trials left. Or four. Or twenty. With guys like you, no wonder the world is in a mess.

EXP.: Let's just say that end play is a problem. In the real world, however, life goes on. Colin could tell his boss that he stuck with C at the last trial because who knows what's to come next, and a working trust should be worth something.

ROW.: Exactly!

Game 11: The Ultimatum Game and the Rewards of Irrationality

The final game we will consider is the ultimatum game. This game takes many forms, but we stick to one first given by Richard Thaler. The rules in this version are as follows.

Two players are given $100 to share between them. The rules specify that one player is the Allocator and the other player is the Recipient. Let's refer to them as Mr. A and Ms. R. At move 1, Mr. A, the Allocator, proposes a sharing division of the $100: X goes to Ms. R and (100 − X) is kept by Mr. A. At move 2, Ms. A either accepts the proposed division or rejects the proposal, in which case both Mr. A and Ms. R get zero. All this is common information.

The amount X that A proposes to give to R will be termed the *ultimatum bid*. So the game boils down to: A offers X to R, who can either accept or reject. A is the proposer, but R has veto power and must ratify A's proposal of X to R.

There is to be absolutely no preplay communication between A and R. The game is to be played just once.

- Prize to be shared: $100.
- Players: Allocator (A); Recipient (R)
- Move 1: A offers allocation of X to R and (100 − X) to A
- Move 2: R accepts X and gives (100 − X) to A—*or* R rejects allocation and both get nothing.
- X = The ultimatum bid (offer) of A to R.
- Rules of play: R has veto power; no preplay communication; single-shot game.

As Allocator, what ultimatum bid (offer) would you make to the Recipient? As Recipient, what is the minimum ultimatum bid you would accept?

After answering the questions above: What proportion of Recipients in similar circumstances would accept nothing less than $50? than $25? than $1?

We'll have Al as the Allocator, and Ree as the Recipient in our debriefing.

REE: I don't want to go first.

AL: O.K. I'll go first. It seems to me that the Allocator has all the power. As Recipient, if I'm offered X, I have a choice of X or nothing. I'd accept any positive amount. If the Allocator is a do-gooder and offers me 5 or 10 bucks, I'll take it.

REE: It's definitely not true that the recipient has no power.

AL: Let me finish. As Allocator, I expect that most Recipients would be happy with any positive amount, and since I'm a nice guy, I would offer the Allocator 5 bucks. I bet he or she would accept it.

REE: Well, here's one Recipient who wouldn't. Can I have a turn?

EXP.: Sure.

REE: In making my decision I first thought about what I would do as Allocator. I think the assignment of roles is just fortuitous. So the obvious *fair* thing to do is to make a 50–50 split; and thus as Allocator, my ultimatum bid would be $50, and I would expect the Recipient to accept this. I had this split in mind—what I would do as Allocator—when I thought of what I would accept as Recipient. I would certainly not accept Al's offer of $5. I consider that an insult. I decided my break point was $30; anything less I would refuse and anything more I would accept. I would, for example, accept $40, but I would feel put upon, because if I were the Allocator I would offer the Recipient the fair share of $50.

AL: I don't understand you at all. Are you saying that if I as the Allocator offer you $10, you would rather get nothing than $10?

REE: In a flash I would reject $10.

AL: But you would be losing $10!

REE: Ah, but I would keep you from getting $90.

AL: Why are you being so vindictive? The Allocator is being nice in offering you $10. You are better off with one cent rather than nothing.

REE: Doesn't fairness count in your opinion?

AL: Sure, but the game gives all the power to the Allocator. If you, as Allocator, offered me $10, I would think what a nice, kind person you are.

REE: I'm not sure you understand the game, Al. The Recipient has the ultimate power of the veto. It's not as if the Allocator gets $100 and offers a consolation prize of one or two dollars to the poor Recipient. In this game the Recipient has the real power to keep a selfish Allocator in line.

EXP.: Ree and Al differ. Their choices would be different and this biases their responses. It is clear that Ree is not trying to maximize only her expected monetary return; she repeatedly raises the question of fairness. She wants to be fair and assumes that others will also want to be fair.

REE: Not only do I want to be fair, but I want to establish the *reputation* that I am fair. When Al, as Allocator, offers an ultimatum bid of $10, not only is he being unfair, but he is jeopardizing his reputation.

AL: How you distort my image! By offering $10, I am being kind, because the Recipient would be happy with $1.

EXP.: Well let's try to sort out the issue of fairness from reputation. Let me ask you to consider the following variation of the ultimatum game. Suppose that the experimenter, say myself, keeps the identities of the Allocator and Recipient completely anonymous. Anonymous from each other and even from the experimenter. This could be achieved in a laboratory with computer consoles hooked up in a network. Now I ask you, Ree, would you as Recipient accept $20 if you didn't know the identity of the Allocator and vice versa? You said before that you would reject $20 if you knew the Allocator's identity and he or she knew yours.

AL: Surely, Ree, you would take $20. You lose nothing by it, and you haven't besmirched your reputation for standing up for fairness because nobody knows you have pocketed just $20. You could use the money to go to a concert. Or better yet, you should take it and give the cash to a charity of your choice.

REE: I'm weakening. Yes, with complete anonymity, I would take $20, I think. But I sure wouldn't take $10.

AL: You continue to surprise me. Why not?

REE: I guess in this case it's not my reputation at stake, but it's *justice*. I'd get more than $10 worth of satisfaction by punishing the Allocator even if he and I remained anonymous. I'll know that somewhere out there I kept someone from getting $90 by exploiting someone else.

AL: What if the amounts are higher. Say we had $1,000 to share. Would you accept $100 as Recipient? What would you do if identities were known? And what would you do in the anonymous case?

REE: With higher values, I would have different tradeoffs among the four components of utility assessment: money, fairness, reputation, and justice. Monetary return would loom larger, I admit. I think I would accept $100 in a $100 and $900 split if the identities are anonymous but not if identities were known.

AL: Well, I would have no trouble. Having thought about it a bit more I wouldn't take $10 from $1,000—whether the identities were known or not—but I would take $50. Maybe, Ree, the difference between us is that I can use the money more than you can.

EXP.: Let me go through another variation. We started by assuming the roles of Allocator and Recipient were chosen by chance, the toss of a coin. Now suppose the roles are selected on the basis of who does better in some external context. For example, who does better on a spelling bee, or who does better in a game of chess, or whose parents are richer. Would this change matters?

REE: Well, if I were rewarded for some past accomplishment, and because of that I was selected as Allocator, then, yes, it would make a difference. There would now be some rationale for me to exploit my good fortune. I have earned it, and now as Allocator I could with clear conscience squeeze the Recipient by offering him or her, say, $10 and keeping $90 for myself.

AL: Finally, I agree with Ree. It would be critical to know how the roles of Allocator and Recipient were chosen in determining questions of fairness, justice, and reputation.

EXP.: Let's take a different tack and move on to equilibria analysis.

Jointly Normative Equilibria Analysis

Let's change gears and employ the strict, standard, game-theoretic assumptions under which the two players have common knowledge about the rules of the ultimatum game. Assume furthermore that each player is ultra-rational and concerned only with his or her expected monetary payoff and has no concern about fairness, reputation, or justice. And all the above is common knowledge. Remember, there is no revenge component (punishing Al, say, for being piggish) in this abstraction. We now examine equilibria analysis in this bona fide game structure. Let's bring back Al and Ree.

EXP.: Do you remember what a strategy is in a game, and what we mean when we say that a pair of strategies is in equilibrium?

AL: Sure, a strategy is a rule or recipe that says what a player will do in every possible situation.

REE: If a principal gave her agent a strategy, the agent would know exactly what to do in the game.

AL: And a pair of strategies is in equilibrium if each player has no motivation to change his or her strategy if the other player holds fixed. In other words, each strategy of the equilibrium pair must be optimal against the other strategy.

EXP.: Great! Now let's put this to the test. Consider the following pair of strategies. The Recipient accepts $10 or more, nothing less, and the Allocator offers $17.72. Is this an equilibrium pair?

AL: Hell, no. A Recipient only interested in money would be better off accepting the $17.72.

REE: And the Allocator, knowing the Recipient's strategy, would be better off making an offer of $10.

EXP.: O.K. Now consider the pair where the Recipient will accept $16.23 or more, nothing less, and the Allocator offers $16.23.

AL: Why those numbers?

EXP.: Why not these numbers? We're just testing what we mean by an equilibrium pair. Well, Al?

AL: I suspect a trick. If the Allocator offers $16.23 then the Recipient would gain nothing by changing her strategy.

REE: And if the Recipient will accept $16.23 or more, but nothing less, then the Allocator would gain nothing by changing his strategy.

EXP.: So?

REE: So I guess that pair is in equilibrium. *But so would any other pair be* where $16.23 is replaced for both of them by any amount between $0 and $50 in the game.

EXP.: Right. How about where $16.23 is replaced by $87.49?

AL: But it doesn't make any sense for the Allocator to offer $87.49 to the Recipient.

EXP.: Let's be orderly here. Keep to the question.

REE: Well, if they both agree to choose $87.49, then I guess there is no motivation for either to change if the other holds fixed.

EXP.: There is simply a huge number of equilibria pairs in this game.

This is an embarrassment to the powerful conceptual notion of equilibria. We'll see that the multiplicity of equilibria pairs occurs often in games, and that game theorists have introduced additional requirements. They concentrate on equilibria that have other special properties.

Let's go back to the equilibrium pair where the Allocator offers $16.23 and the Recipient will accept $16.23 but nothing less. Now consider the dynamics of this situation.

Suppose that for some reason or other the Allocator does not go down the path "offer $16.23" but rather offers her $14.19. In the subgame following this first move by the Allocator, it would not be optimal for the Recipient to stick with her strategy of $16.23 or more. She would want to switch and accept the offer of $14.19.

REE: After the first move of the Allocator, the Recipient in the ensuing subgame would want to accept any positive amount.

EXP: In the vernacular of game theory, the pair of strategies where the Allocator offers the minimum positive amount (say, 1 cent) and the Recipient accepts any positive amount is said to be a "perfect equilibrium" pair. If we continued this line of development, we would come across other games where there are many perfect equilibrium pairs, and the search begins for "perfectly perfect" equilibria. Instead we move on to more practical topics.

Asymmetric Prescriptive/Descriptive Analysis

Now let us all assume the role of an analytically inclined Allocator, A, who must decide his most preferred ultimatum bid, X. Assume A is solely interested in maximizing his expected monetary return and that he is uncertain about the Recipient's, R's, minimum acceptable ultimatum bid (MAUB). If A knew R's MAUB, that's the X he would offer. Not knowing the MAUB, a bid of X would yield an expected monetary value of

$$EMV_A(X) = (100 - X)F(X),$$

where

$$F(X) = \text{Prob}\{MAUB \leq X\}$$

is the cumulative distribution function of the uncertain MAUB value.

Literally thousands of MBA students have played this game. Records have been kept, so that an astute Allocator, playing against a field of MBA students and knowing the empirical results, should offer an ultimatum bid of $35. Furthermore, for this population, while $X = 35$ maximizes $EMV_A(X)$, the optimum slopes off slowly for higher values of X, so that a bid of $50 loses little. But the expected payoff falls more precipitously downward for X bids, on the order of $20. So yes, Ree's ultimatum bid of $30 yields a much higher expected return (against MBA students) than Al's bid of $10. This is because there are a lot more MBAs like Ree than like Al. But—and this is an important but—I suspect that among economics majors, the optimal bid would be far lower, something like $X = \$20$.

How might we think about maximizing $EMV_A(X)$ without having a lot of data in front of us? How might we assess a reasonable $F(X)$? Here's one attempt:
Let

$$a = \text{Prob}\{MAUB = 0+\},$$

meaning that a is the proportion of extreme subjects who would accept any positive amount offered. And let

$$b = \text{Prob}\{MAUB = 50\},$$

meaning that b is the proportion of extreme subjects who would not accept anything below $50. For convenience, let the proportion of the $(1 - a - b)$ nonextremists have X-values that are uniformly distributed in the interval $(0, 50)$. So for this case

$$F(X) = \begin{array}{ll} a & \text{for} \quad X = 0, \\ a + X(1 - a - b)/50) & \text{for} \quad 0 < X < 50, \\ 1 & \text{for} \quad X = 50. \end{array}$$

The cumulative distribution F has a jump of a at $X = 0$, rises linearly to $(1 - b)$ at 50, with a jump at 50 of b. In this case we have, for $0 < X < 50$,

$$EMV_A(X) = (100 - X)(a + X(1 - a - b)/50),$$

which is a quadratic function of X. To find the maximum we can use the calculus and differentiate, or we can complete the square, or we can use the following trick: we find the two root-values of X for which $EMV_A(X) = 0$, and then, recognizing the function to be quadratic, the maximum of X is at the midpoint of these two roots. This program yields

$$EMV_A(X) = 0 \qquad \text{for } X = 100,$$

and

$$\text{for } X = -50a/(1 - a - b).$$

So the optimum

$$X^* = [100 - 50a/(1 - a - b)]/2 = 50 - 25a/(1 - a - b).$$

Now you can find a reasonable optimum ultimatum bid by assessing values for a and b. For the MBA class a reasonable a-value is .2 and a b-value, .4, so that

$$X^* = 50 - 25 (.2/.4) = 37.5.$$

For economics students we might find that $a = .4$ and $b = .2$, so that $X^* = 25$.

Core Concepts

In interactive strategic situations involving two contesting parties, game theory examines how very smart people think hard about what the other is thinking about what they are thinking about and so on. The theory then proposes how each should play, and in that sense the theory is jointly normative.

The chapter has presented a clutch of very simple, two-player, bi-matrix games. The games start out very simply, teaching concepts such as dominance, equilibrium, and iterated dominance. The theory of equilibria in games gets into trouble in many ways. Sometimes there is more than one equilibrium pair and it may not be easy to decide how to choose among these contenders. In some games it would be helpful to have preplay communication between the players with the possibility of making binding agreements; in other games a player may choose not to come to the table for a preliminary chat for fear of being threatened. In some games, a player gains an advantage in learning the choice of the other party before announcing her own; in other games it is advantageous to declare first.

By far the most cited simple game is the social trap, where each player has a dominant strategy but the pair of such strategies leads to miserable payoffs to each. Two naive or "irrational" players fare much better than two "rational" ones. This game structure has a popular accompanying story line that involves two prisoners and is referred to as the prisoner's dilemma game.

When this social dilemma game is repeated a number of times, sophisticated players can achieve jointly desirable results by tacitly cooperating over time even though each can achieve a better payoff in the short run by defecting at a given trial. They develop an operational trust for fear that any deviation will be punished in succeeding rounds. But as expected, havoc ensues when an end-game is anticipated. We'll see in later discussions that complications with end-game phenomena keep on recurring in negotiation contexts.

The ultimatum game is another one of those simple games where subjects are surprised at the behavior of others and grapple with an important question: Just what are we trying to maximize? How to balance concern about self and generosity to others? Much more about this balance in later chapters.

5

Negotiation Analysis

We have now discussed the basics of *individual* prescriptive and descriptive decision making in Chapters 2 and 3, respectively. *Interactive* decision making (noncooperative game theory) was introduced in Chapter 4. A vast array of formal theoretical work underpins the individual and interactive perspectives. We now introduce a third approach: joint decision making, or negotiation analysis, the primary concern of this book.

The Essence of Joint Decision Making

As in our discussion of (noncooperative) interactive decision making, we can extract the essence of the joint decision-making perspective. The essential elements are as follows:

- You and the other individual or individuals can make mutually agreed-upon joint decisions.
- Your payoffs depend either on the consequences of the joint decisions or on each party's go-it-alone alternatives.
- You can reciprocally and directly communicate with each other—about what you want, what you have, what you will do if you don't agree, or anything else. This communication might be honest, or it might not.
- You can be creative in the decisions you make.

This sums up the cooperative focus of joint decision theory. Certainly the components of the theory could be highlighted differently, for instance with less emphasis on the sharing of information and opportunities for creativity. But we have deliberately included these attributes to play up the benefits of cooperation. A cooperative perspective provides a counterbalance to the interactive, game-theoretic perspective, because it focuses on the potential of decisions that are rational when measured against a group, rather than an individual, benchmark. Below we flesh out the basic concepts of joint decision making. To clarify them, we employ game theory as our foil.

Joint Decisions

According to the assumptions of joint (cooperative) decision theory, two or more decision makers have an opportunity to reach a joint agreement. We call these decision entities "parties" rather than "players." Contrast this with game theory's outlook: the players are obliged to make separate decisions that interact to produce joint payoffs. Joint decisions shift the focus from separate interactive actions to group actions. This is a crucial distinction. For instance, if we examine the prisoner's dilemma game from the perspective of joint decision theory, the dilemma dissolves, because group rationality dictates that both players will cooperate (see Game 8 of Chapter 4).

When thinking about joint decisions, our first inclination is to concentrate on the final decision. Can the negotiators close the deal with a joint agreement? The final decision is of central importance, but joint decision opportunities crop up throughout a negotiation, not just at the crunch. Additional instances include joint decisions on what issues to discuss, on what the rules of conduct between the parties will be, on inviting a mediator to take part, on postsettlement settlements, and on ignoring a subset of issues. Joint decisions are sprinkled throughout the negotiation process.

Joint decisions are *not*, by definition, enforceable or binding. In an ideal world, a joint agreement on how to resolve the issues would guarantee the agreed-upon consequences (payoffs) for each party. In reality, a verbal agreement, a handshake, an exchange of assets, or even a legal contract may not be sufficient to make one or more parties honor the joint agreement. An astute negotiator will be wise to consider mechanisms for enforcement, because joint decisions are not necessarily mutually binding.

Joint Payoffs

In game theory, joint payoffs are determined by separate interacting decisions. The game players face a problem of strategic interdependence: each party has partial control over its own consequences and partial control over the consequences of the other party. Payoffs for one cannot be dissociated from choices of the other. This interdependence of payoffs and decisions is an ineradicable feature of game theory. In negotiations (joint decision making), we assume that the joint decision of the collectivity determines the payoffs for each party.

Reciprocal Communication

Joint decision making emphasizes direct communication of interests, aspirations, expectations, beliefs, visions of the future, and so forth. This contrasts starkly with the noncooperative game-theoretic perspective of simultaneous choices and no preplay communication. Negotiators bent on squeezing out joint value through collaborative decision making may wish to exchange their most intimate secrets in the hope of maximizing joint gains. We'll have a great deal to say about this idea in later chapters under the rubric of Full, Open, Truthful Exchange (FOTE).

Let's not go overboard and casually assume that communication is all good news. All too often, bluffing, threats, trickery, exaggeration, concealment, half-truths, and outright lies come to mind when we think of negotiations. Dialogue may not be reciprocated on equal terms. Unscrupulous parties may exploit strategically sensitive information to boost their own payoff. Negotiators may engage in strategic misrepresentation of their own interests in the hope of tricking the other party into making concessions. Of course, these activities are part and parcel of negotiation. But we don't want to lean too far in the negative direction. Positive activities—mutual exchange of internal documents, informal briefings, clear statements of interest, "confessionals" with mediators, joint brainstorming, joint problem solving—are just as much part of negotiation. Negotiations characterized by truthful exchange of information and interests are surprisingly common. When parties in a negotiation jockey around, in some sense they are making separate and interacting moves in an extensive game. So we can often think of the dynamics of a negotiation in the same way. As we have indicated, it's hard to contain any one perspective without involving the others.

Creativity

In the joint decision-making perspective, negotiators can be creative in the actions they take and the decisions they make. Again we can contrast this with the game theory assumption that there exists a predetermined set of strategies and associated payoffs, which are common knowledge among all the players. Each player knows exactly what finite number of strategies are available to him or her and to the other players, and vice versa, and they all know what payoffs are associated with the interaction of different strategies. Joint decision making jettisons the restrictive assumption of common knowledge. It widens the scope of its vision to include the invention of strategies, creation of new alternatives, and increases or decreases in the number of parties. But in mentioning creativity, we are in the theoretical border region between joint decision making and a broader, less tightly defined perspective on negotiation. And to complicate matters even further, pure game theory does not have to remain so pure: assumptions about common knowledge can be relaxed, and once this is done it's hard to distinguish the boundaries between perspectives.

Decision Perspectives for Negotiators

Negotiators need different analytical perspectives to cope with the myriad decision problems they face. Successful negotiators study the same decision problem from different angles to form a deeper understanding of the strengths and weaknesses of their situation. How do decision-making perspectives help negotiators make better decisions?

An *individual decision-making perspective* clarifies the decision structure in terms of alternatives and payoffs. It provides insight through a systematic analysis of a particular problem from a single perspective. A full decision-theoretic perspective allows a negotiator to compare the benefits of a joint agreement with separate or unilateral action. Of course, an individual perspective simplifies the problem by excluding an explicit consideration of the other side's possible decisions. These factors, however, can be folded into the analysis as uncertainties. An individual decision-making perspective helps negotiators decide with whom they should negotiate, assists them in comparing the expected benefits of an ongoing negotiation with the uncertain benefits of alternatives, and provides a theoretically well founded methodology to structure their negotiation problem.

An *interactive decision-making perspective* obliges a negotiator to consider carefully the alternatives, interests, aspirations, and behaviors of the other side. A focus on interactive decisions heightens a negotiator's awareness that his or her payoff is determined not just by his or her own actions, but also by the separate interacting actions of the other negotiators in and surrounding the main negotiations. Will the other party honor what we just agreed to do? How secure is our negotiated contract? How will the other party interpret the purposefully vague language of our final negotiated document? No negotiator can have complete mastery over the final outcome. Thinking strategically about the interaction of separate decisions should help negotiators to understand the underlying threat structure and how they can improve their leverage in a negotiation.

A *joint decision-making perspective* emphasizes the opportunities for cooperation between two parties—and helps them avoid falling into the trap of negotiating solely on the basis of what is individually rational. By adopting a joint decision perspective, negotiators can better conceive how communication will facilitate the drafting of joint agreements to the benefit of both sides. Through cooperation, negotiators might explore agreements based on a process of joint decision making that are mutually superior to disagreements born of separate interacting decisions—the no-agreement state. In highlighting the role of joint decisions, we raise the possibility of a win-win solution to the negotiation problem.

Try as we might, these three decision-making perspectives cannot be melded into an integrated analysis producing clear prescriptions. Negotiations are too complex to be solved with analytical tricks. But a sophisticated negotiator can switch between the different perspectives to probe the complexities of the negotiation from different angles and with different purposes in mind. Which alternative is best? What are the uncertainties attached to each one? What are the other side's priorities? How will the other side react to this proposal? What can we jointly agree that will create shared value? How can we identify and exploit shared synergies?

An adept negotiator can move back and forth between individual, interactive, and joint decision-making perspectives, synthesizing insights along the way, to arrive at well-informed decisions. Experimental evidence shows that, when left to our own devices, we are not much good at negotiating optimal deals. Analytical perspectives can help. But to achieve the best solution that you can, you will need to strike a balance between your analytical endeavors and your cognitive capabilities. Implementing the fruits of analysis relies on your bargaining skills, your powers of persuasion, your nimble thinking, your knowledge of body language, your inventiveness and creativity, your willingness to use credible threats, your skills at drafting complex deals, your coalition-building expertise, your linguistic abilities . . . the list never ends.

The Negotiator's Dilemma

In negotiations, you would like to get a large slice of the pie you jointly create with the other parties. Tactics used to *create* a larger pie (for example, truthful sharing of information) may conflict with tactics designed to *claim* a large slice of the pie. In every negotiation, value creation is inextricably linked to value claiming—this is the negotiator's dilemma.

If Noel attempts to create joint value between himself and Liam, his actions will almost certainly have ramifications for the distribution of value between himself and Liam. Depending on the circumstances, Noel may increase his total payoff and also increase his slice of the pie relative to Liam. But Liam may react to this, and decide to claim more value for himself. This could affect Noel's cooperativeness and, through action and reaction, reduce the pie's overall size. Noel and Liam need advice on how to manage this volatile interdependence so as to maximize their joint payoff and each receive a fair share of the pie.

Any theory of negotiation worth its salt needs to address the tension between creating and claiming value. Practical advice should offer methods both for co-operating and for competing with the other side. In this book we argue that a collaborative approach built around the Full, Open, Truthful Exchange of information, preferences, and so forth is often—but not always—a good practical aid for negotiators to adopt—or at least to think hard about.

Full, Open, Truthful Exchange

In meeting new classes of problems, we will often adopt a jointly normative stance to facilitate clear and straight analytical thinking. We'll investigate cases where the parties agree on an idealized, collaborative style of deliberation in which they try jointly to solve their problem by adopting the style of Full, Open, Truthful Exchange. They keep no secrets from each other—at least as far as the current negotiations are concerned—and they divulge to each other the truth, nothing but the truth, and the whole truth. (There's the rub—the *whole* truth!)

You might think that if this ideal is assumed, there's nothing left to talk about. But this is not the case. Even with the FOTE approach we'll see that there's a lot to discuss and a great number of problems to be resolved. Even when we discard conflictual factors from our analysis, there is still a huge array of advice that can assist cooperatively minded negotiators in reaching mutually beneficial agreements. For instance, in deals between friendly partners, there may be differences in values, in perceptions, in fallback positions, in time preferences, in power. Without advice, the parties involved might not make the most of the potential synergies. Before immersing ourselves in the art of deal making, we should remember that negotiations that begin as deals can, and do, deteriorate into disputes.

Collaborative decision making: Real-world examples.   Consider some real-world examples of decision making assisted by FOTE.

My wife and I (HR) negotiate in the FOTE manner—at least we do most of the time. That may be why we've been happily married for over fifty years.

Business partners may swear that not unlike some married couples, they behave according to the FOTE ideal. It's part of their religion: if ever they start lying to each other, it's time to dissolve their union.

Some students in laboratory-simulated exercises in negotiation start by being maximally deceptive but end up being converts to principled negotiation. They indicate that as businesspersons they will seek out partners who, like themselves, would like to negotiate in the FOTE style. Of course, they had better be careful lest they fall prey to some devious counterpart. FOTE is not advisable in all cases.

Adversaries who hate each other and would not trust each other in real life may nevertheless know so much about each other that it's senseless not to tell the truth during negotiations. These interactions involve "intimate enemies." For instance, A says to B: "Don't waste my time telling me how much you hate me. I know why you hate me and I know that you know why I hate you. You have spies and I have spies. But still we both prefer to be alive than dead, so let's see if we can act professionally in order to find an outcome that we would both prefer to the deadly status quo. I, for one, have nothing to hide in these negotiations, and we have a lot to gain if we trust each other without posturing during them."

Some negotiators may be persuaded to confide the truth, in confidence, to an outside facilitator. They would not negotiate in FOTE style without the facilitator, but with this external helper's assistance, FOTE-style negotiations can take place. Some of the techniques we shall develop could be implemented by such an outside person.

And some negotiators might try to plead with the other side to adopt the FOTE condition because they know that if it can be achieved, they will have a better expected outcome.

Finally, now and then two idealists interact and readily agree to the FOTE approach.

The importance of a polar extreme. In the extreme, idealized case, certain basic concepts such as efficiency and equity, along with the bottom-line values that each side has, gain clarity, simplicity, and a crispness of definition. We'll also see that when we back off from the polar extreme, not all is lost. Parts of FOTE analysis may still be appropriate in giving partisan advice. We will often disengage from FOTE by considering the important special case where the parties tell the truth but not necessarily the whole truth. For example, they may be most reluctant to disclose their bottom lines—their reservation values. We'll refer to this important case as POTE—Partial, Open, Truthful Exchange.

Finally, we mention that it takes two to tango. Idealistically, you may want to negotiate in a FOTE style, but you might not be able to trust the person across the table. In these cases the FOTE approach is unworkable. It also can be true that in some instances undertaking a full, idealized analysis may be daunting; we'll have something to say about pragmatic shortcuts and compromises. FOTE remains a benchmark by which to judge these compromises, however.

Some Organizing Questions

Early in my research on these matters, I had the grandiose idea of devising a taxonomy of negotiations, in which the listing would be reasonably exhaustive and in which overlaps among categories would be rare. This was possible, I found, only after developing a host of abstract constructs—and even then the taxonomy was not very useful. For our purposes here, and to give a flavor of the sweep of topics to be discussed in Parts II–V, a partial classification will be sufficient. We'll do this by identifying the important characteristics of negotiations.

Dispute or Deal?

Think back to the last negotiation you were involved in. What was the context? Haggling at a bazaar? Bargaining for a car? Resolving a disagreement with your teenage daughter? Negotiating the end to an ethnic conflict? Trying to retrieve your passport from a corrupt customs official? Talking down the cost of court fees with your lawyer? Arguing for a salary raise? These sorts of answers are most common—negotiating in a dispute of some kind.

Typically, when people think about past negotiations, they focus on situations involving conflict. In disputes emotions usually run high. Disputants behave competitively and expend most of their energy trying to claim value for themselves. They develop negative perceptions of the other side, and make use of threats and warnings. As a rule, disputing opponents perceive the game to be zero-sum: a loss for me is a gain for you and vice versa. Disputes command a lot of attention in the field of negotiation research. Consequently, we have many ideas and tools of conflict management, conflict resolution, and alternative dispute resolution. But disputes constitute only part of the picture.

Let's return to our opening question but frame it slightly differently: think back to the last *deal-making* negotiation you were involved in. What was the context? Deciding where to go on a much-awaited holiday? Deciding on the architectural plans for your new house? Negotiating over which delicious wines to serve at a feast? Choosing a restaurant for a graduation celebration? Selecting a church for a wedding ceremony? Pondering where to invest the profits of a successful business venture? These examples are distinguished by a good deal of cooperative behavior, quite unlike what you might find in disputes. Surprisingly enough, collaborative, joint problem solving is often ignored in the field of negotiation. As a result, we lack the concepts and tools to extract the maximum potential value out of such deal-making situations.

More than Two Parties?

There is a vast difference between conflicts involving two disputants and those involving more than two. Once three or more conflicting parties are involved, coalitions of disputants may form and act in concert against the others. Game theorists have long made a distinction between two-person games and many-person games, where "many" is interpreted as being greater than two. The United Nations Law of the Sea negotiations is one example of a game with many players: the Group of 77 (in reality some 114 developing nations) is one reasonably stable coalition of players in this game.

In some conflicts the disputing parties are not well specified. Consider a dispute between a developer and a group of disturbed citizens who can organize themselves into negotiating entities but have not yet done so. A group may form, but during negotiations its members may not agree among themselves and may splinter into subgroups, each demanding representation in the negotiations.

At other times, well-specified negotiating parties might jointly decide who else should be invited to join them at the conference table; thus part of the negotiations may be taken up with deciding just who is to negotiate.

Are the Parties Monolithic?

When U.S. ambassador Ellsworth Bunker negotiated the Panama Canal Treaty with his counterpart from Panama, three agreements had to be made: one across the table (United States and Panama), one within the U.S. side, and one within the Panamanian side. Bunker spent much less time negotiating externally than he did internally within the United States, where there were vast differences of opinion—differences among the Department of Defense, the Department of State, the Department of Commerce, the Department of Transportation, and so on. It is a delicate and highly intricate matter to be able to synchronize external and internal negotiations. On the internal side, the president of the United States and his ambassador play a role not unlike that of a mediator, but a mediator with clout.

Far from being exceptional, it is commonly the case that each party to a dispute is not internally monolithic: each party might comprise people whose values differ, perhaps sharply. Even if one side consists of only a single person, that person might still experience internal conflicts. We are not implying that the diversities that exist internally within each team make bargaining more difficult between teams; indeed, the more diffuse the positions are within each side, the easier it might be to achieve external agreement. But we do wish to emphasize how important it is in discussing negotiation to be aware of internal as well as external conflicts.

Sometimes the internal differences will be between agents and a group of principals. Whenever the United States signs a treaty with another nation, the U.S. Senate must ratify it before it becomes binding. Analogously, a union leader might settle on a contract with management, but before it becomes operative the union rank and file must ratify the agreement. Further last-minute concessions might be squeezed out of the other side during this ratification process: "salami" tactics—one slice more. What is even more important, the ratification process might strengthen the side requiring it—but, of course, it might also make negotiations much less flexible and less amicable, and might stiffen the resolve of the other side.

In some circumstances, negotiators themselves may artificially create a ratification requirement. For example, a corporation president, though he has the authority to commit his firm to an agreement, might say to the other negotiator, "Of course, this agreement is acceptable to me, but my board of directors will have to ratify it." Such ploys can adversely affect the atmosphere of the negotiations.

Is the Game Repetitive?

When people haggle in a bazaar-like fashion over one-shot issues such as the price of a used car or the price of a home, each disputant may have a short-run perspective that tempts him to exaggerate his case. Contrast this type of negotiation with those cases in which the bargainers will be bargaining together frequently in the future and in which the atmosphere at the conclusion of one bargaining session will carry over to influence the atmosphere at the next bargaining session. When bargaining is repetitive, each disputant must be particularly concerned about his reputation, and hence repetitive bargaining is often done more cooperatively (and honestly) than single-shot bargaining—luckily for society. But this is not always so: with repetition there is always the possibility that some inadvertent, careless friction can fester and spoil the atmosphere for future bargaining; this is especially true when there are differences in the information available to both sides. With repetition, a negotiator might want to establish a reputation for toughness that is designed for long-term rather than short-term rewards.

Are There Linkage Effects?

When the United States in the 1970s negotiated a contract with the Philippines concerning military base rights, the negotiators had to keep in mind similar contracts and treaties that were pending elsewhere, such as in Spain and Turkey. One negotiation becomes linked with another through the ramifications of precedence. Repetitive games also involve linkages that arise from repetitions with the same players over time.

The U.S. Senate, in discussing the SALT II Treaty, linked these negotiations to others involving defense spending. In grain negotiations with the Soviet Union, the United States threatened to link food to oil.

One must be aware of the intricacies caused by linkages and, to put it more positively, one must use linkage possibilities to break impasses in negotiations. This is not done creatively enough in most disputes.

More than One Issue?

In selling or buying a house, a car, or even a firm, the critical issue is the final price of the transaction. This is the case even in some labor-management disputes in which the wage rate may be the overwhelmingly dominant factor. One side wants a higher settlement value; the other side, a lower settlement value. The sides are in direct conflict. Of course, both might prefer some reasonable settlement to no settlement at all.

In most complicated conflicts there is not one issue to be decided, but several interacting issues. There were virtually hundreds that had to be resolved in the UN Law of the Sea conferences. Some of the issues were economic; others were political; still others involved military considerations. In such settings, each side, in comparing possible final agreements, must carefully examine and thrash out its own value tradeoffs—and one must remember that each side may not be monolithic and that these tradeoffs do not usually involve naturally commensurable units.

The point is that disputants are engaged in a horrendously difficult analytical task in which there is vast room for cooperative behavior. When there are several issues to be jointly determined through negotiation, the negotiating parties have an opportunity to considerably enlarge the pie before cutting it into shares for each side to enjoy. Negotiations rarely are strictly competitive, but the players may behave as if they were competitive; the players might consider themselves to be strictly opposed disputants rather than jointly cooperative problem solvers. We do not live in a zero-sum society—it is not true that what one gains another must necessarily lose. The trouble is that we often act as if this were the case.

The parties may start their negotiations by trying to decide what will be at stake. Often they may need to be flexible; they may want to introduce new issues or eliminate old ones as part of the negotiation process. Thus one issue in the negotiation may be to determine just what issues should be included in the negotiation.

Is an Agreement Required?

If a potential seller and buyer of a house cannot agree on a price, they can break off negotiations. During negotiations, each has a mild threat: he can simply walk away. Contrast this case with that of a city which is negotiating a complex wage settlement with its police force or firemen. By law, a contract must be settled by a given date. True, the parties might delay and miss critical deadlines, but eventually they must settle on an agreement. When contracts have to be made, the parties may be required by law to submit their unsettled cases for mediation and arbitration.

If an agreement is not required—or not required at a particular stage of negotiation—each party must contemplate what might happen if negotiations were to be broken off. In these circumstances, each party faces a complex decision problem under uncertainty, and the negotiator must somehow figure out just how much he must obtain in the negotiations before he would be indifferent between settling for that amount or breaking off negotiations. This phase of analysis—the determination of a minimal return that must be achieved in negotiations—is usually not well carried out in practice.

Even in those cases where, by law, contracts eventually have to be agreed upon, negotiations may be protracted, and at any stage a negotiator might want to think about her rock-bottom conditions for agreement to a contract at that particular point. "If you can't get this much at this time, then break off negotiations until next week"—so go the instructions.

Are Threats Possible?

If the buyer of a house objects to the price offered by the seller, the buyer can threaten to walk away. This is called the *fixed threat* to go back to the status quo ante. Contrast that situation with the case where a party says, "If you do not agree with my offer, not only will I break off relations, but I will take the following actions to hurt you." Certainly threats can influence outcomes, but if used crassly or carelessly they can easily stiffen the opposition. Indeed, it can be demonstrated in laboratory situations that increasing the power of one side (everything else being equal) can empirically result in poorer outcomes *for that side* (and usually for the other side as well). Power is often not used artfully.

This question is related to others presented here, for the divisions among these topics are not distinct. Threats by their very nature tend to link problems, for example, and problems are often linked in order to make threats possible and credible.

Are There Time Constraints or Time-Related Costs?

When the United States negotiated with the North Vietnamese toward the close of the Vietnam War, the two sides met in Paris. The first move in this negotiation game was made by the Vietnamese: they leased a house for a *two-year* period.

The party that negotiates in haste is often at a disadvantage. The penalties incurred in delays may be quite different for the two parties, and this discrepancy can be used to the advantage of one side. It can also be misused by one side to the disadvantage of both sides, as we shall see. In some negotiations, the tactic of one side might be to delay negotiations indefinitely. For example, environmentalists can often discourage a developer through protracted litigation. In a civil liability suit, an insurance company can use delays in bringing a case to court in order to get the plaintiff to accept a more favorable (to the insurance company) out-of-court settlement.

Are the Contracts Binding?

How can the Israelis or the Palestinians be sure that the other side will abide by an agreement after their respective current leaders have left center stage? They can't. Any agreement is risky—but so is no agreement.

In many conflicts within a nation-state, agreements can be signed and actions made legally binding. The courts are there to put muscle into agreements. Contrast this situation with the case of a multinational mining company that is negotiating a joint mining venture with a developing nation. The multinational is to supply the initial capital and know-how, the developing country the physical resources; and if profits are to be reaped, they might agree to share these profits in certain proportions. Indeed, the agreed-upon proportional amounts themselves might be contingent on other factors, including, for example, the size of the cash flows. But suppose the multinational firm is afraid that the developing country might unilaterally break the contract at some later date (for example, by nationalizing). In order to protect itself, the multinational might bargain harder for a quicker payback period—but, alas, this tactic might hasten the very counterreaction that the firm fears. Uncertainty abounds.

Are the Negotiations Private or Public?

It's hard to keep secrets nowadays, at least in the public sector. In negotiations involving many issues, a common tactic is to look for compensating compromises: party A gives in a little on one issue and party B reciprocates, giving in on another issue. When A gives up a little, A might want to exaggerate what it's giving up, while B will minimize what it's getting—all in preparation for a compensating quid pro quo. But now imagine the prime minister of Israel making a concession to the Palestinian leaders and making an exaggerated claim of the importance of his concession. How will this be reviewed by the Knesset once his stance is made public?

Public pronouncements (and leaks to the press) can be artfully employed to bolster the credibility of commitments. The public posture of one side can influence the internal negotiations of the other side.

When negotiating parties are not monolithic or when ratification is required, it is critically important to know just how secret the secret negotiations are. It is not easy to negotiate in a fishbowl surrounded by reporters, who themselves feel conflicting desires to both get at the truth and get a spicy, newsworthy story.

What Are the Group Norms?

What norms of behavior do you expect of the others in your negotiation? Will they tell you what they truly feel? Will they disclose all relevant information? Will they distort facts? Will they threaten? Will they abide by their word? Will they break the law? Certainly, the modes of behavior you expect when discussing a point of disagreement with your spouse or your business partner are different from those you can expect to occur between firms or between countries or between extortionist and victim.

In the chapters that follow, we will dwell at length on the problems of *cooperative antagonists*. Such disputants recognize that they have differences of interest; they would like to find a compromise, but they fully expect that all parties will be primarily worried about their own interests. They do not have malevolent intentions, but neither are they altruistically inclined. They are slightly distrustful of one another; each expects the other to try to make a good case for its own side and to indulge in strategic posturing. The opposing sides are not confident that their adversary will be truthful, but they would like to be truthful themselves, within bounds. They expect that power will be used gracefully, that all parties will abide by the law, and that all joint agreements will be honored.

We will not deal extensively with the problems of *strident antagonists*, who are malevolent, untrustworthy characters. Their promises are suspect, they are frequently double-crossers, and they exploit their power to the full. Sometimes it's not clear whether such a disputant is really a madman or just acting that way. Think of a hijacker, or of an extortionist who is holding an executive's child hostage.

We will also consider the problems of *fully cooperative partners*. Such negotiators might have different needs, values, and opinions, but they are completely open with one another; they expect total honesty, full disclosure, no strategic posturing. They both adopt a FOTE style and this is common knowledge. They think of themselves as a cohesive entity and they sincerely want to do what's right for that entity. This would be true, for instance, of a happily married couple or some fortunate business partners. Only occasionally do teams of scientific advisers or faculties of universities fall into this category.

Our primary subject will be negotiators who conform to the group norm in the middle: that of cooperative antagonists. Sometimes negotiations start in this category and slide toward conflict. One aim of an external helper is to prevent this from happening and to nudge negotiations toward the full-cooperation category.

Can an External Helper Be Used?

Negotiations are affected by the possible availability of external helpers, usually mediators or arbitrators. This is customarily referred to as "third-party" intervention, even when there are more than two disputants. We prefer to use the neutral term "external helpers." For one reason or another, a disputant may say to himself, "If I bargain tough and do not succeed, then I can always submit my case to arbitration." Or, "I'd better be more reasonable—if not, an outsider will be brought in and who knows what will happen." A negotiator must consider if and when to suggest (or to agree with the suggestion of) the involvement of an external helper. Usually this poses a complex decision problem with vast uncertainties. If an external helper does enter the dispute, the negotiator has a new set of tactical options: How much should he reveal? How cooperative should he be? How truthful? The problem can be viewed from the perspective of either negotiator or external helper. We take both perspectives in this book.

Core Concepts

The chapter has discussed the cooperative essence of joint decision making that is the hallmark of negotiation. The parties may jockey about in game-like moves, but eventually they have to adopt a common action or contract that they all have to live with. The negotiator can often identify strategic aspects of his or her negotiation interactions that can be partially analyzed and helped by a decision analytic or game-theoretic perspective.

Most negotiations present a tension between tactics that *create joint value* and tactics that are designed to *claim individual value*. How to balance creating and claiming is the negotiator's dilemma and lurks in the background of most real negotiations.

The book often considers a polar extreme when all parties agree to negotiate in a FOTE manner—that is, with Full, Open, Truthful Exchange—and when this is common knowledge. After better understanding that polar case here, we backed off to consider the case in which the parties tell the truth but perhaps not the whole truth. For example, they might withhold information about their bottom lines. We call this POTE analysis, or Partial, Open, Truthful Exchange. And finally we considered cases where the parties are not to be trusted.

To gain some perspective, we then discussed organizing questions about negotiation that serve as a partial taxonomy.

Two-Party Distributive (Win-Lose) Negotiations

Having laid out a basic introduction to the fields we attempt to synthesize—decision analysis, descriptive decision making, and game theory—we will now look at negotiation. We start with the simplest case—two-party distributive negotiation. By *distributive* we mean those negotiations (or those parts of larger negotiations) concerned with the division of a single good. *Distributive* is opposed to integrative negotiation, where *integrative* means integrating the parties' capabilities and resources to generate more value. Other commentators use *claiming* and *creating* for the same concepts.

Integrative negotiation is making the pie bigger. Distributive negotiation is about getting a bigger piece for oneself. The good at stake is usually money, but could be time (which parent spends tonight with the kids), space (who gets more of the armrest on that long flight), or any other single commodity. Chapter 6 presents a case study of a prototypical negotiation problem. A seller (in this case an institution that owns a halfway house) wishes to sell an asset (its residence, the house) to a buyer (a developer). The seller wants more, the buyer less, and they haggle. It's all a matter of claiming a larger part of a pie of fixed size. Two parties negotiate over one issue—money—but it could be time or any other single commodity.

In Chapter 7 the problem is abstracted, and the discussion addresses such questions as: How to prepare? Who should declare first? Why? How extreme a first offer? What's a reasonable counteroffer? What is the pattern of concessions? And so forth. The chapter discusses a related double-auction game in which the seller and buyer simultaneously offer sealed bids and a transaction takes place if and only if the bids are compatible. This game is analyzed normatively, descriptively, and prescriptively, and we conclude that good old-fashioned haggling might be better.

Chapter 8 complicates the picture by introducing uncertainty, such as the uncertainty of a trial outcome. We analyze the negotiation of an out-of-court settlement in a liability case. The dominating uncertainties are who will win the court battle if the case goes to trial and the size of the jury award if the plaintiff prevails. We examine the problem as an individual decision problem under uncertainty and illustrate and extend the methodology introduced in Chapter 2.

Chapter 9 introduces the complication of time. It starts off with a seller dynamically searching for buyers. When to quit searching is another one of those decision problems under uncertainty. The central case of this chapter is the discussion of a hypothetical strike game, and this is analyzed from the perspective of the individual decision maker and of game theory. Empirical laboratory observations demonstrate the potential for dysfunctional escalatory, vindictive behavior. Strike mechanisms have their problems, but sometimes negotiating parties may wish they had a strike mechanism. Here is an additional opportunity for negotiation.

Chapter 10 considers the case in which many buyers face just one seller. The seller can negotiate sequentially with each potential buyer or engage in a competitive bidding or auction procedure. The chapter compares several different types of auction procedures, and in order to give partisan prescriptive advice to one of our players, we need to further develop each of our supporting building blocks: individual decision analysis, behavorial decision theory, and the theory of games.

We end the chapter by introducing the complexities of combinatorial bidding. There are now several items up for bid, and your valuation of some of these items depends, in part, on what other items you will win. This is the case of the famous FCC Auction in which the government auctioned off hundreds of licenses for the air waves.

6

Elmtree House

The following case study is mostly make-believe; one might speak of it as an "armchair" case. It involves a professor of business—we'll call him Steve—who was quite knowledgeable about finance but not a practitioner of the art and science of negotiation. We'll use the case to get a handle on the distributive part of negotiation—to build vocabulary and analytic concepts and maybe come to a few prescriptive conclusions. I (HR) have used this as a teaching example for quite a while. At critical junctures I stop and ask the students what advice they would give the protagonists. We'll ask you the same questions.

The Problem—and the Opportunity

Steve was on the governing board of Elmtree House, a halfway house for young men and women aged eighteen to twenty-five who needed professional guidance and the support of a sympathetic group to ease their transition from mental institutions back to society. Many in the house had had nervous breakdowns, or were borderline schizophrenics, or were recovering from unfortunate experiences with drugs. Located on the outskirts of Boston in the industrial city of Somerville, Elmtree House accommodated about twenty residents. The neighborhood was in a transitional stage; some said that it would deteriorate further, others that it was on the way up. In any case, it did not provide an ideal recuperative setting because of its noise and disrepair. Although the house was small and quite run down, the lot itself was extensive, consisting of a full acre of ground. Its stand of elm trees, once magnificent, had succumbed to disease.

The governing board, through a subcommittee, had earlier investigated the possibility of moving Elmtree from Somerville to a quieter, semi-residential community. Other suitable houses were located in the nearby cities of Brookline, Medford, and Allston, but the costs were prohibitive and the idea of moving was reluctantly dropped.

Some months later, a Mr. Wilson approached Elmtree's director, Mrs. Peters, who lived in the house with her husband and child. Wilson indicated that his firm, a combined architectural and developmental contractor, might be interested in buying the Elmtree property. This was out of the blue. No public announcement had ever been made that Elmtree House was interested in a move. Mrs. Peters responded that the thought had never occurred to her, but that if the price were right, the governing board might consider it. Wilson gave Mrs. Peters his card and said that he would like to pursue the topic further if there were a chance for a deal.

Even simple-looking decisions can be deceptively complex. For example, should Mrs. Peters let on that Elmtree might welcome the opportunity to move? On the one hand, she would like to affect Wilson's aspiration level (which is positively correlated with his distributive outcome, and thus negatively correlated with Elmtree's), implying that he'll have to pay a high price. On the other hand, she would also like to develop a positive relationship with him, if only for the duration of this negotiation. Revealing her side's desires, especially when it may be against her interest, might inspire trust and raise his degree of concern over the outcome for Elmtree. There is also the risk that Wilson is a more casual buyer, and might easily walk away if discouraged.

What would you do?

The governing board asked Steve to follow up on this promising lead. The other board members included local clergy and prominent individuals in clinical psychology, medicine, and vocational guidance; none besides Steve had any feeling for business negotiations of this kind. Since they fully trusted Steve, they essentially gave him carte blanche to negotiate. Of course, no legal transaction could be consummated without the board's formal approval.

What advice would you give Steve?

Steve sought my advice on how he should approach Mr. Wilson, and we decided that an informal phone call was in order. In the course of the call, Steve accepted an invitation to discuss possibilities over cocktails at a nearby hotel. He decided not to talk about any money matters at that first meeting—just to sound out Wilson and find out what he might have in mind. He insisted, I think rightly, on paying his own bill. I assured him that he also did rightly in not even hinting to Wilson that the governing board was looking for other locations.

On the basis of that first meeting and some outside probing into Wilson's business affiliations, Steve ascertained that Wilson was a legitimate businessman of decent reputation. Steve thought that Wilson's company wanted the Elmtree property as a possible site for a condominium project. Wilson wished to talk money matters right away, but Steve needed a couple of weeks to prepare for negotiations. He used the excuse that he needed the approval of the governing board before he could proceed to serious negotiations.

Are such strategic misrepresentations of the truth an acceptable mode of be-havior? Given that Steve has two weeks to prepare (about fifteen working hours), what should he do?

Students are surprisingly tough in their responses to this case study. They generally suggest that Steve invent all sorts of stories, because such misrepresentations would seem to be in the interests of a good cause and because the students identify with the housing plight of the residents of Elmtree House. This setting was purposely chosen for the case study to stir these emotional feelings.

Preparing for Negotiation

What kind of information should Steve look for in advance of negotiations?

Gathering Information

During the next twelve days, Steve did a number of things. First, he tried to as-certain Elmtree's reservation, or walk-away, price (RP)—that is, the minimum price that Elmtree House, the seller, could accept. The reservation price was dif-ficult to determine, because it depended on the availability of alternative sites to relocate. Steve learned that of the other sites that had previously been located, the one in Brookline was no longer available, but the other two, in Medford and in Allston, were still possibilities—for the right price. Steve talked with the own-ers of those sites and found out that the Medford property could be had for about $175,000 and the Allston property for about $235,000.[1]

Steve decided that Elmtree House would need at least $220,000 before a move to Medford could be undertaken and that it would need $275,000 to jus-tify a move to Allston. These figures took into account the cost of moving, minor repairs, insurance, and a small sum for risk aversion. The Allston site (needing $275,000) was much better than the Medford site (needing $220,000), which in turn was better than the site at Elmtree. So Steve decided that his reservation price would be $220,000. He would take nothing less, and hope to get more—possibly enough more to justify the Allston alternative. This bit of research took about six hours, or a couple of evenings' work.

Meanwhile Steve's wife, Mary, contacted several realtors in the hope of

1. These were not firm figures, but Steve's assessed distributions of these amounts were tightly distributed about these central values; each judgmental distribution had a standard deviation of about $15,000. This means that roughly Steve would give 2-to-1 odds that the actual selling price of the Medford property would be within $15,000 of $175,000 and 19-to-1 odds that the actual selling price would be within $30,000 of $175,000. Analogously for Allston. These are 1975 figures.

finding other locations Elmtree could consider. There were a few nibbles, but nothing definite turned up.

Steve next investigated what Elmtree House would bring if sold on the open market. By examining the sale prices of houses in the vicinity and by talking to local realtors and real estate experts, he learned that the Elmtree property was probably worth only about $125,000. He felt that if sold without Wilson in the picture, the house would go for between $110,000 and $145,000 (with the mean of the distribution halfway in between), and it was just as likely to go below $110,000 as above $145,000. How disappointing! This took another four hours of research time.

What was the story from Wilson's perspective? It was difficult for us to make judgments about the buyer's reservation price—that is, the maximum price that Wilson would be willing to offer before he definitely would break off negotiations, not temporarily for strategic purposes, but permanently. Neither Steve nor I had any expertise in the matter. We went for advice to a number of real estate experts and also queried two contractors in the Boston area. Our experts did not agree with one another, but they all took our question about reservation price seriously; we were convinced that they understood our problem. A lot, we were told, depended on the intention of the developers. How high a structure would they be permitted to build on the site? Were they buying up other land as well? Steve found out that the answer to the latter question was yes. The matter turned out to be much more involved than Steve or I had imagined it would be. After ten hours of his time and five hours of my time, we decided that we were hopelessly vague about our assessment of Wilson's reservation price. Figure 6.1 shows Steve's assessed probability density function (that is, all the possible values, and the relative likelihood of each)—all things considered—of Wilson's RP. As of two days before the start of real negotiations, Steve would have bet even money that Wilson's RP lay in the interval from $250,000 (the lower quartile) to $475,000 (the upper quartile).[2]

How should Steve conduct the talks? Where should he hold them?

Preparing a Strategy

After all this preparation, Steve and I discussed his negotiation strategy. It had already been decided that the meeting would be at a hotel suite to which Wilson's company had access. Steve and I had no objection to this venue; the dining

---

2. One expert thought that there was a reasonable (over 25 percent) probability that Wilson would go as high as $600,000; another thought that the chances of this were minuscule (less than 1 percent). Too bad we couldn't have had them bet with each other and taken a brokerage fee for our entrepreneurial efforts.

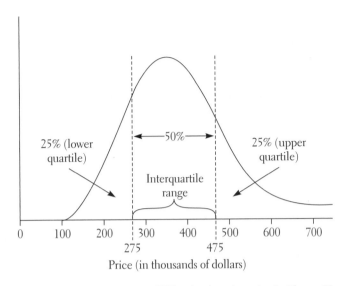

25% (lower
quartile)

←——50%——→

25% (upper
quartile)

Interquartile
range

| 0 | 100 | 200 | 300 | 400 | 500 | 600 | 700 |

275   475

Price (in thousands of dollars)

Figure 6.1.   Steve's probability assessment of Wilson's reservation price for Elmtree House.

room of Elmtree House would have been too hectic, and his own university office inappropriate.

Who should come?

Feeling that he needed someone at the discussions to advise him on legal
details, Steve decided to invite Harry Jones, a Boston lawyer and former member
of Elmtree House's governing board. Jones agreed to participate, and Steve reserved two hours to brief him prior to the meeting.[3]

We also thought it might be a good idea for Steve to bring along Mrs. Peters.
She was the person who was most knowledgeable about Elmtree House, and
perhaps an appeal to Wilson's social conscience might help. It was agreed that
Steve alone would talk about money matters. Mrs. Peters would talk about the
important social role of halfway houses and to argue that it did not make sense
for Elmtree House to move unless a substantial improvement in the surrounding amenities would result: "You know how hard it is on kids to move from one
neighborhood to another. Just think how severe the effects will be on the young
residents of Elmtree House." Mrs. Peters actually did have conflicting feelings
about moving, and it would be easy for her to marshal arguments against a
move.

What should his opening tactics be?

Who should start the bidding first? If Wilson insisted that Steve make the
first offer, what should that be? If Wilson opened with X thousand dollars, what

---

3. One colleague of mine suggested that bringing a lawyer to the initial negotiations might
have hurt Steve's cause: it indicated too much of a desire to do business and to settle details.

should Steve's counteroffer be? How far could this be planned in advance? Were there any obvious traps to be avoided?

Steve and I felt that our probabilistic assessment of Wilson's RP was so broad that it would be easy to make a mistake by having our first offer fall below his true reservation price. But if we started with a wildly high request like $900,000—way over what we would settle for—it might sour the atmosphere.

Steve decided to try to get Wilson to move first; if that did not work and if he was forced to make the first offer, he would use the round figure of $750,000, but he would try to make that offer appear very flexible and soft. Steve thought about opening with an offer of $400,000 and holding firm for a while, but we felt there was a 40 percent chance that this amount would be below Wilson's RP. If Wilson moved first, Steve would not allow him to dwell on his offer but would quickly try to get away from that psychologically low anchor point by promptly retorting with a counteroffer of, say, $750,000.

I told Steve that once two offers are on the table—one for each party—the final point of agreement can reasonably be predicted to fall somewhere close to midway between those two extremes. So if Wilson offered $200,000 and if Steve came back with $400,000, a reasonable bet would be a settlement of $300,000—provided, of course, that that midway figure fell within the zone of possible agreement (ZOPA), the range between Steve's (the seller's) true RP and Wilson's (the buyer's) true RP. For starters, Steve thought that it would be nice if he could get $350,000 from Wilson, but of course Steve realized that his own RP was still $220,000.

We talked about the role of time. What if Wilson's most recent offer was above $220,000? Should Steve be willing to bluff and make a show of walking away from the bargaining table even though the offer is better than his reservation value? If Wilson didn't call back with a better offer, Steve could say that the Elmtree board had reconsidered. I reminded Steve that there is no objective formula for this. He would be confronted with a standard decision problem under uncertainty, and his assessment of Wilson's RP could be better evaluated after sounding out Wilson than it could be with present information. The danger in breaking off negotiations—and a lot depends on how they're broken off—was that Wilson might have other opportunities to pursue at the same time.

Opening Gambits

As it turned out, the first round of negotiations was, in Steve's view, a disaster, and afterward he wasn't even sure that there would be a second round. Mrs. Peters played her part admirably, but to no avail; it seemed unlikely that Wilson would raise his offer to Elmtree's reservation price. After preliminary pleasantries and some posturing, Wilson said, "Tell me the bare minimum you would accept from us, and I'll see if I can throw in something extra."

How should Steve respond?

Steve expected that gambit, and instead of outright misrepresentation, he responded, "Why don't you tell us the very maximum that you are willing to pay, and we'll see if we can shave off a bit." Luckily, Wilson was amused at that response. He finally made his opening offer at $125,000, but first bolstered it with a lot of facts about what other property was selling for in that section of Somerville. Steve immediately responded that Elmtree House could always sell its property for more money than Wilson was offering, and that it did not have the faintest intention of moving. They would consider moving only if they could relocate in a much more tranquil environment where real estate values were high. Steve claimed that the trouble of moving could be justified only by a sale price of about $600,000, and Mrs. Peters concurred.[4] Steve chose that $600,000 figure keeping in mind that the midpoint between $150,000 and $600,000 was above his aspiration level of $350,000. Wilson retorted that prices like that were out of the question. The two sides jockeyed around a bit and decided to break off, with hints that they might each do a bit more homework.

Where should Steve go from here?

Steve and I talked about how we should reassess our judgmental distribution of Wilson's RP. Steve had the definite impression that the $600,000 figure was really well above Wilson's RP, but I reminded him that Wilson was an expert and that if his RP were above $600,000 he would want to lead Steve to think otherwise. We decided to wait a week and then have Steve tell Wilson that Elmtree's board would be willing to come down to $500,000.[5]

Two days later, however, Steve received a call from Wilson, who said that his conscience was bothering him. He had had a dream about Mrs. Peters and the social good she was bringing to this world, and this had persuaded him that, even though it did not make business sense, he should increase his offer to $250,000. Steve could not contain himself and blurted out his first mistake: "Now that's more like it!" But then he regained his composure and said that he thought that he could get Elmtree's board to come down to $475,000. They agreed to meet again in a couple of days for what they hoped would be a final round of bargaining.

4. A student of mine suggested that during negotiations, obvious modifications could have been made to the exterior of Elmtree House to give the impression that the residents indeed had no intention of moving.

5. A colleague to whom I recounted this story thought that our assessment of Wilson's RP should have been updated during the breaks in the negotiations by going back to the experts we had consulted initially; Steve should have been more aware of information he might have obtained from Wilson that the experts could have used to reassess Wilson's RP.

The Dance of Concessions

Following this phone conversation with Wilson, Steve told me that he had inadvertently led Wilson to believe that his $250,000 offer would suffice. Steve also felt that his offer of $475,000 was coming close to Wilson's RP, because this seemed to be the only reason for Wilson's reference to a "final round of bargaining." We talked further about strategy and we revised our probabilistic assessments of Wilson's reservation value and the likely outcome.

Over the next two days there was more jockeying between the two sides, and Wilson successively yielded from $250,000 to $275,000 to $290,000 and finally to a "firm last offer" of $300,000, whereas Steve went from $475,000 to $425,000 to $400,000, and then—painfully—when Wilson sat fixedly at $300,000, inched down to $350,000. Steve finally broke off by saying that he would have to contact key members of the governing board to see if he could possibly break the $350,000 barrier.

Now, $300,000 not only pierced Steve's RP of $220,000 (needed for the Medford move), but also would make it possible for Elmtree House to buy the more desirable Allston property. It had at that point become a question of "gravy." I asked Steve whether he thought Wilson would go over $300,000 and he responded that although it would take some face-saving maneuver, he thought Wilson could be moved up. The problem was, he felt, that if Wilson were involved in other deals and if one of these should turn out well, Wilson might well decide to wash his hands of Elmtree.

Closing Gambits

What should Steve do to finalize the deal?

Steve did two things next. He first asked Harry Jones to put in place all but the very final touches on a legal agreement for acquiring the Allston property. Jones reported the next day that all was in order, but that it was going to cost $20,000 more than anticipated to do some necessary repair work on the house in order to meet Allston's fire standards. Still, $300,000 would meet those needs. Second, Steve worked with Mrs. Peters to find out what an extra $25,000 or $50,000 would mean to Elmtree House. Mrs. Peters said that half of any extra money should definitely go into the Financial Aid Fund for prospective residents who could not quite afford Elmtree House, and that it could also be used to purchase items on her little list of "necessary luxuries": a color television set, an upright piano, new mattresses and dishes, repair of broken furniture, a large freezer so that she could buy meat in bulk, and so on. Her "little list" became increasingly long as her enthusiasm mounted—but $10,000 to $20,000 would suffice to make a fair dent in it, and as Mrs. Peters talked she became even more excited about those fringes than about the move to Allston. She was all for holding out for $350,000.

The next day Steve called Wilson and explained to him that the members of Elmtree's board were divided about accepting $300,000 (that was actually true). "Would it be possible for your company to yield a bit and do, for free, the equivalent of $30,000 or $40,000 worth of repair work on Elmtree's new property if our deal with you goes through? In that case, we could go with the $300,000 offer." Wilson responded that he was delighted that the board was smart enough to accept his magnanimous offer of $300,000. Steve was speechless. Wilson then explained that his company had a firm policy not to entangle itself with side deals involving free contract work. He didn't blame Steve for trying, but his suggestion was out of the question.

"Well then," Steve responded, "it would surely help us if your company could make a tax-free gift to Elmtree House of, say, $40,000, for Elmtree's Financial Aid Fund for needy residents."

"Now that's an idea! Forty grand is too high, but I'll ask our lawyers if we can contribute twenty grand."

"Twenty-five?"

"Okay—twenty-five."

It turned out that for legal reasons Wilson's company paid a straight $325,000 to Elmtree House, but Wilson had succeeded in finding a good face-saving reason for breaking his "firm last offer" of $300,000.

Lest readers think erroneously that it's always wise to bargain tough, I might suggest another perfectly plausible version of this story: Wilson might have backed out of the deal suddenly, at the time when he made his firm last offer of $300,000 and Steve demanded $350,000. An alternative venture, competitive with the Elmtree deal, might have turned out to be better for Wilson.

Core Concepts

This chapter has presented a case study portraying a prototypical, distributive negotiation. The seller (a nonprofit) wishes to sell an asset (the residence) to a buyer (a developer). The two parties disagree on the price, and they negotiate. Each wants to claim a larger share of a fixed pie. The two parties negotiate over one issue: money in this case, but it could be time or anything else. In this chapter we give partisan advice (asymmetric prescriptive in our lingo) to the seller based on a descriptive assessment of the behavior of the buyer.

In preparing for a distributive negotiation (or the distributive aspects of a larger negotiation), the seller first determines his reservation price, the minimum value he would just be willing to accept. The seller then considers his (probabilistic) perception of the reservation price of the buyer, and may solicit help from experts. The seller next reviews his tactics for the upcoming negotiations: preliminary rhetoric, who should attend, where, opening gambits, who should start, how should the other react, anchoring, and so on.

The typical "dance of concessions," in which the parties pretend to be ready to walk out while assessing the possibility that the other is telling the truth, may be frustrating and inelegant, but it tends to work well in practice. In the laboratory, parties rarely walk away when there is a positive zone of possible agreement (ZOPA).

In this case, after a disappointing round of first-stage posturing, the parties moved toward a final contract. In the closing, the seller introduced nonmonetary options that might have resulted in joint gains (the buyer providing services in kind such as renovation or help with moving) but this was not achieved.

7

Distributive Negotiations: The Basic Problem

With Elmtree House as a basis, we can now simplify and abstract. Later we will begin building our understanding of the complexities.

The Distributive Abstraction

Consider the case in which two bargainers must jointly decide on a determinate value of some continuous variable (like money) that they can mutually adjust. One bargainer wants the value to be high—the higher the better—whereas the other bargainer wants the value to be low—the lower the better. We could label these agents "high aspirer" and "low aspirer," but for our purposes "seller" and "buyer" will be sufficient, even though the context we'll be dealing with is much broader than that consisting of simple business transactions in which there is an actual seller and buyer.

To simplify matters, let's assume that each bargaining agent is monolithic: he or she does not have to convince the members of some constituency that they should ratify the agreement. Let's also assume that the bargaining agents are concerned about this deal only. They won't worry about repetitive plays or linkages to other outstanding problems. Setting precedents, cashing in credits for past favors, and logrolling between problems are not concerns. Time is a more troublesome matter. We will try at first to deemphasize the role of time, or at most to keep it only informally in mind.

The two agents come together to bargain. The setting, the language, the costumes are all irrelevancies for us. We'll assume that the bargainers are honorable people—at least according to the code of ethics of our time—and we will also assume that any contracts made are enforceable and inviolable. No neutral third-party intervenors are present to assist the bargainers. We'll also assume a single-threat environment: at most, any party can threaten only to break off negotiations and revert to the status quo before bargaining. The bargaining milieu can be classified as nonstrident.

Reservation Values

Taking our cues from the Elmtree House illustration, we shall assume that each bargaining party has reflected on the decision problem he or she faces if no contract is made. Each has tried to determine his BATNA, or *Best Alternative to a Negotiated Agreement.*[1] We shall assume that by analyzing the consequences of no agreement, each bargainer establishes the threshold value that he or she needs. The seller has a reservation price, s, that represents the very minimum he will settle for; any final-contract value, x^*, that is less than s represents a situation for the seller that is worse than no agreement.

Buyer's and Seller's Surpluses

If x^* is greater than s, then we can think of $x^* - s$ as the *seller's surplus*. The seller wants to maximize his surplus.[2] The buyer has some reservation price, b, that represents the very maximum she will settle for; any final-contract price, x^*, that is greater than b represents a situation for the buyer that is worse than no agreement. If x^* is less than b, then we can think of $b - x^*$ as the *buyer's surplus* (see Figure 7.1).[3]

Zone of Possible Agreement

If $b < s$—that is, if the maximum price the buyer will settle for is lower than the minimum price the seller will settle for—there is no zone of possible agreement. However, if $s < b$, then the zone of possible agreement (that is, the range of possible solutions for the final contract x^*) is the interval from s to b. Suppose that the final agreement is some value x^*, where x^* is between s and b: the buyer's surplus value is then $b - x^*$ and the seller's surplus value is $x^* - s$. The sum of

---

1. See Fisher, Ury, and Patton (1991) for this term. We will assume that each party considered a robust list of possible alternatives to an agreement with this counterpart, and calculated which one of the alternatives was the most desirable (using something like the decision structure laid out in Chapter 2).

2. In the Elmtree House case, Steve's reservation price, s, was $220,000. The bargainers settled at $x^* = \$325,000$, so Steve, as the seller, had a surplus of $105,000.

3. In the Elmtree House case we were not privy to Wilson's reservation price. Let us suppose that it was $400,000. Then $b = \$400,000$ and the buyer's surplus would have been $75,000.

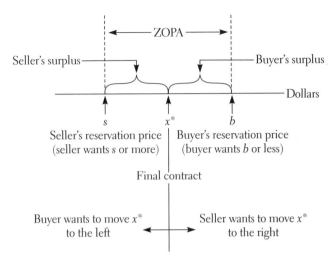

Figure 7.1. The geometry of distributive bargaining. (Note: if $b < s$, there is no zone of possible agreement.)

the surplus values is $b - s$, which is independent of the intervening x^* value. In this sense, the "game"—if we think of the bargaining problem as a game—appears to be constant-sum (in surplus values). But not quite, because if $s < b$ (where a potential zone of possible agreement exists), the parties still might not come to an agreement—they might not agree to settle for a mutually acceptable x^* in the zone of possible agreement. So at most we can only think of this as a quasi-constant-sum game. To make it even more "quasi," the players generally do not know the size of the pie, $b - s$, that they have to divide.

In the abstraction we shall develop, each bargainer knows his or her reservation price, but has only subjectively assessed probabilistic information about the other party's reservation price. Very often in practice the parties have but an imprecise feel for their own reservation price and make no formal attempt to assess a probability distribution of the other party's reservation price.

If we take the asymmetric point of view of one of the bargainers—say, the seller—the seller would be well advised before the negotiations start to ascertain s and to probabilistically assess $\tilde{b}$.[4] During the negotiation, the seller wants to periodically reassess $\tilde{b}$, at least informally; but he also wants to lead the buyer to think that $\tilde{s}$ is higher than it really is. The seller should also be aware that the buyer may be analogously motivated—that is, the buyer wants to make the seller think that b is lower than it really is. To what lengths a player might be willing to

---

4. I use the convention of a tilde to denote an uncertain quantity, or random variable. Thus, the seller knows s but assesses a distribution for $\tilde{b}$; the buyer assesses $\tilde{s}$ but knows b. In the Elmtree House case the seller, Steve, knows that s = \$220,000 and his assessment for the uncertain buyer's reservation, $\tilde{b}$, was depicted in Figure 6.1. Wilson, the buyer, would know b and assess $\tilde{s}$.

go to mislead his or her quasi-adversary—I say "quasi" since we are not discussing a strictly constant-sum game—depends on the culture. In some cultures, it is acceptable to marshal forcefully, but truthfully, all the arguments for one's own side and to avoid giving gratuitous help to the other side. In other cultures it is acceptable to exaggerate or even to bend the truth here and there—but not too much. In still other cultures a really big whopper, if accomplished with flair and humor, is something to brag about and not to hide after the fact, especially if it is successful.

The Streaker: A Simple Role-Playing Exercise

A simple laboratory bargaining problem can be introduced with less than one page of confidential instructions to the seller and one to the buyer.[5] The context is the sale of a used car, the Streaker, and the setting is dated to justify a seller's reservation price of $3,000 and a buyer's reservation price of $5,500. The instructions to each give only the vaguest of hints about the other person's RP. The challenge for a buyer is to get a good deal for herself, and she will be judged in terms of how well she has done in comparison with other buyers in an identical situation; the seller is judged similarly, in comparison with other sellers. This is like a duplicate bridge scoring system.

Players who put themselves in the role of one or the other of these negotiators will naturally ask a number of questions. What analyses should be done? What bargaining ploys seem to work? Should I open first with an offer? If I open first, how extreme should I be? Am I better off giving a reasonable value that would yield me a respectable surplus and remaining firm, or should I start with a more extreme value and pace my concessions with those of the other party? What is a reasonable pattern of concessions?

The Dance of Concessions

A typical pattern of concessions is depicted in Figure 7.2, where s_1, b_1, s_2, b_2, and so on represent the prices successively proposed by the seller and buyer. I call this pattern "the negotiation dance." The seller might open with a value of $7,000 ($s_1$ in the figure); the buyer retorts with $b_1 = \$2,500$; then in succession come $s_2 = \$5,000$ (breaking the buyer's RP), $b_2 = \$3,000$ (breaking the seller's RP), $s_3 = \$4,500$, $b_3 = \$4,000$, and a final-contract price of $x^* = \$4,250$. Would x^* be higher if s_1 were $9,000 instead of $7,000? If so, why not make $s_1 = \$20,000$?

In distributive bargaining, successive offers by the seller are usually mono-

5. I am indebted to John Hammond for this example.

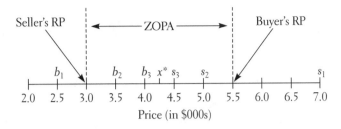

Figure 7.2. The negotiation dance. x^* = final contract price.

tonically decreasing (that is, ratcheting in a single direction), whereas those by the buyer are monotonically increasing. Indeed, one of the principles of good-faith bargaining is that once a concession is made, it is not reversed. The following anecdote depicts an amusing counterexample.[6]

Larry M. gazed somewhat disinterestedly at a briefcase displayed in the window of a luggage store in Mexico City. The proprietor, who spoke English, approached him outside the store and said, "Are you interested in that briefcase?"

"No, I'm just window shopping," Larry replied.

"You can have it for $75. That's a good buy."

Larry had a perfectly acceptable briefcase and said that he was not interested.

"All right, you can have it for $70." Declined.

"How about $65? That's a fantastic buy." Declined.

At this point, Larry became interested. He didn't want the briefcase, but he was curious about how far the shopkeeper would lower his price. So he stayed around, saying nothing.

"I'll sell it for $60. You can't get anything like this at that price in the States." Declined.

"All right, since you're obviously a tourist with a limited budget, just for you I'll give it to you for $55." Declined.

"My final offer: if you promise not to tell anybody, I'll sell it to you for $60."

"Hey, wait a second," interrupted Larry. "You just offered it to me for $55."

"Did I do that? I made a terrible mistake. I shouldn't have done that. But even a mistake must be honored, so for you and only for you I'll sell it for $55."

Larry bought the briefcase for $55.

Empirical Results

John Hammond and I (HR) ran this case with hundreds of business school students at Harvard. More rigorous social scientists would be appalled at our sloppy

6. I am indebted to Philburn Ratoosh for this anecdote.

data collection techniques, and the many ways in which our subjects were influenced by the setting and our lectures. Still, we were looking for broad impressions of how people behave in distributive situations. And what we found seems to match our own anecdotal observations of real negotiators.

Our data yielded a number of interesting findings. First, most pairs of negotiators come to an agreement. The final contracts ranged over the entire zone of possible agreement, from $3,000 to $5,500. A sprinkling (less than 1 percent) of cases were settled out of the zone for a value less than $3,000 or more than $5,500; the subjects in these cases misinterpreted the directions. In some cases, but surprisingly few (around 3 percent), agreement was never achieved.

Second, the average of the final contracts was $4,150 with a standard deviation of 520, indicating a surprising spread of outcomes. The average opening offer of the sellers was $5,250 (standard deviation of 1,160); the average opening offer of the buyers was $2,610 (standard deviation 1,120).

Third, the Boulware strategy of making a reasonable opening and holding firm works sometimes, but more often than not it antagonizes the other party, and many of the no-agreements resulted from this strategy.[7] Advice: don't embarrass your bargaining partner by forcing him or her to make all the concessions.

Fourth, once two offers are on the table (s_1 and b_1), the best prediction of the final contract is the midpoint, $(s_1 + b_1)/2$ — provided that the midpoint falls within the zone of possible agreement. If the midpoint falls outside this zone, then it's hard to predict where the final contract will fall. It is not true that x^* will be near the reservation price that is closer to the midpoint. The reason is that the concessions will have to be lopsided, and it's hard to predict the consequences. Thus if $b_1 = \$2,500$ and $s_1 = \$20,000$, with the midpoint being $11,250, the seller is going to be forced to make huge concessions and x^* might end up closer to $3,000 than to $5,500.

Fifth, from a linear regression analysis it appears that if the buyer's opening bid is held constant, then on the average adding $100 to the opening bid of the seller nets an increase of about $28 to the final contract. If the seller's opening bid is held constant, then on the average subtracting $100 from the opening bid of the buyer nets a decrease of about $15 from the final contract.

Variations

With one group of seventy subjects I ran a variation of the Streaker experiments, with some fascinating but inconclusive results. In the variation, the instructions

---

7. Lemuel Boulware, former vice president of the General Electric Company, rarely made concessions in wage negotiations; he started with what he deemed to be a fair opening offer and held firm. This is commonly referred to as Boulwarism.

to the buyers were the same: as in the original experiments, they still had a reservation price of $5,500. But the instructions to the sellers were altered: they still had to get at least $3,000, but they were told not to try to get as much as possible because of the desirability of later amicable relationships with the buyers. The sellers were told that they would receive a maximum score if they could sell the car for $5,000 and that every dollar above $5,000 would detract from their score; a sale of x dollars above $5,000 would yield them the same satisfaction as x dollars below $5,000. Thus, for example, a selling price of $5,250 would yield the same score as a selling price of $4,750. The buyers, of course, were not aware of these confidential instructions to the sellers.

Surprisingly, the sellers did better playing this variation with benevolent intentions toward the buyer than they did with aggressive intentions to squeeze out as much as possible. In the variation, the average price for the car was $4,570 instead of $4,150. One reason for this might have been that in the original version, the sellers were told only to get more than $3,000 and they did not have any target figure. In the variation, they were told that the best achievable value was $5,000 and this became a target value. Indeed, the sellers' opening offers averaged higher in the variation than in the original exercise ($5,920 versus $5,250). In the variation, the sellers came down faster from high values (above $5,000), but they became more reluctant to reduce their prices as they pierced their $5,000 aspiration level, thus making it seem to the buyers that they were approaching their reservation values.

Some sellers said that they felt some qualms when they let themselves be bargained back from $6,000 to $5,000, even though they got a higher score with the lower price. Some sellers told the buyers that they thought $5,000 was the fair price and that they did not want to get a higher value; but the buyers they were bargaining with tended not to believe them, and these sellers on the average hurt themselves. Additional variations are suggested in Box 7.1.

Analytical Elaboration

Now let's employ the typical mathematician's device: pushing to extreme cases. It might seem that we've already reached the simplest level, but we haven't. Consider the following three special cases.

Case 1. Each Party Knows the Other's Reservation Price

Suppose that the seller and buyer each know their own and their adversary's reservation price. If $b < s$, then there is no zone of agreement: no deal is possible and the parties know it. If $b > s$, then a zone of agreement exists and the par-

---

Box 7.1   Some Distribution Experiments

It would be interesting to run some additional variations. *What do you think would happen?*

1. Give the seller a specified reservation value of $3,000. Hint at a "fair" or "reasonable" value of $5,000, but suggest that getting more would be still better. Let the buyer remain with a reservation value of $5,500.

2. Go back to the first variation in which the seller needs $3,000 and wants $5,000, and in which getting x dollars above $5,000 is like getting x dollars below $5,000. Push the buyer's reservation value below $5,000—to, say, $4,500. It is likely that some sellers will get confused between what they absolutely need ($3,000) and what they aspire to ($5,000).

3. Make the seller's reservation value of $3,000 more vague. Tell the seller, for example, that if the negotiations fall through he will have to sell the car to a dealer, who will offer him one of the three equally likely values: $2,000, $3,000, or $4,000. Since $3,000 is the expected value of the alternative, it should serve as the effective reservation value for the present negotiation; but in this case the seller might bargain more aggressively for values over $3,000.

---

ties have a potential gain of $b - s$ to share. Of course, they get nothing if they can't agree on a sharing rule. Instead of carrying around excess symbols, suppose that $s = \$400$, $b = \$600$, and $b - s = \$200$. How should they share that $200 surplus? The obvious focal point would be in the middle ($100 to each), and that's what happens overwhelmingly in experimental negotiations—provided that some care is taken to balance the environment. Box 7.2 describes several additional experiments, which illustrate some of the dangers of being aggressive.

There is a famous example used by game theorists: How should a rich man and a poor man agree to share $200? The rich man could argue for a $150-to-$50 split in his favor, because it would grieve the poor man more to lose $50 than the rich man to lose $150. Of course, an arbitrator, keeping in mind the needs of the rich man and the poor man, might suggest the reverse apportionment. The rich man could also argue for an even split on the grounds that it would be wrong to mix business and charity: "Why should I be asked to give anything to this poor man? I would rather get my fair share of $100 and give something to a much poorer person."

Instead of dividing up $200, let's introduce another asymmetry by having two bargainers divide up 200 poker chips; as before, no agreement means no chips to either. Suppose further that player A can convert the chips to dollars in equal amounts—one chip equals one dollar—but that player B is given a complicated nonlinear schedule for converting chips to dollars. Figure 7.3 depicts one possible case. If A gets x chips, then B can cash in the remaining $(200 - x)$

Box 7.2   The Zeckhauser Experiments and the Perils of Being Too Aggressive

In one interesting experiment conducted by Richard Zeckhauser, many pairs of subjects were asked to divide $2 between themselves; no agreement meant no money. In the symmetrical version, where both sides had the same instructions, practically all settled on the $1 focal point. In some pairs, one party was secretly prompted to hold out for $1.20 and to hold firm; as expected, the reactions of the opposing parties were also firm—they would rather take nothing than 80 cents. Would this be your preference, too, if you had to share $200 and someone demanded $120?

The subjects were next told to share $2, but they were each penalized 5 cents for every minute it took them to decide on their sharing rule. They quickly jumped to the $1 focal point. Then came an interesting variation: party A was penalized 5 cents per minute of negotiations, whereas party B was penalized 10 cents per minute. Clearly party A had a strategic advantage. But what had become the natural focal point? The surprising thing is that empirically, averaging over many subjects, party A (the stronger party) in this variation did worse—not better, as might have been expected. Once symmetry was destroyed it invited power confrontations, and the seemingly advantaged party A ended up, on the average, worse off than he had been in the symmetric case.

---

chips for an amount in dollars equivalent to the vertical distance above x from the horizontal axis to the negotiation curve. If player A argues that the game is symmetric in chips and that each should get 100 chips, player B would receive $45. If player B argues that the real currency is dollars, not chips, the symmetric solution would give $58 to each: A would get 58 chips, and B would get 142 chips that are convertible to $58. This is analogous to a claim by the rich man that the real currency involved in his negotiation with the poor man should be after-tax dollars; and because he is in a higher tax bracket than the poor man, he should get more than $100 in a "symmetric" split of $200.

Another way of disturbing an apparently symmetric strategic situation is to have different numbers of people on each side of the bargaining situation. A simple case might be one in which party A and party B have to divide $200. No agreement means no money. But now let party A comprise two people (A' and A″) who have agreed to their share, and let B represent one person. At one focal point, $100 could go to party A and $100 to party B; A's $100 could then be split, $50 to A' and $50 to A″. At another focal point, each of the three could get $66.66; each one, after all, has full veto power.

This compendium of possible asymmetries is far from complete, but the examples it presents are instructive: differences in initial endowments or wealth, differences in time-related costs, differences in perceived determination or ag-

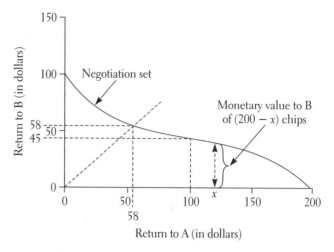

Figure 7.3.   Example of a negotiation set with symmetry in chips but not in conversion to money.

gressiveness, differences in marginal valuations (as in tax brackets), differences in needs, and differences in the number of people comprising each side. There are, of course, many others.

The notion of symmetry and focal points is often associated by bargainers with their notion of "fairness." But one person's symmetry is frequently another's asymmetry, and the discussion of what is symmetric can be divisive. Even in the extremely simple case of two-party distributive bargaining, in which each side knows the other's reservation price and in which a zone of agreement is known to exist, there is still a strong possibility that the players may not agree to an apportionment of the potential surplus $b - s$.

Case 2. One Party Knows the Other's Reservation Price

Suppose that the buyer knows the seller's reservation price (s) as well as his own (b); the seller knows s but has only a probability distribution for b. To be less general, assume that in a laboratory situation s is set at $10 and each party knows this. Next, let b be chosen from a rectangular distribution from $0 to $30—that is, all values in the interval from $0 to $30 are equally likely.[8] Suppose that the chosen value of b is $25, known only to the buyer. How might the players negotiate?

Once the buyer shows an interest in negotiating, the seller can update his

8. This procedure is implemented in experiments by taking thirty-one blank cards; labeling them 0, 1, and so on up to 30; shuffling them; and letting the buyer choose a card at random from the deck. Once the experimenter has shown the card to the buyer, he returns it to the deck.

knowledge about the unknown b. He knows that b is not less than $10. The final determination will depend not only on the bargaining skills of the two contenders but on their levels of obstinacy. The buyer should be able to push the seller down to a value close to $10. The buyer could act as if b were on the order of $14, rather than $25.

In these simple negotiations, in which only a single number b is unknown to the seller, the behavior of the bargainers will depend critically on whether b will become known to the seller after the negotiations are completed. In most real negotiations a reservation price is not just handed to the players: they have to analyze what their alternatives might be if there is no agreement, and uncertainties are usually involved. When inconvenience, transaction costs, and the players' risk aversion are taken into account, it might never be possible, even after the negotiations, for one party to determine the reservation price of the other. Laboratory results depend to a crucial extent on whether true reservation prices are revealed after the termination of the bargain.

Imagine a case in which a business is acquired for a price of $7.2 million. A couple of months after the transaction is completed the seller asks the buyer, "What was the very maximum amount you would have been willing to pay for my firm?" The buyer's reservation price was $12 million, but if she reveals that high number, she might not only make the seller feel miserable but also tarnish her reputation. Of course, there are those who might gleefully and boastfully admit to $12 million. More likely the response of the buyer might be, "You did quite well—I might have gone up to $8 million, but I'm not sure." That's not a truthful response, but it's a kind one. The misrepresentation is not offered for the purpose of squeezing out a few extra dollars—at least not immediately—but in a self-serving way it does enhance the reputation of the buyer. The best alternative is probably a truthful but evasive answer: "Sorry, that's a number that just should not be disclosed." Of course, an analytically minded seller might then muse, "Hmm—she wouldn't use that ploy unless she'd really gotten the better of me."

Suppose that the buyer's reservation price happens to be extremely low, either by chance drawing in a laboratory setting or in a real-world setting because of unexpected exogenous factors. If the buyer reveals this true reservation price—and it may be in her interest to do so—the seller might suspect that this is merely a ploy. The buyer might be better served if she refrained from making such truthful pronouncements, especially if her RP appears to be self-servingly low: the buyer can actually lose credibility by being honest. In one experiment involving successive bargaining rounds with different, independent, randomly drawn reservation prices for each round, a perspicacious buyer who drew an extremely low reservation price in one round decided to make believe that his RP was higher than it actually was; he announced a b' that was *higher* than his observed b. He was willing to lose money in that round in order not to jeopardize his credibility for further rounds of repeated negotiations.

Case 3. Common Probabilistic Information: Game-Theoretic Analysis

The following highly structured bargaining problem might be called the *canonical case of distributive bargaining*. Those who know game theory will recognize it as a formulation based on the work of John Harsanyi (1965).

A seller and a buyer each have a probability distribution, one for the seller's RP and one for the buyer's RP. Both distributions are known to both parties. A random drawing is made to establish the buyer's RP and is shown only to the buyer; a second random drawing is made to establish the seller's RP and is shown only to the seller. The seller and the buyer then negotiate, face to face, and the payoffs are the surplus values that the parties can achieve. If the random values for b and s are such that $b < s$, there is no zone of possible agreement; if $b > s$, there is a zone of possible agreement and the bargainers have to share the excess, $b - s$. They do not know before they start bargaining whether there is an excess and, if so, how large it is. Because each bargainer knows only his or her own reservation price, each has a different probability assessment of the amount of excess to be shared.

Let's dwell on this abstraction a bit. It's critical in the game-theoretic abstraction that there is complete common knowledge about the uncertainties in the problem. Probability assignments are known to both players (that is, are common knowledge), and this effectively rules out subjective probability assessments. In the real-world problem, the seller knows s, but not b, and the seller may have a subjective probability distribution about b that is not shared information with the buyer. Similarly the buyer has a subjective private probability assessment about s.

To be specific and to keep the probabilistic elements simple, let s be drawn from a rectangular distribution from 50 to 150 and let b be drawn from a rectangular distribution from 100 to 200 (see Figure 7.4). All values between 50 and 150 are equally likely for s; all values between 100 and 200 are equally likely for b.

We will assume that the drawings are independent—that the seller's knowledge of the outcome of s does not affect his probabilistic assessments for b, and vice versa.[9] A particular joint drawing can be represented by a point (s, b) in the square shown in Figure 7.5. All points in that square are equally likely outcomes. There is a one-eighth chance that s will be greater than b and that no zone of

---

9. The laboratory procedure can be implemented as follows. The seller has a deck of 101 cards labeled 50, 51, and so on to 150; one of these cards is drawn at random, shown to the seller and the experimenter, and returned to the deck. The buyer has a deck of 101 cards labeled 100, 101, and so on to 200; one of these cards is drawn at random, shown to the buyer and the experimenter, and returned to the deck. The payoffs to the buyer and seller are made in a confidential manner so that each player never knows the other's RP.

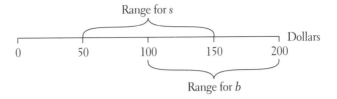

Figure 7.4. Distribution of reservation prices for the canonical case.

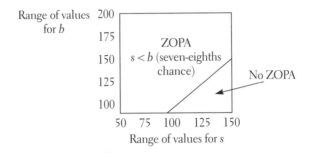

Figure 7.5. Joint representation of equally likely outcomes.

agreement will exist; there is a seven-eighths chance that a zone of agreement will exist.

Subjects are assigned roles and each is given a randomly drawn RP. They negotiate outside any experimental setting and follow no structured rules. They can negotiate face to face or over the phone or write notes to each other. They can make up their own rules, but they *cannot* show their confidential RPs to each other. They are given ample time to negotiate—roughly twenty-four hours, during which they may meet several times, for as little as a few minutes each meeting. They must turn in their negotiation forms at a specified time.

Informal bargaining is surprisingly efficient. The number of actual agreements reached was surprisingly large. One might think that if there were a small zone of agreement—for example, if $s = 110$ and $b = 115$—the parties often would not be able to agree on a final price. Not so. It is true that the smaller the zone, the longer it may take for the parties to locate it, but they almost always come to agreements when agreements are possible. Inefficiencies occur only when there is a zone of agreement and the parties do not come to an agreement. Informal bargaining, without any imposed structure for negotiations and without tight time constraints, leads to more efficient outcomes than do most formal methods. One proposed structured alternative to informal bargaining is the procedure by which both parties reveal information about their reservation prices at the same time. This alternative, though appealing, does not work very well, as we shall see below.

Double Auction: Simultaneous Revelation

In any negotiation experiment there will usually be some bargaining pair who decide to devise rules of their own.[10] A seller says to his adversary, "Let's not waste time. My reservation price is $3,000. What's yours?" What a temptation to a competitive buyer! Let's assume that her reservation price is $5,500. Should she be honest and say so? Is the seller trying to take advantage of her? Perhaps the true reservation price of the seller is really $2,000. According to a commonly proposed symmetric resolution procedure, in what is known as a double auction, the parties simultaneously disclose bid prices: "I'll write down my bid price if you'll write down yours at the same time. If we're compatible, we'll split." Let these disclosed bid values be s' (not necessarily the true value s) for the seller, and b' (not necessarily the true b) for the buyer. If $b' < s'$, then negotiations are broken off; if $s' < b'$, the final contract will be $x^* = (b' + s')/2$, the midpoint between b' and s'. (See Figure 7.6.)

When this simultaneous-revelation procedure was tried, most parties gave truthful revelations: s' equaled s, and b' equaled b. However, in some cases s' was greater than s, and b' less than b; indeed, in some of these cases, there was in fact a zone of possible agreement (s was less than b), *but* the parties did not detect it (s' was greater than b') and an inefficiency resulted.

Suppose that a seller draws a very low s value—say, 60. Should his announced bid value, s', be 60, or a higher value such as 110? Remember that as long as the announced b' is higher than his s', the final-contract price will be midway between these announced values.

In a nonlaboratory, real-world setting, a bargainer may have no way of ever ascertaining the other party's true reservation price. In an experimental setting, however, it's difficult to keep these true reservation prices secret after the fact. Is it "ethically correct" for someone to lie about his or her reservation price when the parties agree to reveal their values simultaneously?[11] Some would say that this behavior is inappropriate, but others would claim that the purpose of laboratory exercises is to provide vicarious experiences: "In real-world settings most people don't even have firm reservation prices. Besides, it's culturally acceptable to exaggerate a bit in your own favor. What's wrong with that? If my adversary did it to me, I wouldn't be angry. I do to others as I expect others to do to me." We'll look closely at this philosophy later.

---

10. This section and the following one are fairly technical and can be skipped by nonmathematically inclined readers.

11. In a debriefing session following one laboratory exercise, a buyer defended her behavior as follows: "My confidential reservation price, b, was 170, and my announced bid, b', was 130. I didn't think of b' as a distortion of the truth but as a strategic bid, not unlike any sealed bid for a contract."

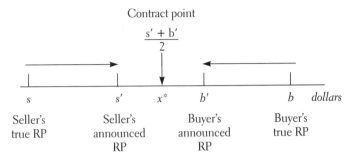

Figure 7.6.   The simultaneous-revelation procedure.

Game-theoretic equilibrium analysis.   Here is a simple exercise. Suppose that the subject playing the role of seller receives a value of s drawn from the interval $50 to $150, and that the subject playing the role of buyer receives a value of b drawn from the interval $100 to $250. All values within these intervals are equally likely. What strategies can the seller devise to determine his value of s' as a function of s (for $50 < s < $150)? Figure 7.7 depicts three such strategies: (1) a *representative strategy*, where, for example, the seller would say $112 if his actual RP were $75; (2) a strategy of *truthful revelation*, where $s' = s$ for all s; and (3) a strategy of *truncated truthful revelation*, where $s' = $100 for all $s < $100 and $s' = s$ for all $s > $100.

Each seller must submit a seller strategy and each buyer must submit a buyer strategy. Each seller is then "scored" by pitting his or her strategy against each buyer's strategy in turn; the seller's score is then his average return—averaged over all s values and over all buyer-adversaries. Buyers are scored analogously.[12]

If s and b are the actual RPs, and if s' and b' are the revealed bid values, the payoffs can be formulated as follows:

$$\text{To seller: } (s' + b')/2 - s \quad \text{if } s' < b'$$
$$0 \quad \text{if } s' > b'$$
$$\text{To buyer: } b - (s' + b')/2 \quad \text{if } s' < b'$$
$$0 \quad \text{if } s' > b'$$

People outsmart themselves. The difference between s and s' can be said to be the amount of exaggeration (or distortion) at s. Subjects in general—even those students who helped me design the game—played it very badly: they exaggerated too much. When truthful revelation strategies, or even truncated truthful revelation strategies (see Figure 7.7) are pitted against each other, the proba-

---

12. This game has been extensively analyzed by Chatterjee and Samuelson (1981).

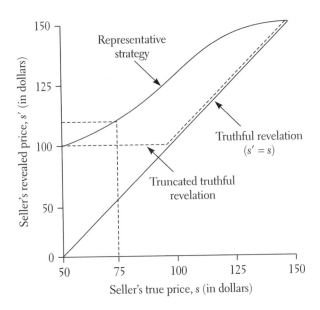

Figure 7.7.   Strategies for the seller in the simultaneous-revelation procedure.

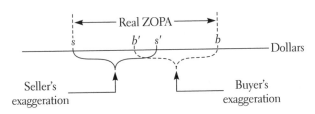

Figure 7.8.   Case in which there is a zone of possible agreement in real but not in revealed values.

bility of getting an (s, b) pair with no zone of agreement is .125 (see Figure 7.5). But averaging over all subject strategy responses and over all (s, b) pairs yielded an extremely large probability, .46, that no zone of agreement (in revealed values) would exist! Thus over one-third of simulated trials resulted in no agreement when in fact a zone of agreement did exist. Not very efficient. However, it should be pointed out that in most cases where the parties failed to come to an agreement when they could have, the collective potential gain of $b - s$ was small. This happened because there was so much exaggeration—so much, in fact, that those subjects who used a truncated truthful strategy did exceptionally well comparatively. They found that a good retort against an extreme exaggeration is (truncated) truth telling. If both parties exaggerate a lot, then the chances for an agreement are very poor (see Figure 7.8).

Thus although the simultaneous-revelation resolution procedure was devised to eliminate the need for haggling, it is obviously not a good substitute.

Consider the seller's equilibrium analysis, as illustrated in Figure 7.9. This figure and Figure 7.10 depict a pair of equilibrium strategies: one for the seller

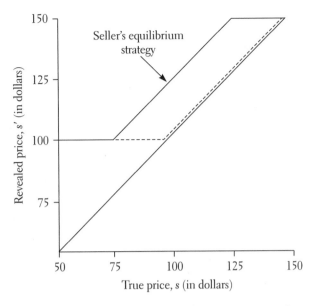

Figure 7.9.   Seller's equilibrium strategy for the simultaneous-revelation procedure.

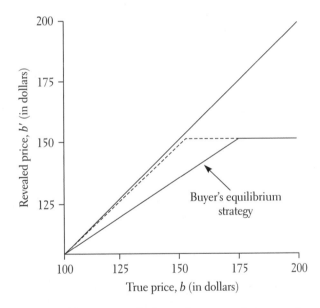

Figure 7.10.   Buyer's equilibrium strategy for the simultaneous-revelation procedure.

and one for the buyer. As long as one party adopts his part of the equilibrium strategy, the other will find it to his advantage to do likewise. But the equilibrium pair is not efficient, because for many (s, b) pairs where there is a zone of agreement, the revealed (s', b') pairs yield no agreement. With a pair of equilibrium strategies in contention, 38 percent of all (s, b) pairs result in no agreement. (The empirical percentage of no agreements was 46.) Two truth tellers do better than two equilibrium strategists. For an analytical elaboration, see Box 7.3.

The simultaneous-revelation resolution procedure is inefficient because it encourages exaggerations; but it's fast and uncomplicated. If time is at a premium or if one is engaged in many such bargaining problems, then this resolution procedure still has merit—especially if the parties can refrain from undue exaggeration.

The more they lie, the more it pays to be honest (and vice-versa). A further result is even more counterintuitive. Many people, when discussing the desirability of lying in a bargaining situation, say something like this: "I wouldn't like to lie against someone who is honest; that would make me the bad guy. But I don't want to be taken advantage of either, so I would lie if I thought the other was lying." Thus the usual rule is, "lie if you think they are lying." The equilibrium analysis shows that this intuitive notion is completely wrong—at least from the point of view of maximizing the return from any one particular negotiation. The more honest the other party is, the less risk there is of destroying a possible deal by misrepresentation. And the more room there is to grab the surplus by misrepresenting. The more honest they are, the more it pays you to lie. Conversely, the more dishonest they are, the greater the risk of reaching no agreement. The payoffs to lying go down, and the payoff to honesty goes up. Apparently, when you are dealing with a crook, the smart thing is to be honest.

Spreading idealistic, collaborative behavior. Two idealistic players can do no better than to truthfully reveal their reservation prices and, if compatible, to split the potential joint gain down the middle. No hassle; no inefficiencies. But they have to trust each other. They could discuss this possibility and brainstorm regarding ways to establish such trust. For example, they could share with each other the analyses of their reservation values. They might agree on a penalty structure if either side is found guilty of acting strategically without Full, Open, Truthful Exchange. Perhaps, over time, a community of like-minded parties can be established, who all agree to negotiate in this manner; and a party found guilty of violating this trust could be expelled from the community. But the obstacles are formidable. Two parties might start trusting each other, but in repeated negotiations over prices a suspicion arises by chance and each side may reluctantly try to protect itself by acting a bit strategically, only to find that trust has been destroyed and becomes hard to reestablish.

Core Concepts

The chapter has examined the abstract distributive bargaining problem (the haggling problem of the bazaar) and introduced the standard vocabulary: seller's and buyer's reservation prices, zone of possible agreement, seller's and buyer's

---

Box 7.3 Analytical Elaboration: Finding an Equilibrium

What can bargainers do when they know about equilibrium strategies but do not have the analytical skills necessary to compute these strategies, or do not have the time to devote to such intricate analyses? Let's take the vantage point of the seller. One simple analysis is to boldly hypothesize a reasonable strategy for the buyer and by trial-and-error figure out a reasonable counterresponse for selected values of s— say, for $s = 60, 80, 100, 120, 140$; these can be compared with a curve for interpolated values of s by inspection. A second simple analysis seeks the best retort against a truncated truthful revelation strategy; this retort distorts the truth more than the equilibrium strategy. Next, one can seek the best retort against the "best retort against a truncated truthful revelation" strategy; this retort distorts the truth *less* than the equilibrium strategy. It can be proved that successive iterations—that is, the best against the best against the best and so on, and finally against the truncated truth—yield a sequence of strategies that converge to the equilibrium strategy, and that these strategies oscillate ever closer and closer around the equilibrium strategy. Two or three stages in that sequence already give a practical approximation of the equilibrium strategy.

---

surplus values. In the standard abstraction, each party knows its own RP but not the other's. Typically there is no common knowledge about reservation values, and thus the standard problem is not in the classical game-theoretic format.

The chapter has discussed the empirical results of the Streaker simulation, focused on the sale of a used car. The data indicate:

- Concessions are monotonic.
- There is widespread variation of outcomes with identical input conditions.
- There is a surprising spread of first offers.
- The Boulware strategy of making a reasonable first offer and holding firm works most of the time, but when it goes sour it goes very sour.
- When two offers are on the table, the natural focal point for further bargaining is the midpoint, provided this midpoint is in the zone of possible agreement.
- The more the first offer is exaggerated—but within reason—the more the bargainer nets.
- It's better to make the second offer because the first offers were not extreme enough.
- Parties who are given enticements to be empathetic (that is, by being given subsidy points for the good scores of the other) fare better even without counting these supplements.

The chapter then considered three special cases of the distributive bargaining problem: (1) RPs are known common knowledge; (2) the seller knows the RP of the buyer but not vice versa; (3) the two RPs are drawn from probability distributions that are common knowledge to both. In case 3 the chapter examined a double-auction scheme where the parties simultaneously reveal their RPs (or exaggerations thereof), and if the RPs are compatible they split the difference; if they are not compatible, there is no deal. When parties play this double-auction game they exaggerate too much, and statistically the procedure results in a lot of no agreements even though joint gains are possible. The game-theoretic equilibrium pairs are also not Pareto efficient—that is, potential trades with positive results for each go unfulfilled. The more one side exaggerates, the more the other side should tell the truth; and the more the one side tells the truth, the more the other side has an incentive to exaggerate. Two truth tellers do better on the average than two exaggerators.

8

Introducing Complexities: Uncertainty

The first part of this chapter develops a real (disguised) case in which a plaintiff sues a defendant represented by an insurance company. It's mostly about the distribution of money, and therefore it falls squarely in this part of the book. The negotiations take place ahead of the trial, and the parties negotiate "in the shadow of the law."[1] The Best Alternative to a Negotiated Agreement is the same for all parties: go to trial. We spend considerable time on this case for several reasons: (a) it's prototypical; (b) it is analyzed from the perspective of each of the parties as an individual decision problem under uncertainty, with methodology that builds on the preliminary material of Chapter 2; (c) it adds to our catalogue of behavioral decision making introduced in Chapter 3; and (d) it illustrates how descriptive and prescriptive notions interact. The decision analysis methods are abstract, and it may be difficult to see how they could be of use to ordinary people. So we present the core of the chapter as a dialogue between an analytical consultant, the plaintiff, and the plaintiff's lawyer.

Case Study: The Sorensen Chevrolet File

Background

Debra Anderson, a young married student of nineteen, picked up her automobile from the repair shop of Sorensen Chevrolet not realizing that her left front headlight was inoperative, perhaps through the negligence of the shop. On a misty, rainy evening with poor visibility, driving alone in a no-pass zone, she "peeked out" from behind a truck and had a frightful head-on collision. She was

---

1. For an interesting discussion of this theme, see Mnookin and Kornhauser (1979).

left permanently disfigured and partially disabled. She had allegedly been travel-ing at 60 miles per hour in a 50-mile-per-hour zone.

The accident occurred in October 1968, and two years later (not an unrea-sonable length of time) her lawyer, Sam Miller, brought suit against Sorensen Chevrolet for $1,633,000. Sorensen Chevrolet was insured with a company we shall call Universal General Insurance (UGI), under a policy that included pro-tection of up to $500,000 per person for bodily injury caused by faulty repairs.

The case extended over more than four years and comprised more than seven hundred pages in UGI's files. The successive steps involved in the suit il-lustrate what we call the "negotiation dance." In this case it's not a pas de deux, but a pas de trois with the lawyer for the plaintiff, the representative of UGI, and, in a lesser role, the lawyer for Sorensen Chevrolet as principals. A greatly abbre-viated guide to the main events of this particular negotiation dance is given in Table 8.1.

According to the original Harvard Business School case study, "UGI policy required a claims supervisor within thirty days after initial notification to esti-mate the amount for which the case would be settled, the so-called reserve. This amount was treated as the amount of loss for accounting purposes until modified or until the claim was actually settled. Regulatory authorities required that a part of UGI's assets be earmarked for settling the case. If additional information sub-stantially altered the estimated settlement amount, reserves were to be modified accordingly. The reserve first set in the Sorensen Chevrolet case when the suit was brought was $10,000." That reserve was set aside in November 1970 (see Ta-ble 8.1). On March 12, 1972, Mr. Miller, the lawyer-negotiator for the plaintiff, wrote to Mr. Bidder, the lawyer-negotiator for UGI, saying: "I am aware of the fact that the Defendant, Sorensen Chevrolet, Inc., has liability coverage with the Universal General Insurance Company in the amount of only $500,000. While I think the settlement value of this case is above that $500,000 figure, I will at this time on behalf of the Plaintiff offer to settle this case for the insurance limits available (that is, $500,000), reserving the right to withdraw this offer at any time." Indeed, Miller argued in the same letter that it was "very probable that the jury would return a verdict in the approximate amount of $1,000,000 to $1,200,000."

As one might expect, Sorensen was extremely afraid that the case would go to court and that the jury would award the plaintiff an amount greater than Sorensen's insurance would cover. Sorensen urged UGI to settle at $500,000. Moreover, it hired counsel to pressure UGI to settle out of court, threatening to sue UGI for bargaining in bad faith if the jury awarded an amount in excess of their insurance coverage. UGI was not impressed.

Let's imagine that it's now the eve of the trial and that one round of negotia-tions remains. What type of analyses might help each of the protagonists?

Table 8.1   The negotiation dance: The Sorensen Chevrolet file

| Date | Event | Universal General Insurance | | |
		Reserve (in dollars)	Offer (in dollars)	Plaintiff's demand (in dollars)
October 1968	Accident occurs			
October 1970	Suit brought against Sorensen for $1,633,000			
November 1970		10,000		
November 1970–March 1972	UGI investigates			
March 1972	Demand for out-of-court settlement; Sorensen urges UGI to accept			500,000
April 1972	UGI wins summary	50,000		
December 1972	judgment that there is no legal basis for trial; plaintiff appeals			
February 1973			25,000	
September 1973				400,000
October 1973			50,000	
December 1973	Appellate court reverses summary judgment; case to be tried by jury			
January 1974				500,000
		300,000		
February 1974		500,000		
			200,000	
March 1974			250,000	
				400,000
				350,000
June 1974–December 1974	Large award in similar case; lawyer for plaintiff loses a different case; lawyer for plaintiff preparing rock-bottom settlement			

Decision Analysis by the Defendant

First of all, it appears that Sorensen can't do much except reiterate its position that UGI should settle out of court for an amount less than $500,000 or else be sued for bad faith. Surprisingly, at the last moment before the scheduled trial, Sorensen actually offered to pay a modest amount ($25,000 for openers) of the out-of-court settlement figure. Thus, if UGI agreed with the plaintiff to settle for $350,000, UGI's actual cost would be $350,000 minus x, which would be Sorensen's contribution. From Sorensen's perspective, the higher the value of x, the higher the probability that UGI would agree to settle out of court. Sorensen's decision analysis would thus center on the question of how high an x it could afford. That maximum value would be Sorensen's reservation price in bargaining with UGI.

In a formal analysis, Sorensen must assess: (1) the chance of a settlement out of court without a Sorensen contribution; (2) the chance of a settlement out of court with a Sorensen contribution of x; (3) if there were no settlement out of court, the chance that the plaintiff might win a jury trial; (4) if the plaintiff were to win, the chance that the jury award might be above $500,000; and (5) if the jury award were above $500,000, the chance of winning a bargaining-in-bad-faith case against UGI and the chance of settling that case out of court for various amounts as a function of the jury award to Mrs. Anderson. All these assessments would have to be processed to yield, for each contribution x, a specified, random distribution of out-of-pocket payments by Sorensen. From there, Sorensen could make an unaided choice of x (that is, select the best—or the least bad—lottery) or could compute an optimal choice by first assessing its utility function (reflecting attitudes toward risk) for money and maximizing expected utility. It could even include, besides monetary outcomes, a secondary component of postdecision regret in its assessment of consequences.

Such formal analyses were not undertaken by Sorensen. Indeed, UGI rejected out of hand any contribution by Sorensen because it would adversely affect UGI's business image; from UGI's vantage point, there was a linkage between this problem and other business affairs.

Decision Analysis by the Defendant's Insurance Company

From UGI's perspective, ignoring all costs to date (up to the end of December 1974), what should their reservation price be in the last stage of pretrial negotiations? In a formal analysis, UGI would need to assess: (1) the chance that the plaintiff might win the court case; (2) if the plaintiff were to win, the probability distribution of the award; and (3) if the award were above $500,000, the uncertainties surrounding a secondary negotiation with Sorensen.

Suppose that Mr. Reilly, vice president of UGI, believes there is a .8 chance that the jury will decide in favor of the plaintiff. Conditional on that finding, let Reilly's estimate of the cumulative probability distribution be as shown in Figure 8.1. Roughly, according to his analysis it's just as likely as not that the award (if given) will fall in the interquartile interval from $275,000 to $550,000; if it falls outside that interval, it is just as likely to be below $275,000 as above $550,000. The judgmental median of the award (if given) is $400,000—that is, the award is

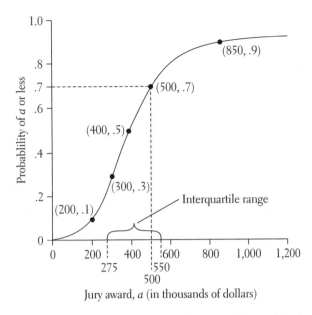

Figure 8.1. Reilly's estimated cumulative distribution of the size of the award (in the event that the plaintiff wins).

just as likely to be below as above $400,000, in the event that one is made. The judgmental probability that an award will be given is .8, and, if one is given, the probability that it will be above $500,000 is .3. The mean (expected value) of Reilly's judgmental distribution is about $360,000, which includes a .2 chance of no payment at all.

Figure 8.2 depicts UGI's decision tree for the last stage of pretrial negotiations. If UGI does not settle out of court and if it loses, the continuum of possible awards is approximated for convenience by five equally likely awards: $200,000, $300,000, $400,000, $500,000, and $850,000. We shall assume that UGI is concerned with three components: an insurance cost (award to plaintiff), a transaction cost (lawyer's fees), and a penalty for linkages to other problems. Note that if UGI fights the case and wins, this linkage penalty is negative. (Some might want to quibble with these assessments. But let's suppose that UGI has reasons for these numbers. In a more sophisticated analysis it is customary to run sensitivity studies, letting the more controversial numbers roam over plausible ranges; we omit these here for brevity's sake.)

If UGI goes down the do-not-settle path, it calculates that it has a .8 chance of losing the court trial. If UGI loses and if the jury grants an award of $850,000, the company will have Sorensen to contend with. This might prove messy, requiring transaction costs, and it would set a bad precedent for UGI to be sued by its own policyholder: sympathy would be on the side of the little guy. All things considered, UGI would rather pay more than its limit than to pay merely

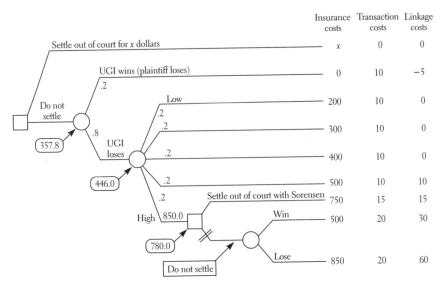

Figure 8.2.   UGI's decision tree for the last stage of pretrial negotiations. Costs are in thousands of dollars.

$500,000 and have its insured liable for the rest. In the decision diagram, UGI assigns a value of $780,000 to the node following an $850,000 award to the plaintiff.

   If UGI chooses not to settle and if it loses, it encounters a five-pronged chance node giving equal probabilities to payoffs of $210,000, $310,000, $410,000, $520,000, and $780,000. The expected value average of these payoffs is $446,000, and that's the value that would be assigned to the UGI node. Finally, the chance node immediately following the do-not-settle branch can be assigned a value of $357,000—or, rounded off, $360,000. Hence UGI from this analysis should want to settle out of court for any value less than $360,000, taking into account future transaction and linkage costs. This analysis uses expected values and makes no allowances for risk aversion—as is roughly appropriate for an insurance company.

   Now let's analyze the problem from the part of the plaintiff.

The Plaintiff's Perspective

It's now four years since the accident, and in the interim Debra Anderson has gotten divorced, has completed her college degree in the social sciences, and is now twenty-three years old. We monitor her deliberations as she discusses her decision alternatives with Jane Coolidge, a management consultant and a friend of a friend, who is helping Debra sort out her feelings. The big question is whether Debra should accept UGI's out-of-court offer of $300,000.

"I feel desperate . . . I never felt this way before, Jane. I just don't know how to decide. I keep on flip-flopping."

"Well, Debbie, isn't that the reason you asked for my help? You probably have never made a financial decision of this magnitude before, and it just isn't easy, even for the pros."

"Boy, I know it. What bothers me most—what keeps me up at night—is that I might decide to go to trial and then lose. Not only would I end up with nothing, but I would always know in my mind that I turned down $300,000 to settle out of court. And if I ever forget how stupid I was, my mom would be at my side telling me that I should have done otherwise."

"Probably a lot of other people would tell you after the fact that you were stupid as well. Debbie, you can't—or rather, you shouldn't—base your decision solely on the feelings of regret that you might suffer afterward."

"I'm being prematurely punished. I haven't made a decision yet and I'm already suffering the anticipation of this feeling you call 'regret.' "

"It's not unreasonable for you to think this way—many decision makers do—but you shouldn't let these thoughts dominate your decision. It's a tricky business deciding what to do. Look Debbie, part of my job is to guide you, given your feelings and priorities, to make a wise choice. We can keep in mind your potential feelings of guilt and regret, but the possible financial payoffs and how likely they are must enter the picture, too."

"My lawyer, Sam, thinks I'm crazy. Life is simple for him. He says that if the other side—he calls them the 'defendants' and our side the 'plaintiffs'—is only willing to pay $300,000, then we shouldn't accept it. He'd say the same thing if they were to offer us $400,000. And he doesn't deny that. I think he just loves to go to court, to jerk around the defendants. With tears in his eyes he'll talk about the scars on my face and how they will irreparably harm my social life and deny me the opportunities that a normal twenty-three-year-old single woman should enjoy in life. He's a poker player, but I'm scared stiff. I don't relish getting up on the stand and being cross-examined. I'll fall apart, I know it. And I have things to hide that have nothing to do with the case.

"Sam says I suffer from excessive pangs of guilt. Deep down I think I'm partially responsible for the accident, but Sam thinks that's nonsense. He says the other guy—the defendant—was outrageously and egregiously—he loves that word—at fault. He tries to pump me up."

"O.K. Debbie, part of our analysis of your case will try to include your potential anxieties and your feeling of remorse or regret if you go to trial and lose. But also your elation if you win. However, a wise decision must depend on three things."

As Jane talks, she writes on her pad, large and clear enough for Debra to see.

1. Your chances of winning the trial and, if you do, the chances of possible jury awards.
2. Your need for money. (She remarks, "For example, going from $0 to $100,000 is far more important to you than going from $400,000 to $500,000, because you have far greater need for that first $100,000.)
3. Your psychological attitudes. (She notes: aversion to risk, anxiety about being cross-examined, worry about suffering regret.)

"Debbie, there's one point I need to clarify. You say your decision is between settling now for $300,000 or going to trial. But isn't there a third alternative? What happens if you wait? I know the trial comes up in two weeks, but on the courtroom steps maybe the other side will offer you $350,000. I understand that Sam has told the other side that you would settle for $450,0000. What if he came down to $400,000?"

"According to Sam, he knows the lawyer on the other side very well and when he says his 'last and final offer' is $300,000, that's it. He has a reputation of never changing. Sam is convinced that the choice is between $300,000 and the uncertain payoff of the trial. Period."

Jane thinks silently to herself: perhaps Sam is right; but I still think there is a chance that Sam is wrong. If it turns out that Debbie should accept $300,000, then I am going to suggest that she not accept that $300,000 until the morning of the trial. And if it turns out that Debbie decides not to accept $300,000, I'm going to have to prepare her to have some number in mind that she would accept at the last minute before the trial. Is it $350,000? Or $400,000 . . . We'll have to see.

"Debbie, does Sam, your lawyer, know about the meeting with me?" asks Jane.

"No. I don't know if I should have told him."

"Debbie, I think I can help you and Sam think this through. If you want to continue, remember, I'm representing you, not Sam, and my analysis may lead the two of you in different directions."

"How come?"

"Well, your need for money and your psychological attitudes differ from his. What's right for you may not be his preferred alternative. But remember, *you* are the decision maker. Sam, and now I, if you want me to continue, are working for you. I hope to help *you* make up *your* mind, to help you strengthen *your* convictions, to help you feel secure in the appropriateness of *your* decision. I can't guarantee that the outcome will be right after the fact. No one can reliably predict what the jury will do; that's the plain hard fact about uncertainty. But the decision process will be right.

"If you go ahead, I suggest that you tell Sam about me and arrange a meeting for the three of us."

Debra agrees. Later, she talks to Sam, who is at first furious, then calms down and reluctantly agrees but remains skeptical. "How can an outsider know what's right? And she's not even a lawyer!"

Debra's decision tree. The next day Debra and Jane show up at Sam's office. Jane has prepared some charts that she hopes will facilitate the joint meeting.

Introductions are made. Sam is polite but cool. Debra is obviously nervous—even a bit hyper. Jane is very much in control of herself and takes command of the meeting.

"I suppose I should start because I'm the one who suggested this meeting. A good outcome of this conference should result in Debra making up her mind whether or not to go to court and to feel comfortable that she is making an appropriate decision."

"That's a tall order," interjects Sam a bit sarcastically.

Jane ignores Sam's remark and continues, "Here's the problem as I see it." Jane passes around a diagram (Figure 8.3) labeled "Describing Debra's Decision Problem as a Decision Tree."

"Start at the left," Jane continues. "At the beginning, at the point that I have marked with the number one, Debbie must decide to go to court or to settle out of court. If she decides to settle, follow the lower branch; there is no uncertainty. But if she goes to court, the upper branch, she faces two uncertainties:

- Will she win or lose (at branch point 2)?
- And if she wins, how much will she get (at branch point 3)?"

Sam interjects, "Where do the numbers $200,000 and $1,000,000 come from?"

Debbie intervenes, "When Jane asked me how much did I think we would get from a jury, I told her that I couldn't pin you down to a figure, Sam, but you said somewhere between $200,000 and $1 million."

"Well, it could go over $1 million. But I agree that's not likely. O.K., I'm with you, Jane. What about that $210,000, at the bottom, and what is this about psyche and lost time?"

"I understand that also," says Debbie excitedly. "If I accept the offer of $300,000, I must give you, Sam, 30 percent, that is, $90,000, and I keep $210,000. Jane and I also talked about my psyche: my loss of sleep, my anxieties about being cross-examined, and my possible feelings of deep regret if we go to trial and lose. That's a factor in my decision as well as the money."

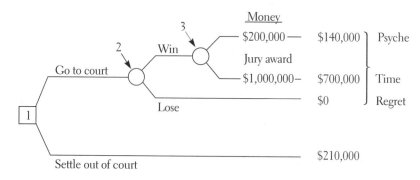

Figure 8.3. Describing Debra's problem as a decision tree.

Sam interjects, "And what about the nagging feeling you would suffer in taking $210,000, knowing that you might have gotten close to a million?"

"Yeah, that too is included in the 'psyche.' Am I right, Jane?"

"You're right, Debbie," Jane replies.

Being specific about uncertainties. Jane now addresses Sam. "You have had a lot of experience in cases like this, so I'd like to try to quantify your knowledge about the likelihood that Debbie will win and about the size of the jury award."

"I warn you Jane, I don't think the way you do. I don't think you can put a number on everything," Sam opines.

"Well, let's see, Sam, if I can probe a bit and draw out what you know so that Debbie can take into account your expertise."

"Am I the witness and you the lawyer? Shouldn't I take an oath to tell the truth?"

"Look, Sam, we need you. Try to suspend judgment until you see what's involved. Now what about the chances of winning the trial?"

"Well, I told Debbie that there's a pretty good chance of winning."

Debbie interrupts, "Yeah, and he told me also that we had more than a fighting chance or, as I remember it, not an unreasonable chance."

"Yeah, that's the way I feel. I'm basing these judgments on how similar our case is to others; in what way it differs; on my knowledge of the judge; on my litigation skills. A lot has gone into that judgment that we have a pretty good chance of winning."

Jane turns to Debbie. "How do you interpret that, Debbie? Give me some number."

"Oh, I would say that Sam thinks the probability of our winning is on the order of 20 or 30 percent."

"That's not what I said! When I say there's a 'pretty good chance,' I mean a higher number."

Jane shifts from Debbie to Sam. "How much higher? More than 50:50?"

"Sure, that's what I meant by a pretty good chance. More than 50 percent."

"How much more?"

"Oh, I don't know how you can put a number on it. It certainly isn't 90 percent. In jury trials you can never be that sure. O.K., if you force me, it's somewhere between 60 and 80 percent."

"Would you say that 70 percent is reasonable, or high, or low?"

"It's a good estimate, as close as we can get."

Next, Jane probes Sam's knowledge about the possibilities of a jury award. After an hour or so of give and take, Jane has prepared a table (see Table 8.2) that summarizes his judgments. She and Sam have divided the possibilities of an award from $200,000 to $1,000,000 into four equally likely ranges that they call "low, medium, high, and very high." For each interval, they decided on a single representative value: (in thousands of dollars) 300 for low, 470 for medium, 610 for high, and 800 for very high. They did this, Jane explained, to capture the essence of the jury award uncertainty, since it would be impossible to consider every possible outcome.

We pick up the conversation with Debbie: "Let me see if I understand this table. Based on Sam's judgments, you two think that if I were to win the trial, it is equally likely that I would receive from the jury 300, 470, 610, or 800 thousand bucks. Hmm, not bad."

"That's exactly right, Deb, but money isn't everything. We also have to factor in your time, your anxieties along the way, your apprehension in being cross-examined, and what your mom is going to say if you lose after turning down a sure thing." So Jane works another hour with Debbie and then prepares the following table (Table 8.3).

"Again let me see if I understand," Debbie asks somewhat hesitatingly. "The numbers in the last column are meant to be the net returns to me incorporating what I pay Sam and reflecting my psychological concerns, like anxiety, regret, and so on, as well as any inconvenience I may sustain. It's the fourth column that I feel is a bit tricky. The negative numbers in column 4, as Jane has said, are the amounts of money I would be willing to pay a genie to wave his magic wand and rid my mind of all that psychological stuff—like feelings of anxiety and of regret. It's the amount I would spend to get rid of the bad vibes. And the +20,000 includes my delight or elation if we win a very high award."

Jane resumes, "That's right. It will be helpful now if we place all these numbers on the decision tree so you can see at a glance what's going on (Figure 8.4)."

Debra's strength of preference analysis. "Well, Debbie, looking at the decision tree, is it clear what you should do?"

Table 8.2   Probabilities for jury award if plaintiff wins

Range	Name	Probability	Representative value
From $200,000 to $410,000	Low	0.25	$300,000
From $410,000 to $550,000	Medium	0.25	$470,000
From $550,000 to $700,000	High	0.25	$610,000
From $700,000 to $1,000,000	Very high	0.25	$800,000

Table 8.3   Determining the net equivalent dollar consequences for Debra

Outcome	Gross award	Deduction of attorney's fees	Adjustment for time, anxiety, and regret	Net evaluation
Win				
Low	$300,000	−90,000	−30,000	$180,000
Medium	470,000	−141,000	−19,000	310,000
High	610,000	−183,000	−12,000	415,000
Very High	800,000	−240,000	20,000	580,000
Lose	0	0	−30,000	−30,000
Settle	300,000	−90,000	0	210,000

"Hell, no! Should it be clear? I'm just as confused. Even more so, now that I understand what's at stake. I do admit you teased some information out of Sam that was a bit of a surprise to me, but not enough to help me decide comfortably."

"Well, Debbie, it's now your turn to answer a few more questions. I want to learn something about your need for money, your worries, and your attitudes toward risk—which I'll explain shortly. O.K.? I can do this alone with you, Debbie, if you like."

"No, no. I don't want to have any secrets from Sam."

"I'm going to ask you something about your assets."

"That's ridiculous. I don't have any. I'm really bankrupt since I have no way to pay my debts. It drives me crazy."

"Oh, but you do have assets. All you have to do is accept the pending offer and you will get $210,000."

"O.K., in that sense you're right. But I thought that the whole exercise was about whether I should accept that offer."

"Of course that's the issue. But it will help me—and I suspect you—if you dwell a bit on just what you would do with that $210,000. How would it change your life?"

"I've thought about that a lot. First, I would pay off my debts. That includes

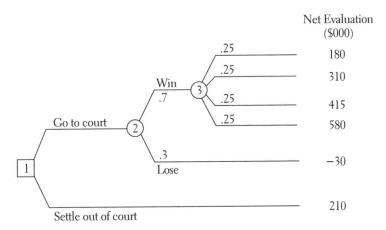

Figure 8.4. Debra's decision tree for final settlement.

$50,000 in student loans for my worthless bachelor's degree. I'd spend another $25,000 for the surgery I still need. My health insurance won't cover that. Should I continue?"

"Please, please. Go on."

"I'd pay my taxes and get the IRS off my back—that's another $8,000—and then I'd have some pure cosmetic surgery to make these damn scars less visible. Then if there is any money left over, I'd get a used car and get out of my dreadful apartment. I would, I suppose, be able to afford an apartment for $1200 a month rent instead of the $950 I'm now paying. Pretty morbid, isn't it?"

"How is your job, Debbie? And how much do you make?"

"It's a dead-end job. I'm a college graduate salesperson earning about $18,000 a year. I hate meeting people who stare at my disfigured face."

"If you paid off your debts and had some money left over, would you keep your job?"

"That's a complicated story. I certainly would have to take a leave of absence for the surgery, and I would love to go back to school and learn something that would help me get a better job."

"And how about marriage and raising a family?"

"I'm not sure anyone I would want would want me."

"Let me summarize," Jane continues. "If you go to trial and lose—this is not the most likely outcome, but it has a probability of .3 on the decision tree—your life would be pretty bad: you would have your debts; you could not afford some of the things that would make you happier; you would have to remain in your present job and keep your present apartment."

Debbie interrupts. "And how about the humiliation of the lost trial and my guilt—I think you called it regret—for not accepting the sure thing of $210,000. I'm in pretty bad shape now and I'd be in far worse shape if I lost the trial."

"But who says you're going to lose the trial? It's most likely you'll win the trial and get about a half million," Sam barks.

Jane continues. "If you netted a lot more money from the trial, what would you do with it? How would it change your lifestyle? How much happier would you be?"

"A lot. If I had more money, I would do all the things I said I would do if I had that $210,000, plus some other things. I could get a condo rather than spending money on rent, buy a better car, take a trip to Europe and other places, and, yes, I would go to graduate school and not worry so much about earning money. I would spend a week in New York going to some shows and maybe buy some fancy clothes. Ah, dreaming is nice."

"Debbie, it sounds as if going from zero to $210,000 is very important to you. You would get lots of satisfaction. How about the next $210,000? Going from $210,000 to $420,000. How important would that be?"

"Well not nearly as important as the first $210,000."

"How much more would you have to get above the first $210,000 so that you would get roughly the same satisfaction as the first 210K?

"Pretty damn high. I think I'd have to go up to a million. Yeah, going from $0 to $210,000 would be like going from $210,000 to $1 million."

Sam can't contain himself. "You can't be serious, Debbie. The gap downward is only a bit over $200,000 and upward it's about $800,000!"

"Yeah, but if it's downward I'm ruined; and if it's up, I'll be richer—not a big deal for me."

"You just don't have any imagination, Debbie."

"I don't need any imagination; I know what it means to overdraw my checking account. I get the picture, Jane. When you force me to think about my attitudes toward risk and the probabilities involved, I now not only know that I should accept that $210,000 offer, but I have some conviction in my decision. When I think of the intrinsic value of money to me, the peace of mind I'll get, it's not worth the gamble to me. I also see why taking the gamble and going to court appears better to Sam than accepting the offer. The value of money to Sam is different than the value to me."

In the end Debbie decided she would rather have the certainty of getting $300,000 than risk going to trial.

Debra's utility analysis.   Let's back up. Suppose we're at the start of the out-of-court negotiations, and Jane is trying to help Debbie and Sam find their plaintiff's reservation value, RV. Following the prescriptions of Chapter 3, Jane would consider the risk profile of the alternative "go to trial." Jane would decompose the problem and develop the uncertainty analysis leading to Figure 8.5.

The certainty monetary (gross amount) equivalent for the lottery associated with going to trial is computed by a utility analysis which we now shall describe.

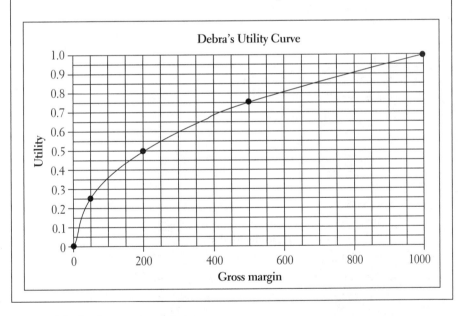

Outcome	Probability	Gross Amount ($000s)	Net Evaluation ($000s)	Utility
Lose	0.3	0	−30	0
Win; Low	0.175	300	180	0.6
Win; Medium	0.175	470	310	0.72
Win; High	0.175	610	415	0.81
Win; Very High	0.175	800	580	0.9
Total	1			
Exp. Value =		381.5	250.875	

EUV = 0.53025
Cert. Equiv. = 235
Risk Premium = 146.5

Debra's Basic Utility Inputs

Gross Amount ($000s)	Utility
0	0
50	0.25
200	0.5
500	0.75
1000	1

Debra's Utility Curve

Figure 8.5. Jane's uncertainty analysis.

The idea, as explained in Chapter 3, is to get Debbie's reactions to a few simple fifty-fifty lotteries and from these to draw her utility curve for net evaluations.

Following Chapter 3, we ask Debbie her midutility value for a fifty-fifty lottery between a gross award of $0 and $1,000,000. We encourage her, in responding, to keep in mind what her net evaluations for those gross awards would be. She responds $200K, which means that if she had $200K in hand she would just be willing to swap that for an equal chance at $0K and $1,000K. Hence if we normalize her utilities by assigning 0 to $0K and 1 to $1,000K, we would assign .5 to $200K. Continuing, we ascertain that her midutility value between $0K and $200K is $50K, so that $50K would have a utility of .25. And since Debbie feels that $500K is her midutility value between $200K and $1,000K, it follows she should assign a utility of .75 to $500.

Next we graph these five input points as shown in the accompanying figure and from the resulting graph we read that Debbie's utilities for the gross awards of $300K, $470K, $610K, and $800K are as shown in Figure 8.5. The expected utility value is the sum product

$$.3 \times 0 + .175 \times .6 + .175 \times .72 + .175 \times .81 + .175 \times .90 = .53.$$

To find the certainty equivalent, we must find the gross award that has a utility of .53, and from the graph we see this is $235K. Since the expected gross award is $381.5K, Debbie's risk premium—the amount of expected gain she is willing to sacrifice to avoid the risk—is $381.5K − $235K, or $146.5K, much too much for Sam, but appropriate for Debbie. If Jane had gotten into the act at the beginning of the out-of-court deliberation, she would have established a reservation gross award value of $235K.

Generalizing: Dealing with Uncertainty and Risk

Let's imagine that you, the decision maker, have to evaluate the worth to you of a particular alternative with uncertain consequences. What do you want to know about it? You certainly will want to know the possible consequences and their relative plausibilities; then you will want to know how well each of these consequences performs on a set of prespecified fundamental objectives. Then, without any further ado, you may decide. That's the way most people do it—without breaking the decision apart into its separate components. That's the way Sam, Debra's lawyer, wanted to do it. "Oh, I would say on the basis of my experience, all things considered, that the court case is worth $450,000."

Jane, Debra's adviser, had a different approach. She wanted Sam's expert opinion about the chances of the different consequences, but she wanted Debra to decide what her alternatives were worth to her, taking into account her judgments about the value that different consequences would have for her life.

That's Debra's province and not Sam's. By breaking apart the problem, Jane was able to use Sam's expertise in Sam's domain.

This is an important point that generalizes. Often by breaking apart a problem into separate parts, the decision maker can tap the experience and deeper knowledge that various experts have on parts of the problem. The business entrepreneur, in deciding whether to launch a new product, will go to the market analyst for judgments about the potential market, but will go to different experts for advice and opinions on financial matters and still others on matters of manufacturing. The business entrepreneur, after gathering these opinions on the constituent components of his or her problem, acts as a synthesizer of all this information to decide what to do overall.

There are two other advantages to this approach: it improves communications and reduces biases in roles. Sam was not used to giving odds or probabilities; he preferred the use of more protective vocabulary. So there was slippage in communications. Quantifying his knowledge about what might occur forced him to be more clear. In addition, Sam wanted to go to court. He wanted to litigate and therefore had a conflict of interest in giving advice to Debra. Focusing on her values clarified the conflict for her. Similar dynamics are evident when you get advice from your surgeon about how to treat an illness or when a CEO gets advice from her production manager.

Your probabilities should reflect your best judgments. But this doesn't mean that they should be pulled out of the air. The more your probabilities are based on hard, objective data and logic, the better off you are. Debra, not being an expert in legal matters, used Sam, her lawyer, as an expert. He had a good deal of experience with similar, and not so similar, court cases. He had to consider the specifics of her case in order to interpret that data—to stress some of his experience as fitting and some as not quite fitting. Of course, he had to be questioned by Jane to communicate his judgments to Debra using the language of probabilities.

Speak the Language of Probability

People are notoriously vague when describing uncertainty. They use words like "pretty good chance," "more than a fighting chance," and "not an unreasonable chance," as Debra's lawyer Sam did in describing her chances of winning her trial. Debra took these words to mean a probability of 20 to 30 percent; it turned out that Sam meant 60 to 80 percent. That degree of miscommunication is easily enough to flip most decisions from one alternative to another. (And it's not unusual; researchers in one study found that "somewhat likely" meant as much as 92 percent to one person and as little as 20 percent to another.) Thus words alone won't do it; they're too imprecise.

The solution? Use the more precise language of probabilities to describe your uncertainties. Probability is a foreign language for many people, but it's not hard to learn. In fact, you use it and act upon it every day without giving it a second thought. For example, you decide what coat to wear on the basis of a morning forecast of "20 percent chance of rain today."

How Real People Act

In laboratory experiments dealing with this case, student-subjects were assigned roles in this last stage of pretrial negotiations. They were given identical information and were asked to assess, first, the probability that the plaintiff would win the trial and, second, the conditional distribution of awards. The plaintiffs' median assessment that they would win their case was .75; the defendants' median assessment that the plaintiff would win was .55. And the set of conditional distributions of the size of the award (if given) as assessed by the plaintiffs was displaced to the right from the corresponding set as assessed by the defendants. Note especially that each party tended to view its own chances in court as better than the other side viewed them. When, as a control, some subjects were asked to assess probabilities before they were assigned their roles, their median assessment fell in the middle. It has been noted in other contexts as well that subjects bias their probability assessments according to the roles they play. Furthermore, in this case the displacement was in the direction of *decreasing*, or even of eliminating, the zone of possible agreement when calculations were based on expected values. Even so, many civil liability cases are settled out of court. We suspect that the reasons for this are primarily the strong risk aversion and potential postdecision regret that push the plaintiff's reservation price downward.

Debra and her lawyer, Sam, probably don't realize it, but they too have an inherent conflict of interest, although the lawyer's incentive structure is designed to motivate him to get as much as possible for his client. Suppose that Debra has a choice between $275,000 for certain, or taking her chances with a jury. Most plaintiffs in Debra's position are probably far more risk averse than their lawyers.

As Daniel Kahneman and Amos Tversky (1979) convincingly demonstrated, the paradigm of expected utility maximizing is not a very good prediction of actual behavior in laboratory experiments—most subjects strongly prefer a certain positive reward to an uncertain reward far in excess of what can be explained by the standard theory of expected utility maximization. Partly this is the avoidance of anticipated postdecision regret.

To be sure, risk aversion and avoidance of anticipated regret will also affect the plaintiff's lawyer—but to a considerably lesser degree. One might speculate that the reservation prices of plaintiffs in civil liability suits would tend to be lower than their lawyers'. If we were to push back the time frame of our analysis from just before the trial to a much earlier stage in the negotiations, the discrepancy in attitudes between the plaintiff and her lawyer would be even deeper: she doubtless would suffer more continuing anxiety than would her lawyer, and she would probably have a greater need for money at an earlier rather than a later date. This would tend to make her reservation price lower than her lawyer's. The lawyer, for his part, would have to consider the great deal of time involved in handling a court case; but this might be offset by the possible advantages to his career and reputation. Of course, all these concerns to the lawyer are irrelevant for his client, and herein lies a possible conflict.

The insurance company, on the other hand, is far less risk averse; and insofar as there is a certainty effect such as Kahneman and Tversky describe, it goes the other way: the choice of a definite, certain *negative* amount is less appealing than a gamble with the same expected value. But one shouldn't make too much of this from the insurance company's point of view. They should think in terms of expected values—but allowances should be made for transaction and linkage costs.[2]

It would appear that plaintiffs in civil liability cases are often exploited: when 90–95 percent of cases are settled out of court, a clear bias seems to exist in favor of the big guys. Not only do they have a better "probabilistic feel" for courtroom realities, but they can unemotionally afford to play long-run averages, and time works to their advantage. Imagine the feelings of continuing anxiety that are experienced by a plaintiff in a protracted four-year, out-of-court negotiation.

But before automatically taking the side of the risk-averse, regret-prone, overanxious victim, think of the reverse exploitation of the big guy by juries who sympathize with the victim—even if she is partially to blame. The jury might reason that after all, doesn't everyone occasionally engage in imprudent excesses? Furthermore, although the cost to an insurance company is passed on to its policyholders, the difference between an award of $500,000 and an award of $1 million is a matter of pennies to those statistically anonymous, faceless multitudes. So even if a case goes to court, it will likely end up as a balance of inequities.

Core Concepts

This chapter has developed the two-party distributive negotiation in which the parties are not given their reservation prices on a silver platter, but have to ana-

---

2. At least the top management of the insurance company will think in terms of expected values—an agent out in the field negotiating the case might be more risk averse. This generally occurs throughout hierarchical business firms.

lyze their alternatives to negotiation; and some of these alternatives may involve critical uncertainties. The methodology for accomplishing this is decision analysis—that is, individual, prescriptive decision making, first introduced in Chapter 2. Thus this chapter is in large part a continuation of the theory developed in that earlier chapter. We added the decision-tree technique, which has much more general applicability than its limited use in distributive bargaining. Another methodological feature we have highlighted is the intimate interrelation between descriptive and prescriptive analysis in the judgmental elicitations of probabilities and utilities. Using numerical probabilities pushes decision makers to work harder on evaluating risk, and it tends to reduce misunderstanding between advisers and advisees. It's important that the "advisee" ponder deeply before responding to hypothetical questions that could affect her lifestyle. We attempt to elucidate these interconnections by once again employing a dialogue form of exposition.

We have developed these methodological concepts in terms of a case study from tort law. There is a plaintiff who sues a defendant, and in the shadow of the law they negotiate for an out-of-court settlement. The best alternative for each is to go to trial, but what will happen at the trial is far from certain; who will win and the jury award if the plaintiff prevails are both unknown. The case is rich enough to develop side issues dealing with differences of opinion between the principals and their agents, a recurring problem in negotiation analysis.

9

Introducing Complexities: Time

Game theory often assumes that the players' actions are simultaneous and instantaneous. But real negotiators don't only have to decide *what* to do—they often have to choose *when*. Should I make an offer now, or wait a while? Should I accept their offer, reject it and make a counteroffer, or simply wait for something better? Does the passage of time put me in a better position, or worse?

The Timing of Concessions

In negotiations conducted in laboratory settings, subjects show an almost uncanny ability to detect even small zones of agreement—but the smaller the zone and the further it is from the range they anticipated, the longer it usually takes them to agree on a solution. As a corollary to this, we can surmise that the bargainer who is willing to wait longer, to probe more patiently, to appear less eager for a settlement will be more successful.

Richard Zeckhauser once conducted a negotiation experiment in which Israeli subjects played against American subjects. He found that the Israelis did better, because they were less impatient to arrive at a negotiated settlement. The Israelis even asked Zeckhauser how firm he considered the deadline that he imposed on the length of the negotiations. When 8:00 P.M. was the deadline for an all-day negotiation, a lot depended, in their minds, on whether Zeckhauser would accept a settlement executed at 8:02 P.M.

Many Americans are uncomfortable with long pauses in the give and take of negotiations. They feel obliged to say something, anything, to get the negotiations rolling. However, it's not what is said in negotiations that counts, but what *isn't* said. Very often the strategic essence of a negotiation exercise is merely a waiting game with self-imposed penalties (embarrassment) for delays.

It is true that during negotiations real penalties may be incurred by one side or the other with the passage of time. And many unskillful negotiators place a dysfunctional premium on speed. Their concerns are not fear that the other side will opt out, or that a totally unexpected event will intervene, or even that a quick response is a polite response, but rather a psychological uneasiness about wasting time. Certainly time is valuable, and sometimes one should be willing to trade money against time. But most people are far too impatient to see a deal consummated.

Let's look at a situation where the question is: should I deal with my counterpart, or wait for someone else to come along?

Sequential Search

Let's suppose that a seller has a single item to sell—say, a summer house—and that she has only a broad probabilistic assessment of what buyers would be willing to pay. She knows that she would rather not sell the house for less than $150,000, and she has a month in which to find a buyer before she has to leave for an assignment abroad. There are no realtors involved, and she advertises in the appropriate places: "Secluded summer place on beautiful, pristine lake. Asking price $225,000, but not firm." She is then approached by a stream of buyers and haggles a bit with each. The first buyer starts at $120,000 but quickly goes up to $135,000, and the seller feels that *maybe* he could be induced to raise his offer to $150,000. A couple of days later a second buyer offers $160,000. Should she wait? The second buyer intimates that he's looking elsewhere and that if he's not approached soon, he may find something else in the interim. The seller gambles. A third buyer shows up and makes a tentative offer of $170,000. Already twelve of the thirty days have expired.

What are some of the uncertainties that the seller faces? First, she doesn't know how many buyers will show up.[1] Second, she doesn't know the distribution of the prices that the buyers might be willing to pay. Third, she doesn't know whether, if she passes a buyer by, she can resume negotiations with that buyer at a later stage.

In this sequential decision problem, the seller is probing the market and thereby constantly revising her beliefs about the intensity of interest of buyers and the distribution of reservation prices for buyers. Such decision models have been formalized—for example, by Zvi Livne (1979)—and dynamic programming algorithms have been devised to generate numerical solutions. If enough simplifying assumptions are made—such as a fixed number of buyers, an unambiguous determination of each buyer's reservation price, no possibilities of going back to a bypassed buyer—then analytical solutions can also be derived. These

---

1. This could perhaps be modeled as a Poisson process.

Box 9.1   "Select the Best Candidate" Problem

Here's one version of the "Select the Most Beautiful Woman" problem. Ernest is given the task of picking the best of one hundred candidates. If two candidates are presented to him, he can unambiguously determine who is better, but he can't say anything quantitatively about how much better. The one hundred candidates are to be presented to Ernest in a randomized sequential order. If he passes a candidate by and does not declare him the best, then he can't go back. Suppose he passes the first by. There's already a one-in-a-hundred chance that this one was the best and that Ernest has failed in his quest to find *the* best candidate. There's no reward to him if he identifies the second best. The second candidate now presents himself and Ernest compares him with the first. If he is not as good, obviously Ernest will not select him; but even if the second candidate is better than the first, he still might want to pass that person by and let him be the standard for judging the ones to come. If Ernest lets x candidates go by, the best of the first x will represent a level against which to judge the remainder. How many candidates should he let go by before choosing? What chance does he have of picking the best?

   Solution: The answer is that Ernest should let 38 percent go by (mathematically speaking, this proportion is the reciprocal of the magic number e) and should pick the next candidate who is better than the preceding thirty-eight. If he follows this plan his chances of selecting *the* best candidate are also 38 percent (or $1/e$). This is a remarkable result. One would suspect at first that achieving as high a probability as .38 would be impossible. (See the appendix to this chapter for mathematical proofs.)

---

models can be thought of as generalizations of a problem that is a classic in the decision theory literature, with an unfortunately sexist name, "Select the Most Beautiful Woman" (see Box 9.1).[2] We will use more neutral language and call it "Select the Best Candidate."

   In the case where a seller is faced with an uncertain number of sequential buyers, it is likely that once a buyer breaks the seller's overall reservation price (that is, the analogue of the $150,000 in our example), the seller will get impatient because by waiting she's trading in a desirable certainty for a potentially desirable uncertainty, and the possibility of postdecision regret looms large. In such situations most people are overcautious, in the sense that if they were inter-

---

   2. Enlightened criticism of the name led successive commentators to rename it the "Select the Best Secretary" problem, "Select the Best Candidate" problem, and "Select the Highest Ordinal Value" problem. While we do not wish to offend, we find the original name more memorable than "Select the Highest Ordinal Value." Perhaps "Select the Best Candidate" is most appropriate. We hope that in the intervening decades the notion that women are limited to being beauties or secretaries has become obsolete. If you prefer, by all means think of it as "Select the Most Sensitive Boyfriend," or whatever other name makes it inoffensive and intriguing. The analysis won't change.

ested in maximizing their expected return, they would probably take more chances.

The Strike Game

Another time-related problem: does the balance of power shift as time goes by? Should I wait for more favorable conditions, or should I make hay while the sun shines? And if the passage of time hurts my counterpart, can I leverage that into an advantageous settlement now?

Although 90–95 percent of civil liability suits are settled out of court, it is the consideration of what might happen in court that determines the zone of agreement for pretrial negotiations. Most labor-management contracts are settled without the bruising penalties of a strike, but it is the possibility of a strike that often makes men and women more reasonable in prestrike negotiations. If two sophisticated bargaining parties have to decide on a wage rate, if both parties feel strongly that their side is in the right, and if neither party can walk away from the conflict, then the waiting game is helped considerably by imposing fines on any delay. The strike accomplishes this.

The Simulated Strike Game

In one experiment, subjects were asked to play the role of management or of the union in a highly structured wage negotiation. In the simulation, management is instructed to hold out for a basic wage of $24.00 per hour, and the union for $25.00 per hour. Equally good arguments could be made for either figure. The issue that the negotiators must decide is the increment x (in dollars) between 0.00 and 1.00 that management will pay the union. Management wants $x = 0$ and the union wants $x = 1.00$. The situation is asymmetric, however, because the net current value to the union (in wages, fringe benefits, and strategic bargaining position for the future) of $x = 1.00$ is $40 million. To simplify, the union payoff (in millions of dollars) for a settlement of x is $40x$. The managers, in contrast, confront a different set of realities. They have to worry about their current inventory condition, their competitive position, and so on. The cost to them (in millions of dollars) of a settlement of amount x would be $50x$.

There is another asymmetry: the costs of a strike. Such costs escalate slowly at first, but each successive day costs more incrementally (the daily cost of a strike goes up quadratically for each side, but with different coefficients—see Table 9.1). To terminate the game cleanly, the rules specify that the union may strike for at most twenty days before its treasury is exhausted. Each party knows

Table 9.1   The costs of a strike: Costs to each party (in $000s)

Day of strike	Management	Union
0	0	0
1	1,150	550
2	2,600	1,220
3	4,350	1,950
4	6,400	2,800
5	8,750	3,650
6	11,400	3,750
7	14,350	5,950
8	17,600	7,200
9	21,150	8,550
10	25,000	10,000
11	29,150	11,550
12	33,600	13,200
13	38,350	14,950
14	43,400	16,800
15	48,750	18,750
16	54,440	20,800
17	60,350	22,950
18	66,600	25,200
19	73,150	27,550
20	80,000	30,000

its own and the other's strike costs. The union negotiator does not have to obtain ratification of the final agreement.

The aims of each side are clearly specified: the bottom line for the union is to maximize its take of $40x$ less its strike costs; the bottom line for the management is to minimize its total costs of $50x$ plus its strike costs. Linkages to other problems or to similar wage contracts at a later stage were to be considered already accounted for in the payoff numbers.[3]

The subjects can negotiate in any way they like before the strike deadline, but once the strike begins, the negotiations become highly stylized. At the termination of each day of the strike, after that day's penalties have been imposed, each side simultaneously submits a settlement offer. Let's denote these offers of management and the union by x_m and x_u, respectively. If management's current offer x_m is less than the union's current offer of x_u, then no settlement occurs at

---

3. Of course, there might be times when management would welcome a strike (for example, to reduce heavy inventories) and other times when management or the union would want a strike to teach the other side a lesson for future bargaining. That was not the case in this exercise. All linkage concerns were meant to be captured in the payoffs given to the subjects. Their aims were simply to get favorable scores for themselves in the game.

the end of that day and the clock moves ahead one day; if management's offer, x_m, is as large as the union's offer, x_u, then a settlement is reached at the midpoint $(x_m + x_u)/2$, and the game terminates.

If, for example, the bargaining parties settle at $x = .40$ at the end of the fifth day of the strike, both parties would have fared better if they had settled at $x = .40$ after four days of strike—and still better after three days, after two days, after one day, and with no strike. Any settlement with a strike cannot be jointly efficient, because there are joint gains to be had with the same settlement and no strike. But with no strike it is impossible to improve the payoff for one protagonist without penalizing the other protagonist. It is in this sense that the jointly efficient set of outcomes is characterized by the simple no-strike condition. Yet despite this obvious characterization, the subjects did strike—and frequently. Remember that each subject was "scored" not against his or her bargaining adversary but against how other subjects did playing a similar role.

The simulated strike game is formulated as a bona fide game. That is, all the information about the game is common information.

Empirical Results

Outcomes were widely distributed. About 10 percent of the subjects settled with no strike; another 10 percent settled only when the union ran out of money after twenty days; about 40 percent settled in one to three days, when the daily cost of the strike was still small; and the remaining 40 percent were sprinkled over the remaining days—more than three and less than twenty. The vast number of settlement values fell between .40 and .60, congregating around the obvious focal point of $x = .50$.

The above outcomes were obtained using subjects who were business school students. Middle managers did a bit worse, senior managers still worse, and young presidents of companies even worse than that. Here "worse" is meant in terms of average payoffs. Of course, the results may have been an artifact of the scoring system, since only with the students did the scores have a real impact, being used as a factor in determining course grades. The students wanted to do well over all the games they played and did not want to do badly in even one game. Still, it is often true that more experienced men and women of the world feel adamant about their insights and thus become less flexible.

Consider two behavioral anomalies that were exhibited in the game. The first anomaly occurred among pairs that took the full twenty days to reach an agreement. These protagonists appeared to have different perspectives on the asymmetries of the situation. Each offered a position and held firm, waiting for the other to admit that he had been unreasonable. Or else one side was embarrassed into making what he viewed as such unduly large concessions that he became angry, going so far as to subsequently act against his own interests simply in order to get revenge upon his adversary. Meanwhile, the adversary might have felt that her behavior had been quite reasonable, given the way she viewed the asymmetries of the problem. In these cases, during the game there was a shift in the payoff functions: a new psychological component reflecting malevolent attitudes had been added to the monetary component, and this added component became dominant.

The second behavioral anomaly that occurred can be exemplified by the concession pattern shown in Table 9.2. After the union held fixed at the apparent focal point of .50, management slowed down its concession rate and dug in its heels at .42. There ensued a slow pattern of reciprocal concessions, culminating in an agreement after day nine of $x = .45$. Remember that the costs of the strike had been mounting daily at an increasing rate. For example, on day five, management offered .42 and the union .48. The midpoint between these numbers is .45. But the protagonists did not reach agreement until day nine; and on days seven, eight, and nine, management spent on strike costs a total of .065 + .071 + .077, or .213 in equivalent wage concessions. On day six, management should have said .45—or better yet .49. The union spent on strike costs on days seven, eight, and nine a total of .09375 in equivalent wage concessions. On day six, the union should have said .45—or better yet .43. By grudgingly making minuscule concessions, each side incurred substantial strike expenses. Why did they do this? One reason might be their expectations about fairness—each side believed that the adversary should be the one to make the concessions. Another explanation is that each knows its own intention to stand firm, but imagines that the other side is about to crack and give in.

Does this happen in the real world? It certainly does.

Table 9.2   Concession pattern in the management-union strike game

Day of strike	Offer made		Daily incremental cost of another day's strike, evaluated in terms of equivalent wage concessions	
	Management	Union	Management	Union
1	.30	.60	.029	.01375
2	.35	.50	.035	.01625
3	.40	.50	.041	.01875
4	.42	.50	.047	.02125
5	.42	.48	.053	.02375
6	.43	.47	.059	.02625
7	.44	.47	.065	.02875
8	.44	.46	.071	.03125
9	.45	.45	.077	.03375

The Escalation Game

Analysis of the strike game is complicated, because at the close of each day of the strike the parties must decide not just whether or not they should concede, but how much to concede. There is a simpler game that is equally fascinating, involving merely the decision of whether or not to concede at any particular stage. It's called the "escalation game" or the "both-pay ascending auction" (see Shubik 1971).

The Problem

For example, two bidders vie for a prize that they value equally in dollar terms.[4] To be perfectly unambiguous about it, let's say that the prize is a $10 bill. The bidders in ascending order cry out their bids. The top bidder wins the $10 and pays the auctioneer his top bid. But now comes the hook: *the second-highest bidder must also pay to the auctioneer the amount of his or her highest bid.* So if the first player bids $7 and the second bids $8, the first can quit and end up with a loss of $7 (the other side netting $2) or he can escalate to $9 with a potential of netting $1 and causing the other side to lose $8—unless, of course, the other side also escalates.

A coin is tossed to decide who will start the bidding. The designated starter can refuse to play (giving the "follower" a profit of $10) or he can bid $1. The follower can escalate to $2. From there on the starter escalates to odd amounts, the follower to even amounts, each in $2 increments. No collusion is allowed between the two bidders or else the game is trivialized: they merely agree that the starter will refuse to bid, and then they share the $10 equally. This may be excellent strategy in the real world, but in this case it misses the point of the game.

Time out to think. What would you do? As odd player would you start? How far would you escalate your bid?

How Real Players Behave

I (HR) remember playing this game in 1958 with two business school colleagues at a faculty luncheon. The opener bid $1; the follower responded hesitatingly with $2; and they continued, still somewhat hesitatingly, up to $5 and $6. There was laughter when the players realized that already I, as the auctioneer, was

---

4. Three or more bidders can start the game—but since eventually the action will come down to just two, it saves time to start off that way.

making money. In quick succession came $7, $8, $9. There was a pause, and the follower said, "Ten dollars," with a note of finality to it. The starter then wanted to clarify a point: "Could I bid $11?" I said there was no reason why not.

Rather quickly the bids escalated to $16. Another pause for clarification. "Must we pay you with the money we have in our pockets?" I assured them, to their amusement, that I trusted them and would take a check.

The bidding resumed. At $25 there was another pause for clarification. "Is this for real?"

"Of course!" I answered. "Wouldn't you have taken my ten dollars if you had won with a bid of $3?"

The bidding continued with a perceptible change of mood: the players showed signs of anger. The 10 dollar bill had become the least of their objectives; each was now intent on winning out over the other. They became entrapped.

When the bidding reached $31, I became uncomfortable and intervened, persuading them that the game had gone far enough and that I'd be satisfied with collecting $20 from each of them. They agreed, with a certain amount of annoyance. It wasn't that they minded losing $20 each to me—but they were irked that I hadn't let them finish the game.[5]

Similarly disturbing results, with much higher final payments, have been obtained by other experimenters.

Prescriptive Analysis

In the literature, the escalation game is sometimes called the "entrapment game" or the "sucker's game." Many subjects who agree to bid in this game know that it can be a trap—that it is often a game that is best avoided. There is, though, a psychological catch: if it makes sense for you not to play, it makes equally good sense for the other bidder not to play—so maybe, after all, you should play. And 'round and 'round this line of reasoning goes. Any rationalization you can make for yourself you can make for the other player, and maybe therefore you should have $n + 1$ thoughts.

After some discussion, subjects realize that if, for example, they plan at the outset to escalate up to a maximum of, say, eight dollars, then they should do it with gusto. If you are a player, there's no use hesitating at the six-dollar level, because this hesitation will encourage your adversary to think that you will finally

---

5. I once played a $1.00 version of this game with some students, forgetting to tell them that the first bid had to start at 10 cents. One student smugly announced 90 cents as a starter, feeling certain that it did not make sense for the other party to escalate to $1.00. To his surprise, the movement upward was vigorous.

back down if he goes to seven dollars. After a little reflection, the best strategy becomes clear: bid aggressively up to a maximum cutoff value and then quit.

Subjects are then asked to think hard about the maximum cutoff values they
would choose as starter and as follower. Each understands fully that his or her
strategy will be pitted against every other person's strategy and that each will be
scored according to the average of these payoffs. Suppose, for example, that a
subject indicated he would bid, as a starter, up to a maximum of five dollars. He
would win nine dollars against each adversary who, as a follower, did not bid at
all; he would win seven dollars against those followers whose maximum cutoff
was two dollars; he would win five dollars against those whose maximum cutoff
was four dollars; and he would lose five dollars to all whose maximum cutoff was
six dollars or more.

How can a player analyze what his maximum bid should be? If he knows
the proportion of subjects for each of the maximum cutoff values, then he can
easily compute an optimum strategy. But how can he assess such a distribution?
He might want to think sequentially and conditionally. For example, he might
ask himself: of every hundred subjects who are "alive" at five dollars (that is, who
have escalated to five dollars), how many would not increase their bid to seven?
If he thinks that more than 20 percent of those alive at five dollars would not go
to seven, then he should definitely increase his bid to six. Of those alive at seven
dollars, how, many would not go to nine? And so on.

In our experiments, using subjects who had not played the game before but
who had been briefed about the possibilities of escalating beyond ten dollars and
the reasons for it, a starter would have been wise to escalate aggressively to a
maximum of thirteen dollars, and a follower to a maximum of fourteen dollars.
This would have been good strategy against the empirical mix of the strategies of
subjects.[6] Once the subjects had played the game and seen the results, they realized that a lot of bidders who had used high cutoff maximums had fared well on
average. When given an opportunity to replay the game, many nonbidders became bidders, and there was a tendency for cutoff maximums to escalate. At this
point, it would have been wise not to bid, or to bid low. Upon repetition, the results vacillated and became more blurred.[7] While the subjects' experience did
not lead to a clear, elegant solution, perhaps game theory could. See Box 9.2.

---

6. A lot of people find this game confusing. How can it be wise to bid up to thirteen dollars for
a ten-dollar prize? The hope, of course, is that many adversaries will quit well below ten dollars.
Some will be alive in the bidding at ten or twelve dollars, but a large proportion of them will quit at
those points, making it profitable for a player to stay in until thirteen dollars. Why not quit at, say,
thirty-three dollars? Because there might be a few obstinate souls who will stick around after ten dollars, and even a few who will have astronomically high quitting values.

7. It would be interesting to try the following variation. Start off with a standard escalation
game for ten dollars and choose some pair of bargainers who have escalated their way to high values,
such as twenty-three dollars and twenty-four dollars, with the game still in progress. With no previous

---

Box 9.2   Analytical Elaboration: Finding an Equilibrium

The natural question (for a game theorist, anyway) is: what about equilibrium strategies? To make sense of this, one has to formalize the end-stage game. In the mathematical abstraction, players can simply escalate indefinitely. We could impose a random stopping rule, but instead let's look for an equilibrium pair among so-called invariant strategies. From an expected monetary valuer's perspective, if the bidding has progressed to x dollars and a player is contemplating raising his bid to $(x + 1)$ dollars, then he has already lost $(x - 1)$ dollars, assuming $x > 1$. Ignoring sunk costs (that is, those that have already been incurred), given that he is alive to raise his bid to $(x + 1)$ dollars, he might want to quit, with probability p that is constant for all $x > 1$. This is what is called an *invariant strategy*: after $x > 1$ a player can quit at any stage, with probability p. If his adversary announces a quitting p that is greater than .20, then the player should stay in the game; if his adversary announces a quitting p that is less than .20, then he should pull out; if the announced p is equal to .20, then he could either pull out, stay in, or likewise play $p = .20$. The pair of strategies according to which (after the game is started) both parties quit at any bid with probability .20 can be said to be in equilibrium.

A second way to make the escalation game into a bona fide game that can be analyzed in terms of equilibria theory is to start with specific capital endowments for each of the bidders. Suppose, for example, the following holds:

1. Odd has a starting capital of $4 and Even has $7 (and this is common knowledge): the equilibrium analysis shows that Odd should not start the bidding.

2. Odd starts with $25 and Even with $18: the equilibrium analysis shows that Odd should bid $1 and Even not bid at all.

In general the equilibrium analysis indicates that the player with the largest initial capital wins the bidding right at the start. Hint: Just work backward.

Behavioral Insights

The both-pay ascending auction is an interesting variation of a regular auction — a variation that's of more than academic interest. Although subjects are fascinated with the game, they at first don't see its relevance to the real world. It takes a while to realize that the game is an accurate reflection of what may occur in arms races, for example, or in wars such as those in Vietnam, Angola, and Eritrea. Gradually, elements of the game become increasingly recognizable in

---

hint, let the experimenter propose a rules change: the even bidder is told that he can deescalate to, say, twenty-two dollars. The odd bidder, who has announced twenty-three dollars, can now quit and collect ten dollars for a net loss of thirteen dollars (the other would have to pay twenty-two dollars) or deescalate to twenty-one dollars. And so on, backward. It would still remain a both-pay auction. What would happen as they approached ten dollars? As they pierced ten dollars in their downward journey, life would become especially precarious.

real-world situations, and it can thus be used to teach—albeit in an artificial set-ting—some valuable lessons.

First, if you are representing some group or constituency, it may be hard for you to explain that sunk costs are sometimes best abandoned; once engaged in the negotiations, you may be forced to stay in longer than you want.

Second, if you are challenged to negotiate and you consistently refuse, then a lot of ripe plums will be plucked by someone else.

Third, if you decide to engage up to a certain level, do it with gusto.

Fourth, if critics on your side make it difficult to proceed with gusto—as would be the case in a democracy—then apparent internal misgivings on your part will encourage the other side to escalate further.

Fifth, the leader who engages in an escalation game, probes the other side, and then withdraws as a loser should not to be hastily criticized. It might be a case of good (ex ante) decision with a bad (ex post) outcome.

Sixth, if you are forced to play, avoid announcing a deterministic strategy. If you announce a high cutoff maximum to impress the other side, remember the effect on your own team; if you ask permission to escalate but with a low cutoff maximum, then you're encouraging the other side to go just one step further. Maybe the best thing for you to do is to act naturally confused, somewhat unpre-dictable.

Last and most important, beware of escalation games—they're treacherous. Think about how you can collude to get out of them.

To tie all of this to negotiations: remember that a strike game may be a par-ticularly vicious form of a both-pay ascending auction game.

The Virtual Strike

The threat of a strike forces the contending parties to get serious. If they don't agree by the deadline, both sides start losing money and this focuses their minds. The trouble is that the strike may impose intolerable penalties on outsiders, the consuming public.

During World War II the navy intervened and forced the Jenkins Company valve plant to continue production past a strike date. Both labor and manage-ment were forced to put monies in an escrow account until an agreement was adopted. At the time, this was referred to as a *statutory strike*. The idea is "to im-pose losses on the disputing parties commensurate with those which would have been incurred in an actual strike and yet maintain normal production at the same time" (Mills 1975). The money contributed to the escrow account could be partially returned to the disputing parties after a settlement.

This idea of the statutory strike was seriously discussed in the 1950s and 1960s by labor economists (for example, Sumner Slichter, Robert Livernash, and James Healy), but was dropped by them because of the difficulty of getting disputing parties to agree to such a mechanism. It came to be known as the *no-strike strike*. Inexcusably, not knowing about these earlier deliberations, I resurrected the idea during the National Football League strike in 1989. David Lax and I wrote an Op-Ed piece suggesting a pseudostrike, which others dubbed a *virtual* strike. Instead of canceling games, the idea was that play would continue as scheduled but the players would receive only expenses and management would have to put all profits into an escrow account. Our suggestion was that, until a settlement was reached, an increasing proportion of the contributions to this fund would be siphoned off into sponsoring athletic activities in the inner cities.

The problem with the no-strike strike is how to structure balanced penalties on the contending parties. These penalty schedules would either have to be negotiated or perhaps imposed by an arbitrator. I apologize belatedly for reinventing an old idea, but I still think this idea should not be summarily dismissed.

Creation of Strike Mechanisms

There are times when there is no existing strike mechanism and the parties may wish there were one. So why not have the parties create artificial deadlines with escalating penalties for breaking these deadlines? They may voluntarily and jointly

- determine a deadline for the completion of negotiations;
- agree on penalty schedules to be assessed against each of them in case they cannot settle by that deadline; and
- agree to forfeit part of any escrow account they amass.

The New York State Budget Deliberations

In 1999 the journalist Elizabeth Kolbert noted that the New York State legislature's "primary responsibility is to put together the state budget each spring. For fifteen years in a row, however, through good times and bad, through surpluses and deficits, through Republican administrations and Democratic ones, the legislature has failed to pass a budget by the start of the state's fiscal year, on April 1st. Expectations had been beaten down so far that when, last April [1998], lawmakers approved a budget only fourteen days after the legal deadline it was considered virtually on time. In 1996, they were a hundred and four days late, and in 1997 a hundred and twenty-six" (Kolbert 1999).

In a variant of the no-strike strike the legislators voted that they themselves would not be paid whenever the budget deadline was passed. It was also, in part, a penance: they voted to stop paying themselves for as long as the budget was delayed. Their pay would go into an escrow account which would then be paid out in full once a budget was enacted. As we might expect, this mechanism had little bite and was mostly ignored. It was a minor inconvenience, and perhaps at worst some legislators were forced to borrow money until the day of agreement. But it would not be hard to add some bite by diminishing the escrow account by systematically giving away some of it for charitable purposes. With that modification, we suspect, this escrow mechanism just might change the hearts and minds of the legislators and cause them to be accommodating in their budget negotiations. (The penalty could be psychologically amplified by donating part of the escrow account to a cause that all would find reprehensible—like to a neo-Nazi party.)

Core Concepts

This chapter has dealt with the question—"should I act now, or later?"—in a variety of contexts. In the prototypical distributive negotiation problem a seller wants a higher price and the buyer a lower price. Each party starts with a reservation price that is assessed by consideration of the Best Alternative to a Negotiated Agreement (BATNA). The complication of this chapter arises when a negotiating party—let's say the seller—is learning sequentially over time information about the alternatives. When seller SSS is negotiating with buyer BBB, his RV depends on what she might get with buyers BBB' and BBB". She is learning about the outside while negotiating on the inside. How much information should be collected? It's a problem of sequential search. The problem is analytically complex but it brings to mind a gem of a problem known as "Select the Most Beautiful Woman," but more appropriately called "Select the Best Candidate." The best solution is to look at .38—or more precisely, $1/e$—of the total pool of candidates and then pick the first one that is better than all the ones so far. Even this approach finds the best one only .38 of the time.

The chapter then examined a highly stylized bona fide game based on a typical contract negotiation between labor (the seller of services) and management (the buyer of services) with imposed penalties for time delays after a strike deadline. The game has been played extensively with student-subjects and the discussion is designed to advance our understanding of noncooperative game theory and of behavioral decision making. Many players end up with terrible scored outcomes because they become entrapped: why should I make the next concession? This game motivates another one with an even simpler structure called variously the both-pay auction or the escalation game. A prize of $10 is offered the high bidder, but in this variation the second-highest bidder also pays the auctioneer. It's a dramatic, simple game to play with a punch line that carries over into many real-world problems. *Don't get entrapped* seems to be the lesson learned, but this is too simplistic. If that approach makes sense to you, it also makes sense to the other contending players, and if they refrain from getting en-trapped, it may then be to your advantage to become engaged.

Appendix A9: Mathematical Proofs for the "Select the Best Candidate Problem"

The mathematical analysis of the problem "Select the Best Candidate," which results in the surprising answer of $1/e$ for both the strategy of how many candi-dates should be allowed to go by to set up the standard and the probability of get-ting the best candidate, is accessible for those of you who know the calculus. It goes as follows.

Divide the universal sequence of all candidates into two groups; the first group—called the standard-setting group—consisting of a proportion t of candi-dates, is used only to identify the best in that group; we then sequentially observe the candidates in the second group—called the selection group—and choose the first candidate who beats the best in the standard-setting group as our poten-tial winner.

Let $P(t)$ represent the probability of finding the best candidate when the standard-setting group comprises a proportion of t candidates. This procedure will result in the choice of the best, provided that:

- the second best falls in the first segment and the first best in the second segment; this has the chance of $t(1 - t)$; or
- the third best falls in the first segment and the first and second best in the second segment, and the first comes before the second; this has the proba-bility of $t(1 - t)^2/2$; or
- the fourth best falls in the first segment and the first, second, and third best in the second segment, and the first comes before the other two; this has the probability of $t(1 - t)^3/3$; or
- and so on.

Hence we have

$$P(t) = t(1 - t) + t(1 - t)^2/2 + t(1 - t)^3/3 + \ldots \tag{A9.1}$$

$$= t[(1 - t) + (1 - t)^2/2 + (1 - t)^3/3 + \ldots]. \tag{A9.2}$$

We show below that the series in parentheses on the right-hand side of (A9.2) is $-\ln(t)$ so that

$$P(t) = -t \ln(t). \tag{A9.3}$$

To find the optimal proportion t for the standard-setting section, we differentiate (A9.3) and set it equal to 0, yielding

$$0 = P'(t) = -\ln(t) - t/t = -\ln(t) - 1 \qquad \text{(A9.4)}$$

or

$$\ln(t) = -1, \quad \text{or} \quad t = e^{-1}, \text{ as to be shown.}$$

In addition

$$P(e^{-1}) = -e^{-1}\ln(e^{-1}) = e^{-1}.$$

Now for the proof going from (A9.2) to (A9.3).
Let

$$S(x) = 1 + x + x^2 + x^3 + \ldots \qquad \text{for } 0 < x < 1. \qquad \text{(A9.5)}$$

Then multiplying both sides of (A9.5) by x, we get

$$xS(x) = x + x^2 + x^3 + \ldots \qquad \text{(A9.6)}$$

and subtracting (A9.6) from (A9.5) we have

$$S(x) - xS(x) = (1 - x)S(x) = 1,$$

or

$$(1 - x)^{-1} = S(x) = 1 + x + x^2 + x^3 + \ldots \qquad \text{(A9.7)}$$

Taking the integral of both sides of (A9.7), we get

$$-\ln(1 - x) = x + x^2/2 + x^3/3 + \ldots \qquad \text{(A9.8)}$$

Letting $x = 1 - t$, we establish the step going from (A9.2) to (A9.3) as to be shown.

10

Auctions and Bids

Intertwined Negotiations

In introducing different auction mechanisms and examining some of the basic problems faced by bidders and auctioneers, we consider auctions as a special type of negotiation and analyze them primarily from a noncooperative, game-theoretic perspective. We relax some of the game-theoretic assumptions, however, and offer advice from an asymmetric prescriptive/descriptive viewpoint. Many of the concepts we use, such as common knowledge and equilibrium analysis, were explained in Chapter 4. It may be useful to look back there from time to time if you don't fully grasp the essentials of the analysis in this chapter.

How do auctions relate to negotiations and games? We treat auctions as a special class of intertwined distributive negotiations involving three or more competing parties. To illustrate, let's assume an auction is taking place in a sales room with one seller and two buyers, Bebe and Cedric. Imagine that bidder Bebe values the prize as worth X. The auctioneer wants to claim as much of X as possible—and if possible get Bebe to pay more than X. If Cedric left the room, Bebe and the auctioneer would be in a distributive negotiation. Whatever Bebe—or the winning bidder—pays (loses), the auctioneer gains (wins). There's no opportunity for joint gains.

In addition to the zero-sum bidder/auctioneer relationship, there is a competitive dynamic between the bidders. Bebe is competing against Cedric to win the prize. This competitive element of the game has a powerful impact on bidders' behavior and often leads them to ignore the distributive value of the game with the auctioneer.

Because this book is concerned primarily with negotiations, it's important for you to see how negotiations—whether distributive or integrative—can be transformed into auctions if another party participates in the game. Imagine, for instance, how the Elmtree House negotiation would have proceeded if three real estate developers had all been eager to buy the property. By adding more parties to compete for a unique prize, a distributive negotiation can be transformed into an auction.

For example, consider the CEOs of two leisure companies, Mr. Winthrop and Ms. Mather, negotiating over potential synergies that would result from giving Mather's members access to Winthrop's golf courses and rowing facilities. Mather and Winthrop pursue an integrative negotiation, and Winthrop thinks he can get a bundle of assets worth X from Mather. Enter Mr. Adams, CEO of a third company, who is very keen on making a deal with Winthrop, because if Mather gets the deal, Adams's business will be adversely affected. So Winthrop starts negotiating separately with each. In the negotiations with Adams, Adams makes an offer that is worth 2X to Winthrop. Winthrop sees that is it possible to make a mint on this deal by stoking the competition between the two aspiring partners. In a meeting with Mather, Winthrop lets it slip that Adams has made a very attractive offer worth 2X to him and wonders out loud whether Mather can improve on this.

In this example, Winthrop has engineered a negotiated auction with the two hopefuls bidding against each other to win the deal. Winthrop uses Mather as his BATNA in the negotiations with Adams, and uses Adams as the BATNA in his negotiations with Mather. Hence the negotiations are intertwined. Let's not forget that the actual negotiations between Winthrop and Mather, and Winthrop and Adams, are pursued on an integrative basis. So strictly speaking, this is not an auction, because the players can change the value of the prize—but it has many of the trappings of an auction.

Auction Mechanisms

In this chapter we discuss different mechanisms that the seller can employ. We compare five types of auctions and bids:

- The open, ascending, outcry auction (English auction)
- The open, descending, outcry auction (Dutch auction)
- The competitive sealed bid
- High bidder wins at the second-high price (the Vickrey or philatelist auction)
- The reciprocal bid or buy-sell bid

Our objective is twofold: to help auctioneers and bidders choose which games they might want to play, and to help bidders make smart decisions once the action starts. Our advice to bidders will be tailored to the characteristics of different auction mechanisms. In crafting our prescriptions, we alternate between two contrasting analytical perspectives. We ask: how (prescriptively) should a buyer bid, given reasonable (descriptive) assumptions about the behavior of other buyers? We then contrast this asymmetric prescriptive/descriptive approach with a symmetrically normative approach. In this perspective we ask: how does game theory, which posits complete rationality of all players, suggest bidders should behave in auctions? As we have done before in dealing with abstract, game-theoretic concepts, we will use a dialogue format to make the discussion more concrete.

Open, Ascending, Outcry Auction: English Auction

Imagine the auctioneer delivering the following patter:

> What do I hear for this beautiful chair? I hope you have all inspected it. You notice that it is just slightly used; it has a beautiful flower pattern and it would go beautifully in someone's living room or den. Now come on, let's start the bidding briskly. I hear the first offer of $15 from that lady over there. That's a ridiculously low bid. Now come on, let's push the bidding up. Oh, I hear $25 from that gentleman; $30 from that lady back there. $35? Do I hear $35? I hear $35 from this gentleman in the third row. Come on, this is still way underbid; $40; I hear $45, $48, I hear $48 going once; I hear $50; I hear $50 going once; I hear $50 going twice; I hear $53; $53 going once, $53 going twice, sold for $53 to that gentleman in the third row.

Let's look at this auction again. Suppose the gentleman in the third row who entered the bidding at $35 and was the high bidder at $53 is named Jay. We ask Jay: "How did you think about the problem? What did you do?"

JAY: First, before the auction started, I inspected the chair, and I knew I could use it in my study. I figured that I would bid up to $60 for the chair.

EXPOSITOR: Where did that $60 figure come from?

JAY: I knew I couldn't get it for that price at a second-hand store, and I was in the market for such a chair.

EXP.: If bidding continued to $60 would you have preferred to get the chair at $60, or not?

JAY: That's tough, but yes, I would have felt comfortable paying $60.

EXP.: How about $80?

JAY: No way. At $70 I would just be at my break point.

EXP.: So instead of saying that you would have bid up to $60, you should have said that you would have gone up higher if absolutely necessary. Now comes an important definition. We'll say that Jay's

reservation value is $70 for the item. He would be indifferent be-
tween getting and not getting the chair at $70. Jay, suppose you in-
spected the chair and couldn't be present at the auction. You had
to tell Ann, over here, to bid for you. What instructions would you
give her?

JAY: I would tell her to start her bidding around $25 or $30 and not to
seem too eager. Bid slowly, and go up to $70 but not a penny more.
I'd tell her to try to get it as far below $70 as possible and to in-
crease her bids in small increments.

EXP.: You got the chair for $53. So your bidder's surplus was $70 −
$53 or $17. Why did you say, Jay, that your agent, Ann, should not
be too eager?

JAY: Well, there is some psychology here. There are other bidders who
might be tempted to raise their prices if the auction is proceeding
briskly. They may not be sure of the price they are willing to pay.

EXP.: Exactly. Their reservation values, RVs, may be soft, and the bidding process itself might influence their RVs. I have a story for you.
When my wife and I were first married, we went to an auction and
bid on a used rocking chair. My wife had in mind to nurse our expected baby in the rocker. We agreed, before the auction started,
that our reservation value would be $20. She would not bid higher
than $20. We were penurious graduate students at that time. Dur-
ing the auction, the auctioneer rapidly got up to $18 and my wife
raised it to $20; and the auctioneer said $20 going once, $20 going
twice; and my wife said $22. The auctioneer then smiled and said,
"Lady you're bidding against yourself." Little did he know that she
was also breaking our RV.

––––––––––––

The Babylonian Wife Auction.   The Babylonian Wife Auction, a fascinating
example, goes back to the writings of Herodotus. We feel uncomfortable relating
this sexist practice to you, but it's the first documentation we know of concerning the use of auctions, and the process is a fascinating one. The rule in this an-
cient land was that all women reaching a certain age had to be auctioned off as
wives of the men of the land. The women were presented sequentially, and the
bidding proceeded in an open, ascending auction for a given woman. The high
bidder married that woman.

One interesting aspect of this appalling event is the fact that the money collected by the auctioneer did not go to the family of the woman sold. It went into
a fund to be used by the auctioneer representing the town. It was used for the
following purpose. Suppose a woman came to the block who was not very desirable, and the swains were not willing to bid for that woman in an ascending auc-

tion. Then the auction proceeded in a descending fashion. If none of the men wished to pay a positive amount for her, the auction continued in the reverse direction: the woman was offered along with a dowry of the equivalent of 100 drachmas to the audience. If no one took the woman plus 100 drachmas, then the monetary value was increased from 100 drachmas to 200 to 300 to 400 and so on until someone was willing to take that woman together with the monetary compensation that would go along with her. That man would have the obligations of a husband to that woman. The money collected for the more desirable women went into the treasury in order to help provide the dowries of less desirable women. The Babylonian auction thus represented a kind of socialization of the value of beauty, which is unfairly distributed by fate or nature. What it meant for the relationship between men and women is, of course, another story.

Variations of the wife auction are possible. You could reverse the roles and let the women select the men. We don't know whether that has ever happened. But you could also reverse the roles to look at what happens if you have a group of women and a group of men that have to be paired off. The women have preferences, the men have preferences. Can you concoct a scheme that will be responsive to both sets of preferences? That's the famous marriage problem, which has been very well researched (Roth and Sotomayor 1990). In India and Pakistan, where arranged marriages still exist, a similar market can be found; sometimes both the female and the male partners have veto rights.

Sometimes auctions can be influenced by those participating in them. See Box 10.1.

Open, Descending, Outcry Auction: Dutch Auction

We now auction off another chair to illustrate the Dutch auction.

> We have this beautiful chair to auction off. Do I hear $100? I don't hear $100. Do I hear $95? Do I hear $90? Do I hear $85? Do I hear $80? Do I hear $75? Do I hear $70? I don't hear $70. Do I hear $65? Do I hear $60? Do I hear $55? I hear $55, sold for $55. The chair goes to the gentleman back there for $55.

Let's examine this Dutch, or open, descending auction, which is one of the most common. It's called the Dutch auction, presumably because tulips in Holland are auctioned off using this procedure. It's very quick. In produce and cattle markets, most items are sold this way: the auctioneer comes down sequentially from a top price and the first person who raises his hand obtains the prize or the item. Indeed this can be mechanized. The bidders sit in designated seats, with a set of buttons on the desk in front of them. Up on the wall is a huge clock. The

---

Box 10.1   Duveen and the Houdon Busts

In the book by S. N. Behrman about the famous art collector Duveen, Behrman tells about the power of Duveen to influence the price of art objects. When Duveen wanted to bid on an art object he would very rarely bid himself; he would use proxy bidders to bid for him. The story goes that Duveen owned, in his galleries in New York, ten Houdon busts, which were not selling very well at the price Duveen was asking. The figures that I am going to give may not be quite accurate but they will tell the story. Let's suppose these ten Houdon busts were languishing on his shelves at a sale price of, say, $15,000. There was an auction of a Houdon bust in England, and Duveen showed up in person and entered the bidding. A rarity. He also posted a stooge of his and they in turn jacked up the bidding on the Houdon bust. Duveen bid $25,000; his stooge bid $30,000; Duveen $40,000; stooge $50,000. It went up to some value close to $60,000. The information spread like wildfire in the art world, and Houdon busts became a very valuable commodity. By that means Duveen was able to unload the ten Houdon busts that he had in his galleries for a handsome profit.

---

handle of the clock points north at a designated high value, and then the clock handle moves gradually clockwise, indicating the reduction of the price. Each individual bidder sitting in a designated chair has to decide when he or she wants to stop the clock by pressing a button, which automatically records the price of the transaction and who pressed the button at which chair; this is automated in a very efficient procedure. By that means many items can be auctioned off efficiently and swiftly.

A bidder's decision analysis.   Now let's examine some of the strategic issues involved in the Dutch auction. Imagine that you are one of the bidders, and you're bidding for a widget. You have inspected the widget and investigated what other possibilities you have for purchasing a similar item, and on the basis of that preliminary investigation, you have determined a reservation value for the item of $800. That's your RV. If you get the item for less than $800, you will have what you consider a consumer's (or bidder's) surplus. If you get the item for a value higher than $800 you will be unhappy. You are trying to get the item for as little as possible. Now imagine that you're sitting at your chair and the clock proceeds from a high of $1,500, lowering gradually. As the handle moves in region A, before the $800 mark, there is no point in your pressing the button and stopping the proceedings (see Figure 10.1). When it hits $800 you are at your RV, and as the handle moves into region B you become more and more anxious. The longer you wait, there is the excruciating possibility that somebody else might push the button.

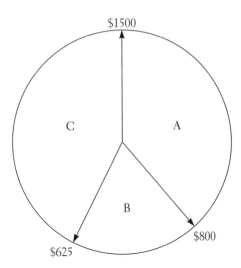

Figure 10.1.   The Dutch auction with an RV of $800 and a bid of $625.

Let's suppose that you have analyzed this, and you've decided to push the button at $625 as long as somebody else does not press the button before then. In that period in B, from $800 to $625, your anxiety is building. You're hoping that nobody will press that button before $625. If you get the item for $625, you achieve a profit, or consumer's surplus, of $175—that is, $800 minus the $625. Of course, you cannot know whether you could have gotten the item if you had held off further; instead of getting the item for $625, you might have held off to $500 or $400. But the longer you wait, the greater the chance that somebody else will push his button. As time goes by, the size of the surplus *and* the risk that someone else will snatch it both increase. How are we going to analyze this problem?

Certainly you should do your RV analysis and arrive at your own break-even price, but then what should your decision price be? When should you press your button? This should depend upon your probabilistic assessment of what will be the maximum execution price of the other buyers in the bidding game.

More about the Dutch auction later. See Box 10.2.

Competitive Sealed Bids

Our next topic is the competitive sealed bid. Each of several bidders submits a sealed-bid value for a given prize. As an example, the government might auction off the rights to drill on a given parcel of government land, and the bidders might be various oil companies or syndicates of companies who are bidding for the rights to drill. Each syndicate submits a sealed bid; the bids are opened up si-

---

Box 10.2   Which Would You Prefer?

If you were the auctioneer or the seller, from your vantage point would you be better off auctioning off an item in the open, ascending (English) auction or the open, descending (Dutch) auction? Of course one issue might be the time it takes to auction off a lot of objects, but let's suppose that we ignore the problem of time. Are you better off with an English auction or a Dutch auction?

---

multaneously and the government awards the contract for the rights to drill to the highest bidder.

In some situations, it's the minimum value of the bid that will win the prize. A typical example: the government wants to award a contract for building a certain project; let's suppose that it's an airplane, a dam, or a highway. Various contracting firms seek the project or the contract; each submits a competitive sealed bid. The bid that promises the lowest cost to the government wins the contract. Rather than mixing up these two examples where the maximum bid is desirable or the minimum bid is desirable, we will concentrate on examples where the maximum bid is desirable. So we will look at problems of competitive sealed bidding where the highest bid wins the prize and pays the auctioneer the amount of the highest bid.

The general case: Private valuations and fuzzy knowledge.   We'll start this analysis by adopting an asymmetric prescriptive/descriptive approach from the bidder's perspective. Because auctions can take place under many different conditions of beliefs, information, and uncertainty, our analyses focus on distinctive cases. Here we begin with the "general case" in which the item up for auction (for example, a garden gnome) is valued differently by the bidders. Moreover, any single bidder has only fuzzy knowledge about what the others' valuations of the item are and what their bidding strategy will be.

Let's assume you are bidding against ten other bidders, say, for a large oak antique desk that would look great in your study. Some people might think that the desk is a monster, but you have just the right place for it. You have shopped for desks like this one, and the prices were about $1,000. This one is even nicer. How do you think about this problem?

Let's suppose you start off with your own reservation value analysis. You ask, "What's the maximum value I would be willing to spend for the desk, all things considered?" If it were an open, ascending auction, and you had to give your instructions to your friend or agent to bid for you, up to what maximum value would you have him or her bid? This is not an easy question. The answer depends on your needs, desires, other opportunities, resale value, and so on. But suppose now you complete your analysis and your break-even value is $800. What bid should you submit?

MBOO analysis.   What is your principal uncertainty? If you had a crystal ball, what would you want to know? You would want to know the maximum bid of others—what we call the MBOO value and pronounced "Maboo." If you knew that the MBOO value was $575, you would submit a bid of $576. But unfortunately you don't know the MBOO value.

You are confronted now with a classical decision problem under uncertainty that can be depicted as a decision tree (see Figure 10.2). At move 1 you must submit some bid. I show just a few branches and extend the tree following the branch: submit a bid of $600.

Following your bid, you come to an uncertainty, or chance, node: you will be told whether you win or lose, that is, whether the MBOO value is below or above your bid. If you win, your consumer surplus or payoff is your RV minus your bid. For the $600 branch you stand to win $800 − $600, or $200, if the maximum bid of others, the MBOO value, is below $600. If the MBOO value is above $600 you are left with a consumer surplus of zero dollars.

With a bid of $600 your expected monetary value is $200 times your subjective probability value of winning, or $200 times your subjective probability that the uncertain MBOO will be below $600.

Let's assume initially that your aim is to choose a bid to maximize your expected monetary value: payoff times probability of winning.

Graphing the MBOO analysis.   For any bid B, let $F(B)$ be your assessed probability that you will win the bid—that is, that the MBOO value will be below B. See Figure 10.3. Notice, for example, that the probability that the MBOO value is less than 700 is .5.

The graph shows the implications of choosing the bid of $500, as an illustration. The potential profits or winnings are $800 − $500, depicted on the horizontal axis. A bid of $500 has a .3 chance of winning; that's the vertical distance from the horizontal axis to the cumulative probability curve at B = $500. The area of the depicted shaded rectangle gives the expected monetary value with a bid of $500. It multiplies potential winnings by the probability of winning.

If you want to choose the bid to maximize your expected monetary value, a reasonable aim, then you maximize the expression

$$(RV − B) \times F(B),$$

which is the area depicted in Figure 10.3.

Look what happens if you bid too low, as in the bid of B_1 in Figure 10.4. The dotted area, depicting the expected monetary value of profit times the prob-

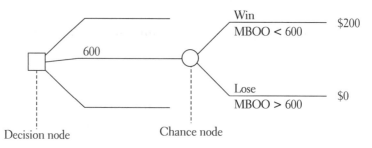

Figure 10.2.   Decision tree for sealed-bid auction.

$F(B) = \text{Prob}(\text{MBOO} \leq B)$

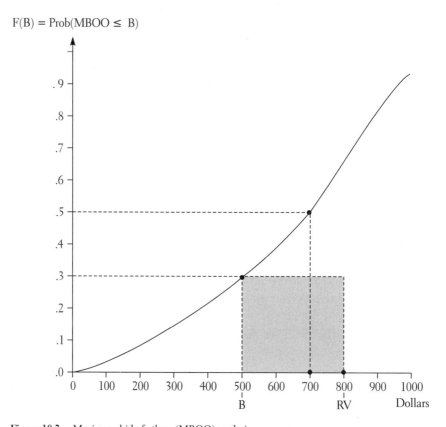

Figure 10.3.   Maximum bid of others (MBOO) analysis.

ability of winning, is small because the vertical rise of the rectangle is small. A bid of B_2 yielding the slashed area is too high, because the horizontal dimension of the rectangle, the profitability, is too small.

If the whole analysis depends so critically on subjective judgments, why not intuitively pick out a bid value without analysis? That's what most people do. Let me see if I can suggest an answer: this process depends on judgment in an or-

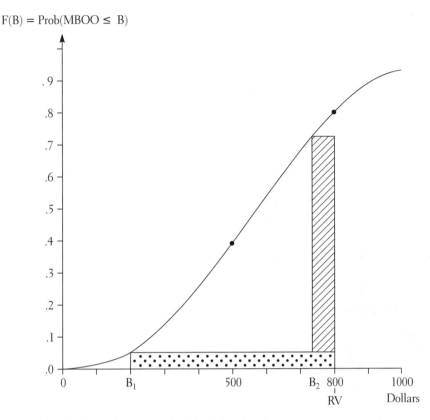

$F(B) = Prob(MBOO \leq B)$

Figure 10.4. MBOO analysis showing bids too high and too low.

derly way—in a way that is "thinkable" at each stage. It decomposes the problem into two separate parts: RV analysis and MBOO analysis.

The Philatelist or Vickrey Auction (High Bidder Wins at Second-High Price)

In this next type of competitive bidding process, each bidder submits a sealed bid for the prize. The high bidder, as usual, wins the prize but pays the auctioneer not the highest bid but the second-highest bid. This bidding process is used for bidding for rare stamps and is called the "philatelist auction" or the Vickrey auction.[1] It may seem strange, but it has some surprising and truly remarkable properties. It's well worth your time to understand this strange competitive process.

1. Nobel laureate William Vickrey was particularly fond of this auction, because it generates honesty.

Table 10.1   Bids of luminaries for condo

Bird	$650K
Boggs	$500K
Evans	$700K
Yastrzemski	$860K

Consider this example. A rich sports enthusiast has decided to sell his condo on the Boston waterfront at a bargain price to one of five great Boston athletes. Bids are by invitation only. Four of these greats submitted sealed bids with the understanding that the high bidder would get the condo but have to pay the second high price. Let's look at some hypothetical data, shown in Table 10.1.

Carl Yastrzemski submitted the highest bit at $860K. So he gets the condo but pays the auctioneer not $860 but $700K, the price bid by Dwight Evans.

Now this sports enthusiast, the auctioneer in this instance, is a philanthropist and wants to sell his condo to an athletic luminary. But why should an auctioneer, who wants as much money as he or she can get, be content with getting the second-highest price? Let's see.

Suppose a new bidder enters the scene, Jim Rice. Let's coach Rice on what he should bid. Suppose that Rice and we, as coach to him, do not know the bids of Evans, Wade Boggs, Yaz, or Larry Bird. First we encourage Rice to do an RV analysis. "In an open, ascending, outcry auction, the usual English auction, what would be the maximum value you, Mr. Rice, would be willing to bid? What's your reservation value? If an agent of yours had to bid for you, Mr. Rice, in the usual ascending auction, up to what value would you instruct her to bid?"

Suppose Rice says $800K. We check if this is right. He wouldn't want the condo for $825K, and if he could get the condo for $700K he would enjoy a consumer surplus of $100K.

In this second-high-price auction, what should Rice bid?

If this were the usual competitive sealed-bid auction, Rice would be counseled to do a MBOO analysis. What would he assess the maximum bid of others to be? (We mean "assess" in a probabilistic sense.) He would have to take into consideration the number of other bidders and his perception of what the condo would be worth to Evans, Boggs, Yaz, and Bird. He would have to reflect not only what it's worth to others but how they would bid.

The punch line is that in the second-highest price auction Rice should bid his RV of $800K! Think about it.

- First argue that he should never bid above $800K. [Suppose we consider some bid, like $830K, above his reservation value of $800K. If the maximum bid of the others (MBOO) is above $830K, Rice is just as well off using $800K as $830K. If the MBOO was between $800K and $830K, he certainly would prefer $800K to $830K. If the MBOO was below $800K, then $800K is just as good as $830K. No matter what, $800 is better or just as good as $830K!]

- Next argue that he should never bid below $800K. [Use similar logic as above.]

With this information there is no need for us to do a MBOO analysis for Rice. It doesn't depend on the number of other bidders. Rice's optimum bid does not even depend on whether the other guys bidding understand the subtle implications of this bidding system. Our advice to Rice: just bid the maximum value you think the condo is worth. Forget about the others. No need to do any strategic analysis. Just be honest and simple minded.

This is lovely. Honesty pays for once. You don't have to try to outguess anyone. But what about the auctioneer? Is the auctioneer better off with the competitive bid designated the high-bid price or second-high-bid price? It's complicated, as we shall see.

The Auctioneer's Perspective

Now for the auctioneer's perspective. Let's compare the four auctions we've talked about so far: the English auction, the Dutch auction, the standard competitive bid (where the high bid wins at the high price), and the competitive bid where the high bidder pays at the second-high price.

ANN: Can I interrupt? Before you go on, could you pull together the punch lines for the four schemes we've discussed—especially from the perspective of the auctioneer.

EXP.: It's complicated! It depends on the true distribution of RVs among the players and their perceptions of the RVs of the others. There are theorems that state that if the auction is posed as a bona fide game—that means common knowledge and all relevant uncertainties have probability distributions that are commonly held, and all are risk neutral (or at least if they have utility functions, they are commonly known)—then equilibria analysis says that all procedures give the same yield to the players and auctioneer.

ANN: That may be interesting to you, a game theorist, but what does that mean to Jay and me if the bidding involves subjective elements—like MBOO analysis using the Dutch auction, or competitive fervor in the English auction?

EXP.: It depends. Take a simple case, bidding for a known $100 prize.

That's a commonly known outcome with complete knowledge all around. In the second-high price auction, the bids should all be $100 without any MBOO analysis. But in the standard competitive bid with MBA students, the best winning bid against the empirical mix of bids is $78. So in that case the auctioneer is better off with the second-high auction.

JAY: How about a case where the competitive sealed bid is better than the standard open, ascending (the English auction)—from the perspective of the auctioneer?

EXP.: It depends on a particularly tricky analysis—the auctioneer's guess about the bidders' MBOO analysis of each other. When there are multiple bidders and the auctioneer thinks they are likely to overestimate each other's maximum bid, then he can benefit from their misperceptions by hiding that information inside the sealed bids. The auctioneer might even pretend that there are multiple bidders when there is only one—investment bankers are famous for doing that.

ANN: What about times when the English auction is better?

EXP.: Just the opposite case—when the auctioneer guesses that the bidder's MBOOs are mistakenly low. Imagine some art aficionados. Each thinks she is the only one collecting this particular painter, so she can get it for a steal. Initial bids are very low because they all think the others will be bidding very low. In the outcry auction bidding could go surprisingly high. Outcry auction bidding may also be preferable for the auctioneer when he thinks the bidders are likely to get into a competition involving interests other than the actual value of the item.

Remember my wife who, in the heat of the open, ascending auction, bid against herself and bid higher than her RV. In her defense she said that her RV changed when she saw how vigorous the bidding was. Other people thought the chair was worth near $20 and this reinforced her new evaluation.

JAY: Couldn't this be factored in at the beginning so that maybe her RV should have been on the order of $30, after all?

EXP.: Sure, normatively it should have been. But descriptively it wasn't. You could also imagine two swashbuckling CEOs with big egos bidding against each other at a charity auction. Each will want to show the crowd that he is richer, or more generous, or whatever.

ANN: It seems, at least, that there are only three types and not four, since the sealed bid is analytically like the Dutch auction. Both produce the same MBOO analysis.

EXP.: Nope. Here again, it depends. Go back to the case of the Dutch
    auction in which utility analysis says you should hold out for $625
    before pressing your button. Your RV is $800. Now imagine that
    the auctioneer has reached the downward value of $700.

    Would you rather be sure of getting a $100 surplus or gamble
    on getting $75 more? Now, descriptively, the standard utility theory
    doesn't track too well. A variant of utility theory, called *prospect
    theory*, predicts descriptive behavior better, and that theory might
    predict that many subjects will grow anxious and hit that button at
    the $700 mark.

    Empirically we know that most subjects do not make coherent
    choices. They tend to unduly favor what is certain rather than a lot-
    tery with higher expected value when the certainty value is posi-
    tive. In behavioral decision theory this is called the "Certainty Ef-
    fect." Subjects become enamored of a sure thing.

JAY: Are you merely saying, in another way, that people are risk averse
    and they should use utility theory to cope with their risk aversion?
    This complication could have been included in Ann's analysis
    leading to $625.

EXP.: I am saying a bit more. It's like what happens in the Allais
    paradox discussed in Chapter 3. Most people do not act strictly
    in conformity with expected utility theory. They deviate from it,
    in weakly predictable ways. One way is that they tend to under-
    value lotteries when the alternative is a certainty with a positive
    payoff.

    Since people tend to fall prey to the certainty effect, we would
    expect that most subjects would bid higher in the Dutch auction
    than in the sealed-bid auction. Perhaps sophisticated players would
    not fall afoul of that trap.[2]

JAY: In the (high bid wins at the second-high price) bidding procedure,
    suppose that the bidders are confused, as I suspect I would be, and
    they bid the same as they would in a straight competitive bid situa-
    tion.

ANN: Of course, if the bids are no different, the auctioneer is obviously

---

2. There is an analytical point that needs clarification. We let $F(B)$ be the ex ante probability
that the MBOO value is less than (or equal) to B. In the Dutch auction, if the clock has rotated
downward to, say, $700, and no one as yet has declared, the conditional probability that the MBOO
value will be less than B (for B values less than 700)—given that B < 700—is $F(B)/F(700)$, and now
we want to choose B to maximize $(800 - B) \times F(B)/F(700)$. The $F(700)$ in the denominator does
not affect the optimal B value—it still is at 625.

better off with the standard process. He's better off getting the top bid than the second-highest bid.

EXP.: Exactly. But now consider, for example, the case where each bidder has a high RV but each thinks the others have low RVs. What happens in the standard first-price competitive auction?

ANN: Well, if I had an RV of $1,000 and I thought mistakenly that Jay's RV was $500, and if it was really $750, I might put in a bid of $525, say; and Jay might put in a bid below his true value of $750, so here's a case where the second-high-bid procedure would yield a higher value than the conventional procedure. The auctioneer would come out far worse with the regular competitive bidding procedure.

The different auctions evoke different psychological responses. The psychologies are different in the different auction schemes and as we have seen, it depends.

Special Cases

We have already discussed the general case of private values and fuzzy knowledge. In this section we address two special cases. In the first example we discuss an auction where:

- The prize has a known given value ("certain value" in the game-theoretic terminology).
- The certain value is worth exactly the same to all bidders ("common value").
- All bidders know that the prize is of certain and common value, and each bidder knows that the others know he knows and so on—the heroic assumption of common knowledge.

In the next special case we'll assume the value of the prize is not deterministic but each bidder has the same objective probability distribution about the value of the prize. For example, suppose that the bidders are shown three envelopes: envelope A has $0 in it; envelope B has $100; envelope C has $200. One envelope is chosen at random, in such a manner that each bidder believes that the prize is equally likely to be $0, $100 or $200. If you want to be dramatic, think of these amounts in thousands of dollars. We dub this the common knowledge, probabilistic case with symmetric information.

We could complicate matters even further by assuming one of the bidders has privileged information about which envelope has been drawn.

Special Case 1: Common Knowledge and Common Values

In this scenario, all bidders have common knowledge about the common, certain value of the prize, and to make life even simpler, we'll assume the prize is a crisp new $100 bill. We'll examine the simplest possible case.

You are playing against just one other bidder—bidding for a $100 bill. What bid would you offer in the competitive sealed-bid auction where the high bidder wins at the high price? What would you bid against two other bidders? . . . four other bidders?

Bidding for $100 against one other bidder.   Let's call in Ann and Jay, who have already done the exercise of bidding for the $100 bill.

ANN AND JAY: Hi.

EXP.: What did each of you bid for the $100 bill when you bid against one other player—say, Jay against Ann?

JAY: I bid $99.99.

ANN: I bid $70.

JAY: So I win. Why did you bid $70?

ANN: Well, you won, but you only made a penny. Don't you feel ridiculous that you didn't bid $71?

JAY: Look, if I had a crystal ball, I'd be rich. Rather than playing silly games, I'd be a regular at the art auctions at Sotheby's.

ANN: The aim is not just winning but to make a reasonable profit if you do win.

EXP.: Let me intervene here. In the theory of games, the analyst tries to give prescriptive advice to each of the players. The problem is: can public advice be given to each player in a manner that will encourage each to follow that advice under the assumption that others will also follow the advice? This is called *equilibrium analysis*. Let's examine equilibrium analysis for this $100 auction with two bidders. Suppose I publicly told you, Ann, to bid $70 and you, Jay, to bid $99.99. Is this advice stable?

JAY: Certainly not. If I knew Ann would follow your advice and would bid $70, I would change and bid $70.01.

EXP.: O.K., so bidding advice of $70 to Ann and $99.99 to Jay is not in equilibrium. How about advice of $80 to each?

ANN: That doesn't work. If he held fixed, I'd move to $80.01. Similarly for him.

EXP.: Can you think of any joint advice that would be stable?

JAY: How about telling us both to bid $100?

ANN: Well, if I knew you were going to bid $100 I would have no incentive to bid less and vice versa, and I wouldn't want to bid more, so I guess that advice would be stable, but not very appealing.

EXP.: Granted. How about $99.99 for each? With the understanding that with a tie you would take the profits, in this case a mere penny, and toss a coin to see who gets it.

JAY: Can I bid in mils? Can I bid $99.999?

EXP.: No, let's rule that out. You can't bid in finer gradations than pennies.

ANN: I guess this would also be in equilibrium, and speaking technically, I suppose the equilibrium ($99.99, $99.99) is better than the equilibrium pair ($100, $100). But I still would defend my bid of $70.

EXP.: Why?

ANN: Well I don't think most people would bid $99.99 or $100 even though Jay did.

EXP.: O.K., that's the lead I want. For you, Ann, your RV is $100 and you are not sure what Jay will bid. So you should do a MBOO analysis, just as I explained earlier. I have played this game with hundreds of MBA students. What proportion do you think would bid as Jay did—that is, $99.99 or $100?

JAY: I would guess about half.

ANN: I would guess about a quarter.

EXP: When this experiment was conducted with hundreds, if not thousands, of MBA students untutored in this exercise, the best bid against the empirical mix of sealed bids turned out to be $78. Of course once subjects learn this result, they bid differently. When bidding against two or more bidders the dynamic changes.

Ann, how should you think about this problem when bidding against two or more bidders?

ANN: I would suspect that most bidders would bid higher against two others than one other. And this time, to win, you would have to bid higher than each of two other bidders.

EXP.: Exactly, Ann. There are two effects. First, to win you have to bid higher than the maximum of the two other bids. So you should bid higher. The second effect is that the others may increase their bids as well, so that you should increase your bid even more. Against the empirical mix of untutored subjects, the best bid against two other bidders is about $87. And as the number of bidders increases to five, the optimal bid jumps to the high nineties.

Special Case 2: Uncertain Common Value, Known Objective Probabilities, and Unknown Private Risk Aversions

Suppose our experimenter and auctioneer takes three envelopes. In the first he puts nothing; in the second, $100; in the third, $200. He shuffles the envelopes; chooses one at random; and puts it on the table in front of him. That will be the prize. As far as you are concerned the prize is equally likely to be $0, $100, or $200. Although the prize is uncertain, there is a known common probability distribution for all the bidders. This case is quite common. Imagine oil companies bidding for the right to drill at a given site, and all have a common report about the chance of hitting oil, for instance. That's close.

What would you bid?

JAY: Let me get this straight. If I bid $60, say, and the prize is worth zero do I lose my $60?

EXP.: Of course. You bid $60 for the prize. I collect your $60 and give you the envelope. Too bad it's zero. It could have been $100 or $200.

What's the first piece of analysis you would do in preparation for each of the four types of auctions and bids?

ANN: I guess I would have to search my soul and do an RV analysis. What's the maximum I would bid for this uncertain prize in the ascending English auction?

EXP.: The expected value of the prize is $100. Is that your RV?

ANN: No, absolutely not. I'm a poor student and I'm risk averse. Keeping in mind I could lose my bid, my RV would be about that $60 value Jay just mentioned. No, make that $50. For the English auction and the second-high-price bid the analysis is easy: I bid gradually up to my RV in the English auction and I bid my RV of $50 in the second-high-price auction. But I would have to do a MBOO analysis, either formally or informally, for the Dutch auction and the standard competitive bid.

EXP.: If—and this is a big if—if it were common knowledge that all bidders were risk neutral and each of their RVs were $100, then it would be like bidding for a $100 bill. It would be possible to give game-theoretic, equilibrium advice to all. Game-theoretic advice requires common objective information. Game-theoretic analysis cannot cope easily with the realistic case in which the risk aversions of the bidders are different and unknown to each other and for which there is no objective way of assessing objective probabilities that are common knowledge.

Now let's imagine that the auctioneer is the U.S. government, which wishes to have a competitive sealed bid for oil-drilling rights at a given site. Let's simplify by assuming that the value of the prize is unknown to each. Let's simplify by assuming that the value can be $0 million, $100 million, or $200 million and let's suppose that there are potentially thirty bidders interested in the deal. Now suppose these thirty bidders form into three syndicates. To continue simply, let each syndicate comprise ten members each. No bidder, of course, can belong to more than one syndicate. The proposal is that the three syndicates will bid against one another. From the government's point of view, is this breakdown into syndicates desirable?

JAY: As we saw before with bidding for the $100 prize, the auctioneer would rather have more bidders than less bidders. But it might be too risky for any single company to bid all that money.

ANN: It seems to me that with risk sharing, the RVs of the three syndicates should be much higher than the individual RVs. For any one member of a ten-member syndicate, if they shared equally, the prize values would be $0, $10, and $20 million rather than $0, $100, and $200 million. The syndicates as a totality would be less risk averse.

EXP.: Exactly. So, should the government allow syndicate formation?

JAY: I would, but only if there were several syndicates bidding. I would be tempted to say no to only two syndicates and yes to five, but with three it's a borderline case.

EXP.: Let's summarize. The topic under discussion is what is best for the auctioneer, and we have come up with the penetrating conclusion "It depends." Why is it so complicated? We are engaged in a prescriptive/descriptive analysis: prescriptive for the auctioneer and descriptive for the bidding actors. It would be much easier to assume rationality and common knowledge all around and come to the result that normatively it doesn't matter which type of auction is used. But reality forces us to consider problems that are not bona fide games: common knowledge is lacking; subjective probabilities are relevant; bidders do not conform to the tenets of the theory of subjective expected utility (SEU). It is still true that some insights can be gleaned from empirical observations and laboratory experiments. But I'm afraid the bottom line is: it depends.

———————

Reciprocal Buy-Sell Bids

Some years ago, when the Celtics were the pride of Boston, I (HR) had two tickets to a Friday night game that I couldn't use. So I gave them to two students, Abe and Bobby, who did best in a class of ninety on a negotiation exercise. The following week I saw one of the students, Abe, in the street and asked how he liked the game. He sheepishly said that he hadn't gone, and the story of why is interesting. It turned out that both students had dates for that night, and so they both couldn't use one ticket but two would be great. "So?" I queried. In the end they tossed for it. They used the easiest fair way to resolve their dilemma: they randomized. But in this case this procedure was inefficient in the sense that it left joint gains on the table.

This case was particularly cogent for me, because just about that time I was doing some consulting for two business partners—let me also call them Abe and Bobby—who decided, after working together for ten years, that they couldn't stand each other. Accordingly, they agreed that one would buy the other one out. Each had, so to speak, one ticket and wanted two. This Abe and Bobby also decided to toss a coin to decide who should be the buyer for the other's share and who should be the seller of his share. Then I got into the act and suggested an alternate solution.

Back to sharing the Celtics tickets. Here are some options.

A. They could randomize—that's what they did.

B. They could have agreed before the toss to have the winner of the toss compensate the loser by the official price of the ticket, which was $25 at the time. The true value of the ticket was far higher.

C. They could have haggled in bazaar fashion. After all, they were taking a course in negotiations. They even could have sold the tickets for scalper's prices and shared the proceeds—but they thought they shouldn't do that, having won a coveted prize. They would be embarrassed to tell their professor that neither of them had used the tickets.

D. They could have used what I call the *reciprocal buy-sell offer.* One fellow, say, Abe, would designate a price for which he would either buy or sell at, and let the other guy (Bobby) choose which side of the deal he wanted. For example, Abe announces $40, and Bobby says at that price he'll buy.

E. I discussed Abe and Bobby's problem in class and a modification of D above occurred to me. Instead of one party announcing a reciprocal buy-sell offer let both parties do this simultaneously with a split-difference compromise. So, for example, if Abe announces $30 and Bobby $45, Abe sells and Bobby buys at $37.50. This, the class agreed, was better than tossing a coin.

There's still the question of how truthful one should be in announcing an offer. But that's for a later chapter.

Oh yes, what about my business clients? I initially suggested strategy D and they agreed to it. They decided that they would toss a coin to determine who would be the party to announce the price and who would select the buy or sell option. After some discussion we revised the procedure from D to E and they actually put it into their agreement. Instead of figures like $30 and $45, they were talking about $150 million and $190 million. But it's the honest truth that I came to this proposal after thinking about the Celtics tickets.

Combinatorial Bidding: The FCC Spectrum Auction

We now consider the case where several items or commodities are to be auctioned off and where some bidders have preferences for particular combinations of commodities. An example of this type of auction took place in 1994, when the U.S. Federal Communications Commission (FCC) auctioned off licenses for radio frequencies for portable telephones and computer services. More than 2,500 licenses were up for auction, netting the government over $10 billion. Companies, or consortia of companies, who wished to bid for spectrum license X, had to take into consideration whether they would get contiguous sites Y and Z, since possession of Y and Z would considerably enhance the value of X. How should such a company bid sequentially (or simultaneously) on X, Y, and Z? Should the FCC auction off combinations of spectrum sites?

Privatization of governmental monopolies.   Another example of combinatorial bidding of significant magnitude takes place when a government like Argentina (say) auctions off its stock of publicly held assets (like railroad equipment and rail passageways). Consortia of companies want some of the items up for auction, and the value of any item may depend on what other items they win.

Let's simplify. Consider the case of three bidders, B1, B2, and B3, bidding for two commodities, C1 and C2. The bidders' evaluations (reservation values) are shown in Table 10.2.

Serial bidding.   Let's assume that each player knows its own reservation values, but is somewhat vague about the reservation values of the others. Bidders B1 and B2 have superadditive preferences: the combination of C1 and C2 is worth

Table 10.2 Bidder's reservation values

	Bidders		
	B1	B2	B3
C1 alone	7	8	13
C2 alone	8	7	12
C1 and C2 together	23	21	14

Table 10.3 History of bidding for commodity C1

Round	B1	B2	B3
. . .			
5	7*	6	5
6	7	9*	8
7	—	9	10*
8	—	11*	10
9	—	11	12*
10	—	13*	12
11	—	13**	—

* = Designated bidder is not compelled to increase his or her bid at next round.
** = Designated bidder wins.

more than the sum of the values of each. Bidder B3, in contrast, has subadditive preferences: C1 and C2 together are worth less than the sum of their values. B3 has a strong preference for just one of the two commodities.

Suppose that the commodities are auctioned off in an ascending auction in a succession of rounds. At each round the loser is challenged to opt out of the bidding or raise the current maximum bid. The process continues until there is a winner. Suppose we start with an auction for C1 and the bidding is as shown in Table 10.3.

By round 5, bidder B1 is high bidder at 7, which is the maximum that he would want to go for C1 alone. (See Table 10.2.) At round 6, B2 and B3 push the bidding along. For bidder B2, the reservation value of C1 alone is 8, yet she bids 9. Why? She hopes to get C1, and this is the only way she can get a chance at C2; and, for her, C1 and C2 are superadditive. In rounds 6 to 10 bidder B3 has a clear motivation to go on up. Perhaps B3 might want to opt out at 10 if he thinks he can get C2 at a value less than 9. B2 "wins" C1 at 13. She is out on a limb and will lose unless she gets C2 in the next auction. It looks as if B2 could bid (for C2) up to a maximum of 21 − 13, or 8; but as we shall see, it's more complicated than that.

Now for the bidding of C2 after B2 gets C1: Bidder B1 is no longer a major player. B2 thinks B3 will not go up very high. Is she in for a surprise, as shown in Table 10.4.

Table 10.4   History of bidding for C2 after B2 gets C1 for 13

Round	B1	B2	B3
. . .			
4	—	8*	7
5	—	8	9*
6	—	10*	9
7	—	10	11*
8	—	12*	11
9	—	12**	—

*   = Designated bidder is not compelled to increase his or her bid at next round.
** = Designated bidder wins.

Sold to B2 for 12. Why at round 6 did B2 break her maximum of $21 - 13$, or 8? She figured that if she lost C2 she would be out $13 - 8$, or 5 units, and if she goes up to 10 on C2, she'll lose less. She argues the same way in bidding 12 for C2. She wins both for $13 + 12$, or 25, for a net loss of 4. Moral: Don't go over your reservation value for C1! Nope, that can't be right, because her strategy would have worked if B3 had valued C2 far less.

Combinatorial bidding is tricky. Bidders are motivated to bid strategically, and mistakes will happen. This happened a lot in the FCC auction. In this simple case, perhaps the auctions for C1 and C2 should have taken place simultaneously or perhaps the auctioneer should have asked for a sealed bid for the combination of C1 and C2. But it's not hard to concoct a set of bidders and a set of numbers where reasonable strategic behavior could lead to disaster for some bidders.

Parallel bidding.   In the above case, the commodities C1 and C2 were bid on serially. Suppose they were auctioned off in parallel: both auctions being conducted simultaneously in a sequence of rounds. One rule might be: any bidder who is not high on some commodity must bid higher on some commodity or bow out of the competition. Let's consider the example in Table 10.5.

- At round 7: B1 is hopeful, because C1 and C2 together are worth 23.
- At round 8: B2 drops out, and B1 is overextended on C2.
- At round 9: B1 is again hopeful.
- At round 10: B1 is again overextended on C2.
- At round 12: B1 is now overextended on C1.
- At round 13: Bidding ends with B1 taking a loss.

At the completion of the bidding, and after the auctioneer collects the fees and distributes the commodities, the players may engage in an aftermarket, where they try to negotiate some Pareto improvement—a sort of postsettlement

Table 10.5 History of bidding in parallel

	For C1			For C2			Who must
Round	B1	B2	B3	B1	B2	B3	bid next
. . .							
5	7*	6	5	4	6	8*	B2
6	7	8*	5	4	6	8*	B1
7	9*	8	5	10*	6	8	B2, B3
8	9	—	10*	10*	—	8	
9	11*	—	10	10*	—	8	B3
10	11	—	12*	10*	—	8	
11	12.5*	—	12	10*	—	8	B3
12	12.5*	—	12	10	—	10.5*	
13	12.5*	—	12	10	—	10.5*	

* = Designated bidder is not compelled to increase his or her bid at next round.

settlement. For example B1 has purchased C1 for 12.5, a loss of $(12.5 - 7)$, or 5.5. In an aftermarket, B2 might offer B1 an amount 9 for C1, and offer B3 an amount 11 for C2. This would be a Pareto improvement for all. And so negotiations may start.

Design of the FCC auction. Now imagine the FCC auctioning off 2,500 commodities to hundreds of potential bidders who can form syndicates and bid collectively. Not all 2,500 items can be auctioned off in parallel, so a combination of serial and parallel negotiations took place. A company that might "win" some items in one of the early auctions may be capital constrained at a later auction. Some strategic bidders might bid early on, with the intent not of winning but of raising the winning prices of their competitors so that they will not have adequate capital in later auctions. But some of these strategic ploys may backfire, to the bidders' dismay, and a company may end up with merchandise it doesn't want. In the event, game theorists had a field day.

FOTE analysis. An alternate procedure would be for a neutral intervenor to collect from each party separately its reservation values for commodities alone and in combination with others and then to calculate what a fair allocation would be that would squeeze out favorable returns to the auctioneer but at the same time afford some profit to bidders. Tricky to do with 2,500 wireless spectrum sites but not impossible in this age of super-speed computing. For the simple case in Table 10.2, it seems reasonable that B1 should get both commodities at 22. But it's more complicated than that. Certainly the existence of B3 raises the value to the auctioneer. Shouldn't that dictate that some value should be allotted to B3 and less to the auctioneer?

Core Concepts

We have considered the distributive negotiation problem when a single seller is confronted with several potential buyers. Face-to-face negotiations may be replaced by several types of auctions or competitive bidding mechanisms. We introduced several of these designs and examined them from a game-theoretic, decision analytic, and behavioral perspective and from the orientation of both a bidder and an auctioneer.

In the open, ascending, outcry auction a bidder should (but often doesn't) prepare for the auction by determining her break-even value. There is no need for an analytically inclined bidder to assess a probability distribution of the maximum bid of others—a so-called MBOO analysis—for this case. A MBOO analysis is central, however, to the decision analytic approach—but not the game-theoretic approach—of the Dutch (descending) auction and the competitive, sealed-bid mechanism.

The second-high-price auction (or philatelist or Vickrey auction) has the nice feature that the best strategy is to bid one's break-even value regardless of what other bidders choose to do. This auction does not require a MBOO analysis to act optimally.

The chapter then considered two special cases: (1) reciprocal buy-sell bids that are especially useful when two business partners seek to dissolve their partnership; and (2) combinatorial bids, such as the FCC auction when the federal government auctioned off 2,500 licenses for radio frequencies. The complicating feature is that for many bidders the value of a given lease depends on the other leases that bidder wins.

Additional material about auctions and bids can be found in the *Negotiation Analysis* Supplement. See the end of the Preface.

Two-Party Integrative (Win-Win)
Negotiations

This part of the book, comprising chapters 11 to 16, deals with two-party integrative negotiations. By "integrative negotiations," we mean negotiations where joint gains are a potential result. Part II dealt with two-party distributive negotiations, which involved partitioning a fixed-sized pie. Whereas Part II was mostly about *claiming* tactics, Part III is mostly about *creating* tactics — how to create a bigger pie. But there is a tension between the tactics used to create a larger pie and those used to claim a larger portion of the pie created. How to balance this tension is part of the science and art of negotiation.

Much of negotiation involves the settlement of disputes. In Part III, however, we will for the most part focus on deals, which present opportunities for joint gains (accompanied, of course, by some distributional strains), in contrast to disputes or problems, which have to be resolved. Behaviorally speaking, the negotiations involved in deal making tend to be more collaborative than those involving the settlement of festering disputes. But still, much of the advice given for deals is relevant for disputes as well. Part of this advice is to try to convert a dispute into a deal and to prevent a deal from becoming a dispute.

Chapters 11 and 12 concentrate on preparing for negotiations after the potential for such deal making is already known (we don't examine how we might find promising potential deals). Three phases are involved. In Phase I, the parties deliberate separately and consider their interests, objectives, and visions; their alternatives to the present negotiation under review; and their options. They also gather information about the other side regarding its interests, alternatives, and negotiation style. In Phase II, the parties engage in an informal dialogue in which they feel each other out, selectively and adaptively share information, brainstorm together, try to establish an appropriate relationship, and decide whether they should proceed further. If they see a green light ahead, we suggest that they jointly decide just what must be decided. We strongly suggest that they construct a *template* to guide their further negotiations. The template specifies the issues that have to be decided and possible resolutions for each issue. In Phase III of their preparation, the parties individually and confidentially evaluate the template by clarifying their preferences for different contract outcomes and by exploring their Best Alternative to a Negotiated Agreement (BATNA). Some analytically minded party might even quantitatively score the template and establish a numerical reservation value (RV) that will specify the minimum value that will be acceptable in his or her negotiations. Chapter 11 deals with template construction, Chapter 12 with template evaluation, Chapters 13 and 14 with template analysis, Chapter 15 with real and laboratory behavior, and Chapter 16 with partisan advice to an analytic and collaboratively inclined client who is pitted against a party not so similarly inclined.

Chapter 13 examines an idealization: both parties negotiate agreements using Full, Open, Truthful Exchange (FOTE)—they tell each other the truth and the whole truth. (It was a surprise to find that in real-world, deal-making negotiations, this ideal is often the reality.) The chapter discusses efficient and equitable contract outcomes for the special case in which two parties have to share a set of indivisible items, the so-called fair-division problem. The problem is special in two ways: (1) the ith issue of the template involves what should be done with the ith item, and there are just two possible resolutions (give it to A or to B—no sharing allowed); and (2) "no agreement" is not an alternative—they must agree. This problem, interesting enough in its own right, sets up the analytical agenda of the next chapter on the efficiency and equity of the scored template when some issues have more than two resolutions and where BATNAs and RVs play an important role. Chapter 13 deals with template analysis for the special template of the fair-division problem, and Chapter 14 generalizes this analysis to the general template and to cases where alternatives to negotiations are present. In Chapter 13 we learn that joint gains stem mainly from the differences in players' evaluations of the template, and this insight carries over into 14. The flavor of 14 is still FOTE, and the approach, as for Chapters 11–13, is normative.

Chapter 15 is mostly descriptive and deals with behavioral realities. We completely disengage from FOTE and comment on laboratory behavior with student-subjects. What happens when subjects without any training act as agents for principals who have provided them with a scored template and precise reservation values stating the minimum acceptable contract values? When such simulations were first carried out, it was surprising to see how varied the final contracts were. It's important to examine what seems to work and what doesn't.

Still in Chapter 15, we next consider anomalies, biases, and errors of behavior in real-world negotiation settings. In an interactive setting, misinterpretations beget misinterpretations, and parties may spiral downward in the pursuit of joint harms rather than joint gains. Cultural differences make it harder to establish a constructive negotiation style, even though cross-cultural differences in interests are often the source for potential joint gains.

Chapter 16 has a primarily asymmetric prescriptive/descriptive orientation. It offers partisan coaching advice to an analytically inclined collaborator, suggesting how he might negotiate with another party who is less analytically inclined, who falls short of the FOTE ideal—perhaps falls in the POTE camp—and who claims a share of the pie prematurely and excessively. We suggest how to negotiate with a disputatious, hostile adversary who needs to emote about your client's past despicable behavior. Part of the strategy is to try to foster more collaboration by effectively changing the other's negotiation style.

11

Template Design

Negotiators who neglect preparation do so at their peril. Purposeful preparation is a launching pad for successful negotiation. All too often it is not done. And the preparation that is done is often woefully unsystematic.

In our comments on preparation, we will be giving advice to negotiators with a high tolerance for analytical thinking. But negotiators who are not so inclined should also carefully consider the advice given and explore topics that are not "natural" for their proclivities. The suggestions are relevant even if the other side is not aware of, or chooses to ignore, them. They don't require reciprocal preparation, although as shown in later chapters, they will be most beneficial in generating joint gains when both negotiating parties follow the advice. If I were your personal partisan adviser, much of the advice I would give to you would also be appropriate to give to the other side, which might, in fact, help you.

This chapter can be used as a checklist for preparation. Every situation is different, so there will be room to pick and choose from the preparation advice given here depending on the specifics of your case.

Setting the Stage

Identifying a Potential Deal

We'll ground this discussion by thinking about the following problem. Let's imagine Alan Albert, president of AAA, met with Brian Bartlett, president of BBB, on a cruise, and they realized that they might have an opportunity to do business together. In this instance the goal is to set up a potential *deal*—a joint venture or merger or some weaker form of collaboration—rather than trying to resolve a dispute. The two presidents—enlightened negotiators—each want to get as much value as possible out of the prospective deal. Albert (A henceforth) has asked us to give him and his A-negotiating team advice on how to prepare. Or if you prefer, we're coaching the A team.

To start with, we know that representatives of the A team have already talked very briefly with B team members, and the two sides have ascertained that there may be significant value in joining forces. This project will entail a complex negotiation on lots of issues: production matters, finance, governance, distribution, profit sharing, contingencies, future involvement, repatriation of profits, grievance procedures, and so on. A fairly typical scenario. As advisers to the A team, we want to do the best for our side, but to do so, we may also want to do well for the other side. Part of our primary concern is with helping the two teams jointly create value. Much of our partisan advice to A would also be applicable if we were asked to give joint advice to both parties.

Three Phases of Preparation

How should the A team prepare? In this highly stylized example, we suggest that they divide their preparation into three phases.

- In *Phase I* each side should think alone. Indeed, The A team should straighten out its private thinking for the upcoming negotiation regardless of whether the B team does the same. Surprisingly, though, A's preparation may help B and vice versa. If the advice given to A is also shared with B, it might, as expected, help B, but it might also help A. This is a case of helping A by helping B and is possible in a non-zero sum world. Box 11.1 relates a case where partisan advice to A was inadvertently leaked to B, perhaps to the benefit of A.
- In *Phase II*, the teams should think and plan together informally and do some joint brainstorming, which can be thought of as "dialoguing" or "prenegotiating." The two sides make no tradeoffs, commitments, or arguments about how to divide the pie at this early stage. They should jointly decide on the agenda that they must later consider when they get serious about give-and-take concessions.
- In *Phase III*, to be discussed in the following chapter, we again encourage our side to think alone, but this time more concretely about what it is willing to give up in return for concessions from the other side, and what is the minimum it should achieve before signing an agreement. It would be nice if B decided to do likewise.

Some of our colleagues would prefer putting different labels on these phases. They think of Phase II as the first round of negotiations. Therefore Phase

---

Box 11.1

I'm embarrassed to relate this true incident because I shudder at my ineptness. I had a client, whom I'll call Mr. X, who gave me his fax number early in our work together. After finding out the nature of X's dispute with Y, I sent X, by fax, a detailed memo on how to prepare for negotiations. I was crushed to learn that X's negotiation adversary had access to the same fax machine and had learned about the confidential advice I was giving to my client. Luckily the tone and details of the memo were such that my client was well served by this leak. Indeed, much later on, my client X found out that his adversary, Y, thought that I had purposely drafted the memo to be intercepted by him. I had no such intention in mind and was just plain lucky that my carelessness didn't disadvantage my client.

I is preparation for the first round of negotiation, and Phase III is preparation for the next round. That is one way of thinking about this problem, but we prefer a different perspective. Nothing precludes, of course, having several rounds of prenegotiations with preparations interspersed between them.

Phase I: Preparing Alone

Our first step is to get our A team to privately review the realities of its situation—sources of strength, weaknesses, special endowments, and other matters. The nature of this review will depend on the details of the enterprise. In addition, we suggest that our team would do well to think about:

- The *interests* they wish to fulfill
- Their *visions* for the future down the road
- The *options for agreement* to be generated in the negotiation
- The *comparative advantages* of both sides
- The *alternatives* to the proposed negotiation
- *Objective criteria* they may wish to invoke in support of claims they would like to make
- The key *uncertainties* confronting the proposed merger (or variation thereof)

Every deal or dispute has distinctive features, but in most cases it is still appropriate to ponder these basic, general issues. It is not necessary to examine them in a particular order, and they need not all be considered in equal depth.

Examining Interests

We suggest that the members of the A team think hard about what they want—not *how much* of it, but *what*. Yes, making profits is important, but money is not everything. This step is essentially the same as the second key element (Objectives) of the PrOACT way of thinking described in Chapter 2. Those on the team should focus on reputation, contributions to society, maintaining the environment, leadership, and precedence as well as on responsibilities to stockholders, employees, and creditors. Some matters arise in the context of the negotiations themselves. For example, President A may be concerned not only about doing well for his side—but also about being fair and being perceived as being fair to the other side, about showing that negotiations need not be divisive when carried out in a creative, problem-solving style, and about establishing a good working relationship with the other side that can be a model for future interactions.

In examining A's interests, objectives, and fears, those on the team may find it too limiting to try to be too systematic at first; they may wish to first thrash about a bit. But at some point the team should try to bring these interests into clearer focus and give them direction. To identify and structure objectives, we would coach them in the art of using simple techniques: drawing up a wish list, thinking carefully about a range of options, considering the problems that might be encountered down the line, pondering the consequences of an agreement, and examining the issue from different perspectives.

Now, armed with a comprehensive list of interests, the team arranges them in a practical, useful format. First off we want them to identify the most fundamental interests. To do so we repeatedly ask the "Why?" question. Why are you interested in X or Y or Z? What is fundamental, and what is derived? What is an end, and what is merely a means to an end? We make it clear that neither the negotiation, nor any alternative course of action, is likely to fully satisfy these possibly conflicting objectives.

Envisioning the Future

We contend that when negotiators concentrate mostly on the issues that divide the parties, they often lose sight of possible long-term benefits. In making calculations about whether to start negotiating in the first place or whether to agree to some compromise during negotiations, they should reflect on their long-term visions. For instance, they should reflect on the ramifications of a deal now for the long-run possibilities of future collaborative efforts. The drawbacks of not doing so are apparent in many seemingly intractable disputes between nations (for example, Israel and the Palestinians, Greek and Turkish Cypriots, Indians and Sri Lankans). Not enough effort is made to dream of what could be "if only . . ."

When A and B strike an accord, they are also buying an option for the future (in the financial sense of the term). If things work out well between them, if there are real synergies, and if the relationship remains favorable, the two sides might wish to march together in realizing some of their joint visions; if not, they can bail out. It may sour the current relationship, however, if they dwell too much at this point on procedures to terminate the joint enterprise if need be. This is a bit akin to agreeing on a prenuptial agreement. Not everything has to be tied down. How much should be left implicit or unstated depends on the interactions between A and B and between their respective cultures.

But now comes a critical step that is peculiar to the context of negotiations: what should be shared with the other side? Thorough preparation by A puts a great deal of valuable information about its interests at its negotiators' fingertips. Is it less or more valuable to A's side if its preparation is shared with side B? There is no easy answer. We propose a graded approach: some of A's interests can be openly shared with B; some A may choose not to divulge; and some might be disclosed in an adaptive manner depending on how forthcoming the other team turns out to be. Open disclosure of strategically sensitive information can be a powerful influence in negotiations. The more one side judiciously shows its hand, the more the other side will disclose. The better the relationship, the more the two sides trust each other, the more open they can afford to be, and the more valuable the agreement they are likely to reach.

Cataloguing Resources

The flip side of interests is the ability to satisfy them. Interests are about demand. Making a deal means matching demand with supply. What does A have that might interest B? We advise the A team to reflect on what distinct capabilities it brings to the proposed joint endeavor and to give some preliminary thought (to be subsequently enhanced) about its comparative advantages. (For example, it now looks like A has better skills in production and B in marketing and distribution.)

Taking an inventory of its own resources can also remind the A team that there are things it already controls that it doesn't have to negotiate for—indeed, A may be able to get along quite well without B. So the next step for A to think about is what it will do if there is no deal.

Exploring Alternatives to Agreement

We are at the stage where the two sides have not yet decided to enter into serious negotiations with each other. Before doing so, our side should carefully review our alternatives to negotiating with B. Perhaps A should go it alone (for example, by expanding internally or by using more out-sourcing alternatives), or perhaps our side should negotiate with C, D, or E instead of B. If A does not negotiate with B, it may have a lot of alternatives open to it in addressing its perceived problem or opportunity. A's upcoming, anticipated negotiations with B should act as a trigger to review the nature of the problem it should be addressing. (See Chapter 2.) The better A's alternatives are, the less it makes sense to invest even in preparation for talks with B.

Serious thought should be given to generating more alternatives for two reasons. First, just as in individual decision making, the better the selection of contending alternatives, the better will be the one selected. Second, in negotiating with B, the better A's alternative to B, the more A should expect from the negotiation with B. To find A's Best Alternative to a Negotiated Agreement with B, the A team may have to conduct some intricate decision analysis.

In describing the PrOACT way of thought, we discussed the use of the *How* question (as distinct from the *Why* or *What* questions) in going from objectives to alternatives. For a given goal we can ask: "How might we achieve that objective?" This same bit of advice applies equally well here in the negotiation context, both in generating alternatives to a negotiated settlement with B and, as we shall later observe, in generating creative options for A and B to consider.

Once negotiations get started it's easy to fall in love with making a deal, so it's important for the negotiators to keep their alternatives in mind. We would tell our side that if it has particularly attractive alternatives, it can legitimately seek a larger share of the jointly created pie. We will have much more to say about going from BATNAs to reservation values when we discuss Phase III preparations in the next chapter.

We are now touching on a sensitive issue: how forthcoming, truthful, and open shall we be in disclosing to B our alternatives to negotiating with B? Telling the members of the B team how we could hurt their side is not exactly conducive to establishing a good relationship with them, which may be an important objective of ours. Here is one place where giving partisan advice may differ from giving joint advice.

Generating Options for Agreement

Let's momentarily interrupt our joint-venture negotiations for some light entertainment. A standard, textbook negotiation (see *Getting to Yes*) involves a contract between a performing artist—let's call her Irene Sweetvoice—and the management of a recital house. They haggle over the price of her contract. Sweetvoice's agent says: "My client won't accept a penny less than $30,000!" Management retorts: "But we can't afford more than $20,000."

The pedagogical point of this simulation exercise is that the role-playing subjects get so involved in monetary compensation that they often overlook creative options that can bridge their differences. Such options include (1) Sweetvoice gets $18,000 plus a sliding share of the gross revenue intake above a certain figure—a contingent risk-sharing deal that exploits different probabilistic predictions of Sweetvoice's pulling power; (2) Sweetvoice will be given perquisites that she values much more than the management has to pay, such as flowers, a larger dressing room, admittance of music students to fill empty seats, or targeted advertisements that would appeal to Sweetvoice; (3) Sweetvoice gets an invitation to appear on the local television interview program; or (4) Sweetvoice gets to teach master classes at the local conservatory.

Likewise, executives from enterprise A and enterprise B have to be imaginative. They should think creatively, not only about introducing options that would help satisfy their own fundamental interests, but also about options that would help satisfy the other side's interests at not, they hope, too great a cost to themselves. If the A team can keep the B team happy, it might expect a reciprocating action. But both sides must also think about issues they want to avoid—options that would be so harmful for their side that no compensation would make them worthwhile.

When it comes to individual or joint creativity in generating options, it is hard to be directive. The mind works in mysterious ways. Nevertheless, a systematic review of the objectives (individually and jointly generated), accompanied by the question "How might we achieve that?" is always important.

Brainstorming is a common technique for generating options. Focusing the attention of a group on a creative task is definitely worthwhile—especially since any member of the group may draw inspiration from another. But listening to another's ideas can also distract individuals from formulating their own. It is useful to have members of a group each generate options alone—as homework—before coming together to share and improve on them. Generating options can include the other side as well. We shall return to this topic shortly.

Exploiting differences.   A great deal of value can be created if both sides systematically exploit their differences. If both sides have different probabilistic beliefs or projections about the future, they may be able to make use of contingency contracts. If both sides differ in attitudes toward risk, they may be able to employ financial instruments such as insurance, diversification, hedging, and futures markets to spread risk. If both sides have different financial constraints and different costs of capital, and therefore different intertemporal tradeoffs, they can creatively allocate upfront and downstream costs and benefits. The question is how can one side give the other side what they dearly want at a reasonable cost to themselves, and what quid pro quo should they ask for in return? The teams should be thinking hard, in private, about the options to be considered in the negotiation, and in the prenegotiation stage the teams should do some joint brainstorming on further options they both might want to introduce to exploit differences between them that have become evident.

Invoking Objective Criteria

The A team wants X, the B team wants Y. How should they jointly decide? To convince their negotiating counterparts, A team members will have to argue the merits of X and the inadequacies of Y. But the other side will be doing exactly the same thing in reverse. What arguments would appeal to a neutral outsider, concerned with fairness and with what is ethically right? For instance, the A team could say: "If we were neutral, we believe it would be only fair to use objective criterion Q for the resolution of our disagreement; and if we do that, X is better than Y."

In preparation, we want to think about all the possible standards that we might invoke to lead toward a resolution that we would like. But we also want to think about standards that we might not like so much. What will they think is fair? If there are standards out there in the marketplace that will help them, we should assume that they will find them and use them. We will be better off if we are not sandbagged by their arguments, but are ready to explain why we think they are not fair or why this particular case is different.

The rub, of course, is that there are often multiple objective standards for what is "right." Instead of arguing about X or Y, the parties can negotiate at a meta-level: should we invoke fairness principle Q or R or S? In preparing for negotiations, we advise both sides to be aware of principles Q, R, and S and do some homework about which principles would logically lead to the adoption of X or Y. This may be hard to do without concrete choices in front of us, so we shall have to return repeatedly to this topic in the following chapters.

Gathering Information on the Other Side

We are in delicate territory once again. What does A know about B? What should it try to learn before the next joint meeting? What should it try to learn at the next meeting? How? What information about ourselves are we prepared to share with the other side? Do they have to know our marketing manager is about to leave us? How truthful do we want to be? We don't want to lie, but do we have to disclose all of our secrets? It depends on how much they choose to reveal.

Here are some pertinent questions the A team might want to investigate:

- Why do they want to negotiate with us?
- Why now?

- What's really driving them? That is, what are their interests and values?
- What are their alternatives to negotiating with us? If their best alternatives to negotiating with us are dismal, then we can expect to extract a bigger piece of the pie for ourselves. Do they have constraints that we can exploit?
- What do we know about how they negotiate, their trustworthiness, their ethics, their cultural style?
- Do we know something about the personalities of the other negotiators?

The negotiating teams might want to assign a member of their delegation, or even someone from an outside consulting firm—there are such firms—to gather information about the other side and to prepare a briefing report about them before negotiations begin. These investigations can come from published sources or open inquiries, but also from clandestine (legal, questionable, or illegal) sources. How far should they go?

Imagine that the A team is from America, and is negotiating with a Russian or an Indian firm. Team members should worry about the impact of potentially dysfunctional cultural differences. The same questions are relevant to both sides. What should we know about the idiosyncrasies of their cultural negotiating style? What are some of the stereotypical notions they may have about our style?

Assessing Uncertainties

The deal between A and B is complicated. There are many uncertainties: about the economy, about interest rates, about currency fluctuations, about costs, about future demand, about future environmental regulations, about performance levels, about management compatibilities, about legal matters, and so on. Both the A and the B teams need to acknowledge the existence of these uncertain elements. In anticipation of a joint discussion of them in the next phase of negotiations, our side, A, might do some preliminary homework in addressing the following list of questions.

- What are some key uncertainties of our problem?
- How would we sort them into three categories (very important, mildly important, comparatively unimportant)?
- What are the dependencies between these uncertainties?
- What do we know objectively about them?
- What data, information gathering, and experiments can we conduct to better assess these uncertainties?
- How will information about these uncertainties unfold over time?

- Will we know more about uncertainties X, Y, or Z by the time we have to commit ourselves to action M or N?
- Who are the experts on our team?
- Do we expect that our beliefs about these uncertainties are different from the other side's beliefs?
- What does the other team know about the uncertainties that we do not know?
- What are their track records and biases?
- Are they wishful thinkers or do they try to protect themselves by hedging their assessments?
- How can these differences be exploited by contingent contracts, by risk sharing, or by management interventions such as diversification or hedging policies?
- What joint projects can we collaborate on to get more accurate information?

A team members should think creatively about what issues should be resolved now and what issues could be deferred to a later time and made contingent on occurrences that are not predictable now. They might want to negotiate a process or general principles that will help guide an appointed, joint decision-making committee to resolve future contingencies that are now only partially discernible.

Phase II: Preparing Together (Dialoguing or Prenegotiating)

Suppose that both teams have thought hard about Phase I of their problem. Let's assume that regardless of how well the B team has prepared, our side, A, has done its preliminary homework. It's now time for the two teams of representatives to have an informal dialogue, or prenegotiation.

Planning the Logistics

Before beginning a prenegotiation, a host of practical questions need to be answered. Who will negotiate? How many team members should participate? Where should the meeting take place? Who should pay the expenses? What language(s) should be used? We will skip over these preliminary issues for now, and instead examine the question: What is the desired outcome of such a dialogue?

Setting Goals for the Meeting

Before entering any meeting, it's wise to reflect about what you want to accomplish. For the joint preparation phase, we might recommend these objectives to the A team:

- To ascertain whether we want to proceed further with B—do we see a red, yellow, or green flashing signal for further negotiations?
- To create a favorable ambience.
- To establish a good working relationship with them if we proceed further.
- To share some information and perhaps induce them to reciprocate. Some things we won't share.
- To brainstorm together about interests (objectives), options, uncertainties, visions.
- To discuss gingerly some of our reservations about them and ask about their reservations about us.
- To discuss the processes we may wish to employ in future negotiations.
- To jointly decide on what has to be decided—a so-called template for further give-and-take negotiations. (Much more about this shortly.)
- To avoid premature "claiming"—to concentrate on building up a big pie to share and not to worry too early on about how to share that pie.

Creating the Ambience

In this phase we don't want our negotiators to get into specifics about compromises; we want them to explore what a negotiation might entail. Throughout the initial meetings, care must be taken to prevent the dialogue from spiraling toward mutual, competing tactics that are certain to impede joint creativity. Our negotiators should be prepared in advance with eloquently phrased pleas concerning the joint need to create a larger pie, and these pleas should be sprinkled appropriately into the conversation.

The first joint decision facing the two sides is whether or not they want to engage in serious negotiations. If the answer is affirmative, we suggest they try to build up trust and enhance their abilities to communicate effectively with each other. Ideally, at this delicate stage, we want them to funnel all their energy into *creating joint value*—exploring options, brainstorming without commitment, devising mutually satisfactory alternatives—and minimize (or, even better, ban) tactics for *claiming individual value*.

What happens if one of their negotiators blurts out that his firm needs X, Y, and Z before any deal would be acceptable to them. This is premature, so we suggest that our side respond: "We also have demands of our own, but this is not the time for agreeing to or rejecting each other's demands. Let's record X, Y, and Z as possible options-for-agreement. We'll consider such things when the appropriate time comes for us to make hard compromises." An answer of this sort should help defuse the situation and let the value creating continue.

Sharing Interests

The teams should selectively share some of their interests, aspirations, and visions. This has to be done adaptively. How much A divulges depends on B and vice versa. The two sides should identify key issues that might be stumbling blocks to negotiations. They need to learn about each other's potential hang-ups. They might want to record some joint objectives or a broad framework to guide their ensuing negotiations. By dwelling on their joint interests and objectives, they can once again ask the *how* question: How might we accomplish that? There should be an interplay between discussions of interests, objectives, and the creation of promising options.

Agreeing on a Process

Negotiations usually run more smoothly if the parties discuss the processes they propose to follow. They might want to appoint a joint fact-finding committee or a joint team to investigate key uncertainties that will have an impact on the operation but are not subject to their joint control. They might agree to assign some neutral group (say, a consulting firm) to project financial futures for the joint enterprise, given various sets of input parameters. These collaborative activities allow the teams to learn how to work together and test modes of cooperation in a nonthreatening environment.

Preparing the Template

Identifying the issues to be resolved. We strongly suggest that after establishing the proper ambience and identifying a desire to go to the next stage of negotiation, the two teams jointly design what may be called a *template* (but others call a framework) for negotiation. They have to decide what needs to be decided. The template lists the set of issues that need to be resolved at the later negotiation stage. In many complex international negotiations, the template commonly lists hundreds of issues, suitably partitioned into different sections. For the joint venture between enterprises A and B, the issues listed in the template might be grouped under headings such as finance, location, production, distribution, organization, management, accounting and control, conflict resolution, employment policy, review of operations, and contingency planning, as well as others depending on the specific context.

In very few instances will two parties raise exactly the same issues for negotiation. In our example, B may want to include an issue which is of no apparent consequence to executives from enterprise A—and it is essential that these nonsymmetric issues be included. In fact, the template should include all issues of relevance to either A or B, not just issues that they both agree are of importance. In template construction, adherence to the principle of *reciprocal inclusivity* avoids nefarious claiming tactics and gives both parties an incentive to create as large a pie as possible.

Specifying possible resolutions of each issue.   In addition to listing the issues to be resolved, the A and B teams should list for each issue a set of possible resolutions for that issue. Thus, for example, the template may have twelve financial issues that need to be resolved. Financial issue 1 (F_1), perhaps dealing with initial equity contributions, might have eight possible resolutions, each of which is termed a resolution level; financial issue 2 (F_2) calls for a yes or no resolution; . . . financial issue 12 (F_{12}) might call for a high, medium, or low value. Further down on the template is a set of marketing issues: the first marketing issue (M_1) may have five levels of resolution; M_2 may have nine levels, and so on.

Ultimately, in a later negotiation session, the task will be for both sides to select and agree on a particular resolution level for each of the issues under consideration. We'll call such a joint selection a "contract." The number of possible contracts is the product of the number of possible resolutions of each issue. This can be a huge number. For example, if there were 10 issues and 5 resolutions for each issue, the number of possible contracts would be the multiple of ten fives (that is, 5 to the power of 10), or 9,765,625. This means that a negotiation template is very flexible. A simple template can be used to produce a huge variety of different contracts with dramatically different values for the two parties.

Delegating template construction to a subcommittee.   Generation of a template for negotiation may entail a lot of joint work. Indeed our team A might privately draw up a version of a template earlier, during Phase I. We might choose not to reveal this earlier draft, because ideally the preparation of a template should be a joint enterprise. If our side is too aggressive in putting forward our own ideas, then the other side might not feel a sufficient sense of involvement or ownership in the joint undertaking. But still we suggest that A might want to come into the prenegotiation phase with its own template, hidden in its collective briefcase.

Suppose that in the joint deliberations about the process they should employ for negotiations, the A and B teams decide that they should jointly develop a template listing issues and their potential resolutions. They could do this collectively with all participants, or perhaps they could assign this task to a joint team. In the latter case they should admonish this joint team not to engage in claiming tactics; but despite this admonition, these may occur. The B-team executives might, for example, wish to add some issues that are unimportant to their side except as bargaining chips. Or they might insist on adding additional levels to a list of possible resolutions of an issue because they might not want their preferred alternative to be at an extreme end of the distribution (far from the focal midpoint). Notwithstanding these caveats, the joint task force constructing the template should be instructed to maximize creativity and minimize claiming tactics.

On the basis of the template it might be desirable to craft a framework agreement that includes a fair number of blanks to represent different possible resolutions of the issues to be negotiated. Indeed such a framework agreement might be designed even without first putting together a template. Nevertheless, in complex negotiations it is desirable for the parties to generate a template before they engage in detailed negotiations that involve exchanging concessions. Creating a template for negotiations does not require an atmosphere of Full, Open, Truthful Exchange (FOTE)—an atmosphere in which each side tells not only the truth but all the truth; neither does it require Partial, Open, Truthful Exchange (POTE) in which the parties tell the truth but not all the truth. They may, for example, keep their bottom lines close to their chests. The template contains no explicit information dealing with tradeoffs across issues, strengths of preference for different resolution levels on issues, and an overall reservation value. Using a template may be good discipline even if the negotiators are involved in a dispute (in contrast to a deal) and even if they anticipate engaging in future hardball, vituperative, emotional exchanges.

Let's look at an example of a completed template in an example of a labor-management negotiation.

Case Study: AMPO vs. City

We consider the case of a partially concocted, fictitious contract negotiation between the Associated Metropolitan Police Officers (AMPO) and the City (an amalgam of New Orleans and Atlanta).[1] There is the usual problem of salaries and vacation days for various seniority levels, and the establishment of more officers (corporals and sergeants). As expected, the City always has budgetary restrictions and will oppose these expansions because they will lead to an increase

---

1. For the original version, see Edwards and White (1977).

in the financial package. For this current contract there are three new wrinkles that have to be resolved. There is the festering problem of the number of police officers to be assigned to each police vehicle. As of now the assignment is one officer per vehicle. The police want two officers to be assigned per police car out of concern for both efficiency and safety. The City is dragging its feet about this modification, because, according to them, it will inflate the overall budget by 25 percent. Just last spring fourteen officers refused to solely man police cars, and they were suspended from the force. AMPO wants these officers reinstated with back pay. The City seems adamant. The second issue concerns Police Commissioner Daniels. He is the mayor's man and the police despise him. The third issue concerns what to do about a police review board. The police argue the board is stacked against them, and they want more congenial representation. At the time of the negotiations there were no police on the board, and a vote for censure of an officer had to be unanimous. AMPO wanted the board disbanded; failing this, they wanted to add police officers to the board—but not if the voting rules were changed.

AMPO and the City have agreed that their jointly designed template include the following ten issues.

AMPO/City: Issues to Be Resolved

1. Starting salaries for police officers
2. Maximum salaries for police officers
3. Vacation for officers with less than five years' service
4. Vacation for officers with more than five years' service
5. The status of fourteen officers under suspension
6. The percentage of two-man patrol cars
7. Creation of the rank of corporal
8. Expansion of the number of sergeants
9. The fate of the police commissioner, Mr. Daniels
10. The status of the Police Civilian Review Board

Note: AMPO wants an increase in salaries and more days of vacation whereas City does not *(Issues 1–4)*. Police concerns over the safety of one-man patrols had, prior to the negotiations, resulted in an unauthorized wildcat strike on this issue, and fourteen officers had been suspended. The issue now is whether they should be reinstated, and if so, with or without back pay *(Issue 5)*. Ideally AMPO would like to have two officers to every patrol car; City wants only one *(Issue 6)*. AMPO wants to create the new rank of corporal; City is against it *(Issue 7)*. The same disagreement exists over the creation of new sergeants *(Issue 8)*. AMPO wants to fire Commissioner Daniels *(Issue 9)*. The Police Civilian Review Board is a complicated problem *(Issue 10)*.

Table 11.1 displays the template which exhibits not only the issues to be resolved but the agreed-upon possible resolutions for each issue.

In the case as originally written by Harry Edwards and James White, there was a fascinating wrinkle that raised some ethical issues. AMPO hated Daniels and assumed that City desperately wanted him because he was the mayor's appointee. But AMPO did not know beforehand that the mayor was secretly disgusted with his political appointee and was looking for an excuse to get rid of him. AMPO falsely assumed that the City wanted to retain Daniels. City, by first defending Daniels and then ingenuously backing down, exacted a concession from AMPO.

With the completion of their template, AMPO and City are ready for Phase III, template evaluation, the subject of the next chapter.

Core Concepts

We have begun Part III by offering prescriptive suggestions to party A in preparing for negotiations with party B. We had in mind the parties negotiating a *deal* rather than settling a *dispute*, although there is quite an overlap. Much of the partisan advice we offered A on preparing for negotiation could also be profitably shared with B.

We proposed that preparation be conducted in three phases, the first two of which are covered in this chapter and the third in the next chapter. (Some of the concerns of Phase II are also addressed in Chapter 21, where we examine how to run meetings effectively.)

In Phase I each party prepares alone by thinking hard about its interests (objectives, concerns, needs, fears, wishes), about its visions, about its comparative advantages and disadvantages, about possible creative options it could introduce in negotiating with B, about its alternatives to negotiating a deal with B, about possible synergies, and about what it hopes to get from B and what it can offer to B. Party A should also develop its perception of B's interests, visions, alternatives—but perhaps not in as much detail.

A might also want to identify the key uncertainties of the problem and, depending on the seriousness of the problem, do some preliminary analysis of these uncertainties—for example, decide which uncertainties are important, structure their qualitative interdependencies, indicate what data are available, and who are the experts we can call on to assist us.

Phase II of preparations is to be done jointly. The parties selectively and adaptively share information and insights, do some joint brainstorming, explore potential synergies and compatibilities. Then they must decide whether or not to proceed with negotiations.

If the parties decide to continue, they should then decide on just what has

Table 11.1 AMPO vs. City template

Issue		Resolutions
1. Increase in starting salaries	A	0
	B	$100
	C	$200
	⋮	⋮
	K	$1,000
2. Increase in maximum salary	A	$0–$500
	B	$501–$600
	C	$601–$700
	D	$701–$800
	E	$750+
3. Increase in vacation with less than five years on force	A	0 days
	B	2
	C	3
	D	4
	E	5
	F	6+
4. Increase in vacation with more than five years on force	A	0 days
	B	1
	C	2
	D	3
	E	3+
5. Reinstatement of suspended officers	A	No
	B	Yes, without back pay
	C	Yes, with back pay
6. Two-man patrols	A	Status quo
	B	In high crime areas
	C	At night
	D	Both
7. Create rank of corporal	A	No
	B	1–19
	C	20+
8. Increase number of sergeants	A	0
	B	1
	C	2
	D	3
	⋮	⋮
	J	9
	K	10
	L	10+
9. Commissioner Daniels	A	Fire
	B	Keep
10. Police Civilian Review Board	A	Disband
	B	Add police, no vote change
	C	Add police, change vote
	D	No police, no vote change
	E	No police, change vote

to be decided. It will be very helpful for our developing analysis if they jointly construct a template for further negotiations that (a) lists the issues that they need to resolve, and (b) for each of these issues, lists their potential resolutions.

In Chapters 13 and 14 we pursue an analysis in which both parties want to negotiate in a spirit of Full, Open, Truthful Exchange (FOTE), and then in later chapters we disengage from this ideal. Much of the advice we have offered in this chapter is quite robust even for strident, duplicitous disputes; in those cases we shall have mostly to add some topics to complete our preparations.

12

Template Evaluation

Let's take stock. In advance of negotiations between enterprises A and B, we're giving advice to A. In Phase I we helped our side prepare alone: thinking about interests, searching for alternatives, and so on. In Phase II we suggested that A and B prepare together by selectively sharing interests and visions, inventing options together, and by jointly constructing a template to guide further negotiations. Whether the parties are negotiating a dispute or a deal, we think it is helpful for them to prepare a template that exhibits the issues to be negotiated and possible resolutions for each issue. By issues, we mean anything of relevance to either A or B, not just items that concern both.

The advice is relevant for all parties in a negotiation, whatever the other side is doing. Ideally, A and B have jointly prepared this template. But all is not lost if they fall short of this ideal. The parties may have constructed a partial template, or after preliminary discussions with B, our more analytically inclined side, A, whom we are coaching, could have prepared a template of its own (not yet shared with B).

Now let's turn to Phase III of our preparations. First, working with our A client, we suggest that A prepare a wish list for the resolution of the issues. During negotiations A may have to pull out the list if B insists on proffering one of its own.

During the negotiations, B might say something like: "Listen, as you request, we'll move from level c to level d on issue 5, but only if you will go from level c to b on issue 4, and from h to g on issue 7." If these are the kind of vexing choices our side might have to make, A had better sort out its preference (or value) tradeoffs in preparing to negotiate with B. And we will want A to do this systematically, not just intuitively or impressionistically. Granted, tradeoff analysis is rarely done in practice, but it should be. And even if A does not follow our advice wholeheartedly, partial preparation will help.

Second, we can anticipate that at some later stage in the negotiations, our side will have to decide whether or not to accept the last, final offer of B on the table or to terminate negotiations and seek another alternative. For instance, what should A do if B delivers an ultimatum?—this or else! To make the right choice, the A negotiators should do two additional things: (1) analyze promising alternatives to negotiation and choose the Best Alternative to a Negotiated Agreement, and, on the basis of this BATNA, (2) decide on a bottom line stipulating which contracts are acceptable and which are not. We'll demonstrate how to assess tradeoffs and calculate reservation values in the next two sections.

Preferences and Value Tradeoffs

Within-Issue Qualitative Analysis

We continue to identify with side A. In most situations it is fairly obvious how our side should ordinally rank our preferences for different resolution levels on any given issue. For example, if issue 7 has four resolution levels, it may be clear how to order these according to our preferences. So let us label these four levels by a, b, c, d in ascending order of preference ($a < b < c < d$). So henceforth, level a on any issue is our worst choice, and as we go from a to b and on up our satisfaction improves.

Between-Issue Qualitative Analysis

Not only should we encourage our client, A, to rank by preference the resolution levels within each issue, but we should also have A place the issues themselves in rank order in terms of their relative importance. For instance, in a labor negotiation, is salary more important than number of vacation days? Many individuals feel comfortable ranking the issues to be resolved in accordance with their importance. But when this is done, a mistake is often made. If we are talking about a range of possible salaries that is very small and a range of vacation days that is very large, then the issue of vacation days may be more important than the issue of salary. The decision whether one issue X is more important than another issue Y in preference should not be divorced from the ranges in the levels of resolution for the two issues. Adding a more wide ranging list of possible levels on issue X should raise the importance of X in comparison with the other issues.

    In ranking issues by their importance, the decision aid shown in Table 12.1 is helpful. Imagine a template with 10 issues. In the table, we use the expression xxx to represent symbolically the actual resolution called for.

Table 12.1   Ranking issues by importance

1	2	3	4	5
Issue	What is worst for us	What is best for us	Ranking (1 is best)	Scoring
1	*a.* xxx	*e.* xxx	5	8
2	*a.* xxx	*d.* xxx	7	5
3	*a.* xxx	*e.* xxx	4	10
4	*a.* xxx	*c.* xxx	1	30
5	*a.* xxx	*b.* xxx	10	1
6	*a.* xxx	*e.* xxx	2	20
7	*a.* xxx	*f.* xxx	3	15
8	*a.* xxx	*e.* xxx	9	2
9	*a.* xxx	*d.* xxx	8	3
10	*a.* xxx	*g.* xxx	6	6
Total				100

Looking at the fourth column in the table, we see that A thinks the fourth issue is the most important, in the sense that he gains more incremental satisfaction from pushing issue 4 from its lowest level, *a*, to its highest level, *c*, than in pushing any other issue from its lowest to its highest level. Next in A's preference structure comes control of issue 6 and so on, with the least important being issue 5.

Within-Issue Quantitative Analysis

Suppose, for the sake of argument, that A is looking at issue 2 with four levels and he has already chosen his labeling to reflect that moving from *a* to *b* to *c* to *d* improves his satisfaction. In negotiating with B he may also want to consider his strength of preference for different resolution levels. Let's scale these, giving level *a* the score of 0 (the worst level for A), and level *d* the score of 100 (his most preferred, or best, level). How then can we assist A to scale resolutions *b* and *c*? This is not unlike the teacher giving partial scores for partially correct answers on question 2 (issue 2) in an exam. Without prejudicing the importance of question 2 in comparison to other questions, the teacher might say that answer *b* deserves a little less than half the credit; *c* a lot more than half the credit; *b* is midway between *a* and *c*; . . . and the teacher might finally come up with the scoring system on question 2 as follows:

Level:	a	b	c	d
Score:	0	40	80	100

Between-Issue Quantitative Analysis

Let's suppose we're now working with A on his preferences and tradeoffs. He has already ranked the importance of each issue. But there is also the question of strength of preference. Suppose A ranks issue Z as being more important than issue Y, which is more important than X. In making tradeoffs it may be important to know that Z is only slightly better than Y, which is much more important than X. Again, let's draw an analogy to a teacher who has to grade an exam involving ten questions, say, on the basis of 100 points (an arbitrary figure). The teacher may choose to sprinkle these 100 points among the ten questions, giving one question a potential total of 20 points and another only 5 points because of the relative importance of the two questions.

The fifth column of Table 12.1 indicates for A's preferences that the importance of issue 4 is not quite as good as the combination of issues 6 and 7; he would rather push up 6*a* to 6*e* and 7*a* to 7*f* than to go from 4*a* to 4*c*. However, the importance of 4 is equivalent to the joint importance of 6 and 3 combined. And issue 7 is more important than issues 10 and 2 combined.

Notice that *between* quantitative scoring can take place independently of any *within* qualitative or quantitative analysis.

Continuing with our analogy to grading exams, the teacher might assign partial scores to incorrect answers to questions and add up the scores over all ten questions to determine the overall grade. We can help A do an analogous "additive" scoring that will, among other things, help A to be precise about value tradeoffs across issues. It's "additive" because we add up the scores over the issues.

The Additive Scoring System

Imagine that we have gone through Phases I and II of our prenegotiation preparation. This has allowed A and B to identify ten issues to be resolved, on the basis of which they have constructed a template as shown in the first two columns of Table 12.2. Let's assume that A has done qualitative analyses within issues and has labeled the resolutions so that preferences for A increase with the alphabet; next assume that A has also done quantitative analyses *within* issues and has done qualitative and quantitative analyses *between* issues. A then produces a scored template (shown by the third column of Table 12.2). In doing so, we shouldn't forget that at this early stage, the scored template contains private information for A that may or may not be shared with B as negotiations advance.

Let's interpret some of the results. The fourth issue is the most important for A. He would rather go from *a* to *c* on issue 4 than to go from worst to best on any other issue. Indeed, A would rather go from worst to best on issue 4 than on is-

Table 12.2   Scored template for A

Issue	Resolution	Value to A
1	*a*	0
	b	3
	c	6
	d	7
	e	8*
2	*a*	0
	b	2
	c	4
	d	5*
3	*a*	0
	b	2
	c	3
	d	8
	e	10*
4	*a*	0
	b	18
	c	30*
5	*a*	0
	b	1*
6	*a*	0
	b	5
	c	15
	d	18
	e	20*
7	*a*	0
	b	2
	c	7
	d	10
	e	13
	f	15*
8	*a*	0
	b	1
	c	2*
9	*a*	0
	b	1
	c	2
	d	3*
10	*a*	0
	b	2
	c	4
	d	4.5
	e	5
	f	5.5
	g	6*

Note: The asterisked figures in column 3 sum to 100.

sues 1, 2 and 3 combined (the gain of 30 points on issue 4 is greater than the gain of 8 + 5 + 10 on issues 1, 2 and 3). Also, issue 4 is equally as important as issues 3 and 6 combined (30 = 10 + 20). We hope A's scoring system is now so clear that you can imagine how it might have been constructed. Again think of the teacher grading an exam of 10 questions.

We're assuming that A has quantitatively scored the template. Negotiator B may or may not do the same. If we were giving joint advice rather than partisan advice to A, we would suggest that B also score the template. Indeed, acting as a mediator, we would encourage them both to do so—of course, without disclosure to the other side.

Moving on, let's suppose parties A and B settle on a contract depicted in the first two rows of Table 12.3. Here A's score is the sum of the third-row entries, or 62. Presumably if A were to settle on this contract, it would have had to surpass A's bottom line (that is, A's reservation value, or RV), which we have not yet discussed. Once this additive scoring system is established, it will be easy for the chief negotiator, A, to have an agent negotiate for him or her. It's clear now what A is trying to maximize. Note, however, that an agent for A still does not know A's bottom line or reservation value. We'll have something to say about this later on.

Appropriateness of an Additive Scoring System

To make sure that the additive template is not misused, a short theoretical interlude is in order. Let us return to the case where we sprinkle 100 points of total value over ten issues and add up the partial scores for each player as a function of the agreement that is jointly chosen. When is the additive scoring system legitimate? It certainly would *not* be legitimate if the value we associated with a given resolution on one issue depended on the resolution of another issue (for example: the longer we stay on our vacation, the less I want to stay in an expensive hotel); or if the relative importance of two issues depended on the levels chosen for still other issues. Keeney and Raiffa (1976), Raiffa (1982), and others have shown that an additive scoring system is appropriate if and only if the value tradeoffs between any two issues do not depend on the levels chosen on the remaining issues. But suffice it to say that there will be many cases where strict additivity does not apply. Our purpose now is twofold: (1) to talk about the case of almost-additivity or quasi-additivity, and (2) to point out that not all is lost if we

Table 12.3   Illustrative contract

Issue	1	2	3	4	5	6	7	8	9	10
Level	*c*	*a*	*e*	*c*	*b*	*a*	*d*	*a*	*b*	*c*
Score	6	0	10	30	1	0	10	0	1	4

have to resort to nonadditive scaling. In fact, in the more complex world where additivity does not seem to apply, a little thoughtful analysis can often provide more useful insights than in the simpler additive case.

Cutting Corners

Let's imagine that value tradeoffs between issues 1 and 2 depend on the level of resolution on issue 3, but not on issues 4 to 10. In technical parlance, we might be able to say that the set of issues {1, 2, 3} are preferentially independent of issues {4, . . . , 10}, but there are dependencies within the grouping {1, 2, 3}. We may be able to define a new composite issue X that comprises issues {1, 2, 3} and achieve additivity across the redefined set of issues {X, 4, . . . , 10}. It's like the teacher who refuses to score test questions 1, 2, and 3 separately but rather gives a "gestalt grade" to questions 1, 2, and 3 collectively, and then proceeds as before to add this collective score to the scores on questions 4, 5, . . . , 10. We can think of this as an additive scoring procedure on eight independent questions. The point of all this is that additivity makes life especially simple, and if violated, it can sometimes be artificially reinstated with a trick or two. The trick we just commented on shows how we can accommodate some simple dependencies. There are other tricks to convert nonadditivity to additivity, like redefining the set of issues. For example, if one issue is compensation to beginning workers and another to senior workers, and if there is a concern for balance and equity between the resolutions on these two issues, we might redefine the issues by considering (1) average compensation and (2) the spread of compensation between beginning and older workers. With this redefinition of the issues, it might be more plausible to justify the additivity assumption.

But now let's suppose life is just a bit more complicated, and no matter how hard we try, we cannot justify an additive scoring system. Let's suppose we nevertheless can still associate a total score to each possible contract (that is, to a contract that specifies how each of the ten issues is to be resolved). The recipe or function is no longer additive with simple +'s and −'s, but includes multiplicative aspects (doing well on issue 2 is extra important if we do poorly on issue 1), minimum and maximum requirements on issues, and if-then statements sprinkled all over. Conceptually, the results of analysis without additivity are little different from the results of analysis with the additivity assumption that is glibly assumed in simple textbook examples.

Finally, let us emphasize a point that might have gotten lost in all the details. In Phase I of our analysis for A and B, we talked at length about the desirability of describing the two parties' interests. We suggested asking the "Why?" question: Why are you interested in X? By that means we were able to distinguish between means and ends and to specify a set of evaluative interests (or objectives). Now in doing qualitative and quantitative analyses within and between issues, we should keep in mind our fundamental evaluative interests. Indeed one might want to assign quantitative scores not just at the level of issues and resolution levels, but more fundamentally for the set of evaluative interests. One would then ask how each contract fares on these fundamental interests, thereby scoring final contracts by this measure. This subject gets too complicated for us to explore, but the linkage between interests and the scoring of the template is fundamental even if we choose not to formalize that linkage here.

Finally, before we go on to BATNAs, in scoring a template one might have a creative insight and want to restructure the issues to be resolved or the resolutions for a given issue. If there were no prior consensus with the other side about a common template, then you should modify the template unilaterally. If, however, the template was jointly agreed upon, another joint meeting might be called to consider the modification of the original template. We like to think that template analysis need not inhibit creativity. Ideally there should be an interplay between it and analysis. Make a template, but don't treat it as sacred. Be willing to play around with it.

BATNAs and Reservation Values

Creating Best Alternatives

Some years ago, I (HR) was asked by hospital A for partisan advice in its upcoming negotiations with hospital B. Because of the restructuring of medical care and the role of health maintenance organizations, my client, hospital A, thought it was in their interest to form an alliance with hospital B. How should they negotiate? I asked my clients about the alternatives to negotiation, and they responded that I was hired to establish a plan for negotiating with B and not to worry about other alternatives. I nevertheless suggested that when they negotiated with B, they would want to keep in mind their alternatives, and have some idea of what they would minimally need in order to form an alliance with B. If their opportunities elsewhere were favorable, they could comfortably ask for more in their negotiations with B. Quite reluctantly they talked about joining with hospitals C and D, but it was B that they were interested in. I insisted that they take the exercise seriously and develop alternate plans that did not include B before they opened up formal negotiations with B. I prevailed, and subcommittees were formed to pursue consideration of linkages with C and with D. To cut a protracted story short, they ended up with C rather than B. But even if they had ended up with B, their actions would have been far more credible in the creating and claiming game with B if they knew, and B knew, that they had an attractive alternative with C. Indeed, for a while alternatives B and C were being actively pursued together, each alternative acting as the BATNA for the other.

Let's abstract from this example. Suppose that before firm A enters serious give-and-take negotiations with firm B, it considers its alternatives to negotiating with B. There may be several alternatives, C, D, E, . . . , but for our purposes, two, C and D, will suffice. The trouble is that it may not be clear what will happen if we go down the C or D paths. How can A choose between negotiating with B, C, or D? Which is better? This is the prototypical problem of decision making under uncertainty, which we discussed at some length in Chapter 2. So let's take our cues from that discussion. Many errors are made by not clarifying the *nature of the problem*. The problem is not whether our firm should form an alliance with B, but rather how A should grow its firm. The alternatives are many: by expanding its capabilities from within, by outsourcing some tasks, by acquiring another firm, by being acquired, by mergers, by joint ventures, and so on.

Following the way of thought discussed in Chapter 2, a key step is articulating A's *objectives*. When A specified its interests in negotiating with B, it had already examined some, but probably not all, of its primary objectives. By focusing on the broader problem, A can reexamine its objectives from a broader perspective. Next come the *alternatives*, and here is another place that errors are made by being too narrow. In the example cited above, my client, hospital A, after some cajoling on my part, agreed to develop the alternative of a merger with hospital C. Even then, at first they didn't consider the alternative of simultaneously negotiating with B and C, playing each off against the other.

Finding the Best Alternative to Negotiated Agreement can be a daunting task, especially when lots of interdependent uncertainties are involved. But as pointed out in *Smart Choices* (Hammond, Keeney, and Raiffa 1999), the task sometimes becomes quite obvious with little analysis. For example, there may only be a couple of viable alternatives, and listing the pros and cons of each and comparing them à la Benjamin Franklin's advice (see *Smart Choices*, p. 88) may suffice. But let's not kid ourselves. Sometimes the problem gets fiercely complicated, and just as most firms have to hire lawyers to sort out legal entanglements, so too A may be well advised to get decision analytical help to deal with the intricacies of finding its BATNA. As I said thirty years ago, decision trees can easily become bushy messes if one includes too many uncertainties, with all their dependencies.[1]

Turning a BATNA into a Reservation Value

So let's return to our negotiation. Following our prompting, A has concluded that the best alternative (strategy) for negotiating with B is an agreement with D. Where do these preferences come from? Ideally they should come from a deep

---

1. For those readers who want to pursue this topic in more depth, we suggest chapters 5, 6, and 7 in Hammond, Keeney, and Raiffa (1999), which describe how to grow a decision tree that is informative.

understanding of party A's evaluative interests ("objectives" in decision analysis parlance).

How should this analysis inform A's decisions and behavior when he is negotiating with B? A somehow has to compare, in desirability terms, any specific contract under consideration with B with his best external alternative, D. That's conceptually tough to do, because the evaluation of internal contracts and external alternatives may be in different summary units. This needs clarification.

Suppose we consider the scored template in Table 12.2 and consider the contract in Table 12.3 that achieved a score of 62. Is this contract better than alternative D? If the contracts are scored, as we have assumed they are, then party A can seek the cutoff score or reservation value using the following procedure.

Start off with the absolutely worst contract, with all issues resolved at level *a*. This contract, according to our conventions, has a score of zero. Let's assume that this contract is far worse than A's best alternative, which is alternative D. Now the trick is to gradually increase the resolution levels from *a*'s to *b*'s to *c*'s until we achieve indifference between the resulting contract and the external alternative. For example, in Table 12.2 suppose all resolution levels are *b*'s, resulting in a score of 37. Assume that this contract is clearly worse than D. In this case, sweeten up the contract to achieve indifference. Suppose that this can be achieved by keeping all resolutions at *b* except for issue 7, which is pushed up from *b* to *c*, adding 5 points to the total score of 42. We have thus ascertained that A's reservation value for internal negotiations is 42 points.

This is simple enough in theory but extremely hard to do in practice. By playing around with different combinations of levels, it may become clear that A's reservation value is somewhere between, say, 35 and 45 points. That may be all the accuracy we may need in practice. As with all subjective evaluations, it may be possible to compare D with several contracts whose scores are near the break-even point. What should be going through the mind of chief negotiator A when comparing the external alternative D with a specific internal contract? Somehow everything should be referred back to how well each composite bundle fares in terms of the fundamental interests (objectives) of our decision maker and how important these interests are. All this can be formalized, but we will resist doing so here. The important messages are contained in the conceptual ideas, not the quantitative details.

B's Perspective without Formalization

Let's take a look at B's side. B is acting as most negotiators are wont to do. Imagine that B has also thought somewhat about tradeoffs, but has not quantitatively scored them. Also assume that B has identified a best alternative but has not done this through any formal mechanism. B does not have the luxury of telling his or her agent to reject any contract that scores below some reservation value.

Certainly B will be called upon to make judgments, but these judgments will have to be made during the negotiations, without the guidance of a few well thought out numbers. B will have a hard time delegating responsibility to an agent. The agent will have to report back frequently for instructions comparing one contract with another or ascertaining whether a given contract meets some minimum standard of acceptability, a so-called hurdle rate.

In our discussion of A's reservation value, we assumed that A did not impose separate hurdle requirements on individual issues. In our example, A adopted the summative view that the contract had to achieve a minimum of 42 points. Without a scoring structure, there is a tendency for B to impose reservation restrictions on separate issues, such as: I'll need levels (a or b) on issue 1, *and* (a, b, or c) on issue 3, *and* (a, b, or c) on issue 7. This makes it tough on the negotiations. Team A can provide its negotiating agent with a single RV number: any contract that has an associated score below that value is not acceptable. Team B, on the other hand, for some reason or other, has no way of associating a score that evaluates the overall "goodness" of a contract. B's negotiating agent, instead of being armed with a single RV number, may be provided instead with the specification of a set of hurdle levels on each of several important issues: "You need to get level c or better on issue 3, *and* level d or better on issue 6, *and* level b or better on issue 9." There is no way that a fantastic outcome on issue 8, say, can make a contract tolerable if it falls slightly below the prespecified hurdle level on any issue. This lack of a sufficiently flexible compensatory mechanism forces B to negotiate harder—and that may be good, but sometimes it forces a rigidity that blocks any progress. Ideally, RVs should have an overall summative feature. B may be forced to insist on reservation hurdles on specific issues because there is no adequate vocabulary for thinking about compensatory tradeoffs.

Remember that we are on A's team. There is no guarantee that B will be trying to act reflectively, coherently, and rationally. We may be able to exploit the lack of preparation and decision-making biases on B's side for our own benefit. But the sword cuts both ways. If B has not prepared well, he may set his reservation level way above his BATNA. This is one of those behavioral realities: in the face of the fuzziness of value tradeoffs, most subjects act very cautiously and conservatively and in this case may use an inappropriately high reservation hurdle. Consequently B might claim much more than his situation justifies. It may be in A's interest to coach B on how to be more flexible and suggest ways that B can better align B's BATNA and minimum threshold requirements for an acceptable contract.

Practicalities and Cutting Corners

The scoring of the template, the determination of the BATNA, and the calculation of the associated RV are formidable tasks in practice. Granted, in real-world negotiations these calculations are rarely done. But the development of the ensuing theory will often be considerably simplified if we assume that this quantification has taken place in a nonfuzzy way.

We would like to encourage the practitioner to follow the advice given here on preparation of the template. In practice, however, one might cut corners by not scoring the template. For example, it might suffice to conduct only qualitative analyses within and between issues, perhaps not even rank ordering the importance of the issues but dividing them into three clusters of low-, medium-, and high-importance levels. For some high-importance issues, the practitioner may wish to do a more quantitative within-issue analysis. If, during the negotiations, more detail is needed, then further quantification of the template may be in order. Similarly for BATNA and RV analyses.

In laboratory or classroom simulations, it is often helpful to give each role-playing subject crisp quantitative scores for the template as well as a numerical RV. But it takes a good deal of preparation in practice to reach that ideal.

Monetary Quantification

In some applications money looms so large that instead of scoring the template in terms of desirability units going from 0 to 100, one can value outcomes in terms of monetary equivalences. There are two advantages to doing this. First, a monetary amount is readily interpretable; a payoff of $65,000, say, is more understandable, standing alone, than a desirability value of 62 units. Second, in negotiations, monetary amounts may be directly transferable. For example, party A may sweeten up his, or her, offer to party B by throwing in a side payment of $15,000. Doing so will simplify some subsequent analyses.

But there is sometimes a downside to using money as the scoring mechanism. Going from a monetary amount of −$25,000 to $0 may be far more important than going from $0 to $100,000. A side payment of $15,000 may be far more important when the other party starts at an asset position of −$25,000 rather than at +$150,000. Converting money to desirability values may help to capture these nuances. Thus, for example, if −$25,000 were given a desirability score of 0, and $100,000 a desirability score of 100, then $0, which is closer to the lower end in monetary terms, might still be given a score of 60 units by some subjects.

Core Concepts

Chapter 11 concluded with template construction. This chapter has focused on template evaluation. We continued giving partisan advice to A, but our advice would have been substantially the same if had we been giving simultaneous advice to both parties.

We suggested for each issue that our client rank order its preferences for the different resolutions (within-issue qualitative analysis) and follow this with a rank ordering of the importance of the different issues (between-issue qualitative analysis). In this latter task it is critically important that the evaluator keep in mind the ranges of possible resolutions of the issues being compared.

Ensuing analysis will be facilitated if the qualitative rankings (within and between) are bolstered by having the evaluator provide, in addition, indications of his or her strengths of preferences in the form of within-issue and between-issue quantitative analyses leading to an additive scoring of the template.

We proposed that our client identify its Best Alternative to a Negotiated Agreement with the other side—this usually entails doing some decision analysis—and, assuming the template is scored, determining the minimum acceptable score (reservation value) in its negotiation. The analysis is considerably more difficult if the template is not scored, and this is one reason why it may be worth the effort to do this scoring. When the template is not scored, there is a tendency for negotiators to impose minimal cutoff levels on each issue instead of a summative cutoff level for all issues combined.

13

Template Analysis (I)

Efficiency plays a pivotal role in negotiation theory, and we start this chapter by clarifying what we mean when we use the term. We then introduce the problem of fair division: two parties have to divide a number of indivisible items between themselves. This problem is a good launching pad for more general problems, because the template for the fair-division problem is especially simple and a full efficiency analysis can be made without getting bogged down. In later chapters, we will expand the method to cover more complicated cases. We briefly touch on the question of equity: what is fair? We then examine the fair-division problem as a problem in mathematical programming and introduce the power of SOLVER, an add-on of the spreadsheet Excel.[1] This powerful bit of technology is introduced here (where it is not really needed) because it will be handy in more complicated problems we'll consider in later chapters.

Finding Efficient Contracts

We start with a special case involving a single issue with multiple resolutions—meaning more than two. Our two protagonists are Ann and Bill, who jointly must select one contract (or resolution) from a set of ten possible contracts. We'll label these choices A, B, up to J. We'll assume that Ann and Bill are going to negotiate according to the prescriptions of the FOTE condition—Full, Open, Truthful Exchange. They trust each other, have no secrets, and wish to maintain their good relationship. But notwithstanding all this, they have different tastes, values, and interests. So which resolution, from A to J, should they select? Could they easily reject some contracts? Do other contracts look particularly attractive? How should they proceed in making their choice?

1. No doubt other programs could be used as well—this is the one we know.

Joint Ordinal Rankings

Let's suppose that they start out by ranking the contracts A to J. Ann and Bill's rankings are shown below.

Ann's rankings from worst to best:

$$E, C, B, I, H, A, G, F, J, D$$

Bill's rankings from worst to best:

$$D, I, J, H, F, G, B, A, E, C$$

For Ann, E is the worst and D the best; for Bill, D is the worst and C the best. These are ordinal rankings, with no strength of preferences given. We'll meet some examples shortly where there are hundreds, thousands, and even millions of contracts and it may not be so easy to rank them. But let's assume that Ann and Bill have no trouble ranking the contracts from worst to best.

In Table 13.1, Ann and Bill assign numbers from 1 to 10 to the contracts A to J, where 1 is worst and 10 best. We'll say that contract A has the joint evaluation (6, 8), meaning that Ann ranks A as 6 from the bottom and Bill ranks A as 8 from the bottom. It may seem more natural to rank the best as number 1 rather than number 10, but it will become apparent why this convention has been chosen.

Dominance and the Elimination of Noncontenders

From Figure 13.1, we see that contract H, with a joint evaluation of (5, 4), is a noncontender because it is out-ranked by A (6, 8), by G (7, 6), and by F (8, 5) — check this yourself. We'll say that H is *inefficient* since it is *jointly dominated* by A, G, and F. More formally we say that contract X dominates contract Y if at least one party prefers X to Y and the other party either prefers X to Y or is equally satisfied with both. Contracts E, B, H, and I are inefficient in this formal sense. This means the *efficient frontier* comprises contracts C, A, G, F, J, and D.

Table 13.1 Joint ordinal rankings: Tabular display

	Contract									
	A	B	C	D	E	F	G	H	I	J
Ann's ranks	6	3	2	10	1	8	7	5	4	9
Bill's ranks	8	7	10	1	9	5	6	4	2	3

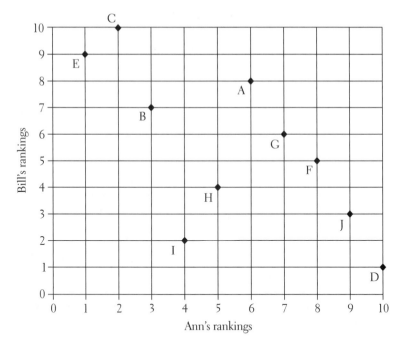

Figure 13.1.   Joint ordinal ranking—graphical display.

Efficiency is an ordinal concept—it does not depend on strengths of preference that are captured by cardinal evaluations. In some circumstances it may be easier to assess cardinal rankings than ordinal rankings. For example, a teacher might grade her students' exams using a cardinal scale from 0 to 100 and then only afterward rank the students (ordinally) according to their cardinal scores.

Observe in Figure 13.1 that along the efficient ridge going from C to D, point G falls in a valley between A and F. This is purely an artifact of the uniform spacing we used in representing contracts in graphical form. We're now going to switch from ordinal to cardinal values that reflect Ann and Bill's *strengths of preference.* You'll see that the efficient frontier will not change, but along the frontier there will be new points that fall in the valleys.

Extreme Efficiency

With a little more work, we can do even better.

Joint cardinal ranking.   We ask Ann and Bill not only to rank the ten contracts ordinally, but to score them cardinally. Ann uses a scale from 0 to 10, putting her worst contract, E, at 0.0 and her best, D, at 10.0. According to her tastes and values, J is not far below D, and J gets a score of 9.5. The gap between J and G is rather wide and between H and I even wider. See Table 13.2.

Table 13.2   Joint cardinal rankings: Tabular display

					Contract					
	A	B	C	D	E	F	G	H	I	J
Ann's score	6	2.5	2	10	0	8	7.8	5.5	3.5	9.5
Bill's score	80	78	96	30	90	55	75	50	32	45

Bill scales the contracts from 0 to 100 but scores his worst contract, D, at 30 and his best, E, at 90. In keeping with the FOTE orientation, we now assume that the information in Table 13.2 is common knowledge. Figure 13.2 plots these points. We see that the efficient frontier remains the same, but now notice that A and F fall in the valleys along the efficient ridge. We'll make the distinction between *efficient* points (C, A, G, F, J, and D) and *extreme-efficient* points (C, G, J, and D).

Why are we so concerned about extreme efficiency? Two compelling reasons.

Randomization.   In Figure 13.2, rather than choosing contract F, which falls in the valley, Ann and Bill might each prefer to toss a coin: heads they take G and tails they take J. This will be called a "mixed" or "randomized" strategy. If the parties are not risk averse, by randomizing between any two points, say Q and R, they can achieve any score on the line connecting Q and R. This is better for both of them.

When we push on with this analysis, we will be introducing complex cases with millions of possible contracts, and it will be hopeless to plot them all. So we will need to have the analytical means at our disposal to identify efficient contracts. Using these methods, it is far easier to find those that are extreme efficient.

The ruler representation.   In Figure 13.3 we depict an example in which contracts A through G lie on the efficient ridge, with C and E falling in the valleys. Imagine a ruler, shown in the figure, moving up and down but maintaining the same slope. Let's start at a location at the lower left.

Now imagine moving the ruler upward (with the same slope) until its lower edge just touches a single point as it is about to leave the field of points—do this yourself. With this slope the ruler rests on extreme-efficient contract D. By changing the slope, the same upward movement will perch the ruler on other extreme-efficient points. Every extreme-efficient point can be a resting point of the ruler for some slope, and any extreme-efficient point can claim the same distinction. For some slopes the ruler might rest on two or more extreme-efficient points.

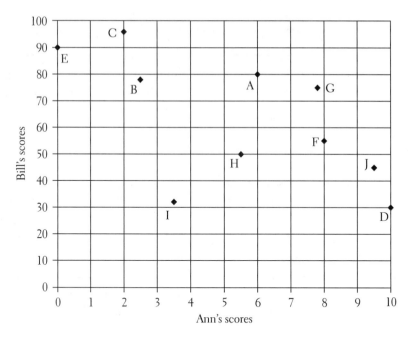

Figure 13.2. Joint cardinal scores—graphical display.

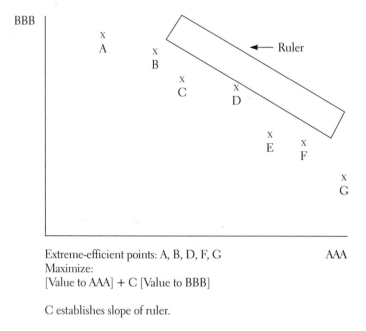

Extreme-efficient points: A, B, D, F, G

Maximize:

[Value to AAA] + C [Value to BBB]

C establishes slope of ruler.

Figure 13.3. The ruler representation: finding extreme-efficient contracts.

Striding forward with our analysis, we now want a technique to help us find extreme-efficient points among millions of potential points. The preceding bit of geometry will provide us with the necessary means. Let's suppose there are a vast number of contracts listed in a column on a computer spreadsheet, and that each is scored for negotiators AAA and BBB. For the ith contract, let a_i and b_i be AAA's and BBB's score, respectively. Each contract corresponds to a point on the AAA and BBB axes—representing their valuation of the contract. For each contract we can compute the weighted sum of the scores:

$$a_i + C \times b_i, \qquad \text{for some positive constant C.} \qquad (13.1)$$

The constant C determines the slope of the ruler, and this simple formula allows us to choose the contract that maximizes the sum of the scores. The point associated with this contract would be an extreme-efficient point. (Why?) [Argue if (1) is maximized by a contract that is not an extreme point, then it would be dominated by some other point whose weighted sum in (1) would be larger.] By changing the parameter C, we essentially tilt the ruler, and then by maximizing we pick up another extreme-efficient point. We'll exploit this piece of geometrical trickery later on in our analysis.

Fair Division

The topic of fair division not only is interesting in its own right, but will be the springboard from which we'll move into the more general case of two-party and multiparty negotiations. We use a case with two parties and many issues, but each issue has just two possible resolutions.

The Fair-Division Problem

The problem is this: Janet and Marty are given twenty items to share, to divide between them. Imagine

- that they have inherited an estate from their recently deceased great aunt; or
- that they are dissolving a partnership, and after distributing easily divisible commodities, like money, they are left with twenty indivisible items to share "equally"; or
- that they are getting a divorce and everything has been resolved except for twenty items of emotional significance to each of them.

The template for this problem has the special characteristic that each issue can only be resolved in one of two ways: the item can go to Janet or to Marty. Let's posit that Janet and Marty, despite their faults, don't lie to each other. In our vocabulary, they will be negotiating under the conditions of Full, Open, Truthful Exchange.

A contract will specify ownership for each of the twenty items. A typical contract might give to Janet items 1, 2, 7, 10, 15, 17, and 19; the rest would go to Marty. There are two possible resolutions for the first item, two for the second, and so forth up to the twentieth. This means that there are a total of:

$$2^{20} = 1,048,576$$

possible contracts. Janet and Marty must jointly select one of these contracts. That's their problem. What are some of the ways they can do it? Think about it.

Some Proposed Solutions

Sequential choice.   An obvious suggestion is that Janet and Marty should choose sequentially. Toss a coin (randomization again) to see who goes first. Suppose it favors Janet. She chooses, then Marty chooses, then Janet, then Marty, and so on. This solution is certainly a fair one, but is it a good one? And what do we mean by a good solution?

Toss a coin.   Another "fair" solution is to give *all* twenty items to either Janet or Marty, depending on the outcome of a single toss of a coin. But this is only fair a priori and is not fair a posteriori when one of the two has everything, the other nothing. The process might seem fair, but the outcome isn't. Janet and Marty would probably prefer to choose sequentially rather than having an all-or-nothing coin toss.

Divide and choose.   As a second serious alternative, they could use the famous divide-and-choose procedure. Toss a coin to determine the divider. Suppose it's Marty. Marty divides the twenty items into two piles and Janet selects which pile she prefers. Again this seems fair. If you were the divider, which strategy would you use? Would you rather be the divider or the chooser? We'll examine questions like this later.

Allocation of points.   Another suggestion is this: Give 100 points each to Janet and Marty. Let them assign their 100 points to the twenty items, giving more points to the items that are more appealing to them. They do this in isolation from each other. Then the allocations of points are mutually disclosed and each

item goes to the party who assigns it more points. Ties can be solved by randomization. Fair? Yes. Good? Perhaps.

FOTE Analysis

Allocating points.   So let's say that Janet and Marty both sprinkle 100 points
over the twenty items as shown in the first three columns of Table 13.3. Janet
gives a whopping 25 points to item 13; but Marty goes one better and gives 30
points to his favorite item, number 8. In Janet's opinion, item 13 (with 25 points)
is slightly better than both items 8 and 15 (worth collectively 14 + 10 points). It's
that simple. The points reflect their tastes, all things considered, but we will not
attempt to describe the thinking behind Janet's and Marty's allocating of points
to items. Once again, imagine that the process is similar to our schoolteacher
grading an exam.

Now for any contract whatsoever that allocates some items to Janet and the
remaining to Marty, we can (with the assistance of a computer) find a score for
Janet and Marty and plot this joint evaluation. If there aren't ties there will be
over a million such points. Our objective is to show how to find all the extreme-
efficient points without having to plot all million points.

Ratios of scores and characterization of extreme-efficient points.   We're now
going to make extensive use of Microsoft Excel, a computer spreadsheet, which
greatly facilitates our task. The fourth column of Table 13.3, under the heading
J/M, displays the ratios of Janet's scores to Marty's scores, item by item. As an example, for item 5, the ratio is 9/7, or 1.29. Intuitively this says that by giving item
5 to Janet, on the margin, Janet picks up 1.29 units for every unit that Marty
gives up. Clearly, high J over M ratios favor Janet, low ones favor Marty.

We assert:

- *For any non-negative constant C, the contract that gives Janet all items with
  J over M ratios greater than C (and only those) yields an extreme-efficient
  contract.*
- *Any extreme-efficient contract is attainable this way for an appropriate non-
  negative C.*

Proof. We argued earlier that for any non-negative constant C, the contract
that maximizes:

$$\text{(Janet's score)} + C \times \text{(Marty's score)} \tag{13.2}$$

is an extreme-efficient contract. (Recall that the constant C determines the slope
of the ruler in Figure 13.3.) Expression (13.2) is the sum of twenty summands,

Table 13.3   Point allocations

Item	Janet	Marty	J/M
1	1	2	0.50
2	1.5	1	1.50
3	8	5	1.60
4	1.5	3	0.50
5	9	7	1.29
6	2	3	0.67
7	3	8	0.38
8	14	30	0.47
9	0.5	1	0.50
10	0	1	0.00
11	7	4	1.75
12	0.5	0.5	1.00
13	25	18	1.39
14	0.5	1	0.50
15	10	5	2.00
16	4.5	3	1.50
17	3	0.5	6.00
18	8	4	2.00
19	0.5	1	0.50
20	0.5	2	0.25
Total	100	100	

one for each issue—the allocation of the item. The contribution to this sum for the ith issue is:

(Janet's score for the ith item) if the ith item is allocated to Janet,

or

$C \times$ (Marty's score for the ith item) if the ith item is allocated to Marty.

Hence to maximize (13.2), we should allocate the ith item to Janet if and only if:

(Janet's score for the ith item) $> C \times$ (Marty's score for the ith item)

or if and only if

$$\frac{(\text{Janet's score for the } i\text{th item})}{(\text{Marty's score for the } i\text{th item})} > C. \tag{13.3}$$

This completes the proof of the assertion.

Finding all extreme-efficient contracts: Dealing from the top.   Table 13.4
sorts the twenty items according to their J over M ratios. Spreadsheets like Excel
love to show off how fast they can do this type of sorting. Observe that any contract that cuts the array into two parts, giving the top part to Janet and the bottom
to Marty, is an extreme-efficient contract. Let's do this systematically. Look at
columns (5) and (6) of Table 13.5. To start with we can give everything to Marty,
yielding Janet 0 and Marty 100. This allocating results from using any constant
C greater than 6. Next we give item 17 to Janet, adding 3 to Janet's accumulated
score and decreasing Marty's score of 100 points by 0.5 units. We get the next
extreme-efficient point by giving Janet item 15 as well as item 17, which yields
13 units to her and 94.5 to Marty. We can keep repeating this transfer of items,
from Marty to Janet's growing portfolio. In doing this we identify all the extreme-
efficient points we're looking for.

Figure 13.4 exhibits the extreme-efficient allocations of the fair-division
problem. The point (42, 77.5), for example, is associated with the allocation to
Janet of items 17, 15, 18, 11, 3, 16, and 2. Adding to this mix item 13 would give
Janet an incremental gain of 1.39 times Marty's decrement, and the addition of
this item to Janet would result in the joint evaluation of (67, 59.5).

Table 13.4   Scores with ratios sorted

Item	Janet	Marty	J/M
17	3	0.5	6.00
15	10	5	2.00
18	8	4	2.00
11	7	4	1.75
3	8	5	1.60
16	4.5	3	1.50
2	1.5	1	1.50
13	25	18	1.39
5	9	7	1.29
12	0.5	0.5	1.00
6	2	3	0.67
4	1.5	3	0.50
9	0.5	1	0.50
14	0.5	1	0.50
1	1	2	0.50
19	0.5	1	0.50
8	14	30	0.47
7	3	8	0.38
20	0.5	2	0.25
10	0	1	0.00
Total	100	100	

Table 13.5   Ratios and extreme-efficient contracts

1	2	3	4	5	6
				Extreme-efficient contracts	
Item	Janet	Marty	J/M	Janet	Marty
				0	100
17	3	0.5	6.00	3	99.5
15	10	5	2.00	13	94.5
18	8	4	2.00	21	90.5
11	7	4	1.75	28	86.5
3	8	5	1.60	36	81.5
16	4.5	3	1.50	40.5	78.5
2	1.5	1	1.50	42	77.5
13	25	18	1.39	67	59.5
5	9	7	1.29	76	52.5
12	0.5	0.5	1.00	76.5	52
6	2	3	0.67	78.5	49
4	1.5	3	0.50	80	46
9	0.5	1	0.50	80.5	45
14	0.5	1	0.50	81	44
1	1	2	0.50	82	42
19	0.5	1	0.50	82.5	41
8	14	30	0.47	96.5	11
7	3	8	0.38	99.5	3
20	0.5	2	0.25	100	1
10	0	1	0.00	100	0
Total	100	100			

What Is Fair?

What would be equitable or fair? We're jumping ahead here a bit to later chapters, but we'll digress briefly. Let's investigate the allocation that maximizes the minimum payoff to either Janet or Marty—in other words, the contract that jointly maximizes the minimum of the two scores.

The maximin solution.   From Table 13.5 we see that if Janet were to receive all items down to and including item 2, then each would get at least 42. If item 13 is added to Janet's bundle, then each could get at least 59.5. This maximizes the minimum among these extreme-efficient points. But we can improve this floor by randomizing: give items 17 down to 2 to Janet and give item 13 to her with some probability X. Janet's expected score is then

$$67X + 42 (1 - X),$$

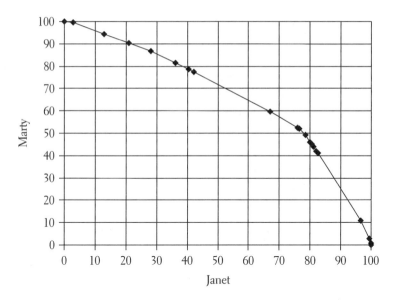

Figure 13.4.   Plot of extreme-efficient contracts.

and Marty's becomes

$$59.5X + 77.5 (1 - X).$$

If we choose X to equate these two expected scores, we get $X = .8256$. So if Janet receives items 17 down to 2 outright and in addition receives item 13 with probability .8256, her expected return before randomization takes place is

$$67 \times .8256 + 42 \times .1744 = 62.64,$$

and Marty's expected return is

$$59.5 \times .8256 + 77.5 \times .1744 = 62.64.$$

Using a randomized strategy, each could get an expected value of 62.64 of their full potential satisfaction. Not a bad result. But there are other compelling alternatives that purport to be fair as well, and we shall meet several of them later in the book.

Let's keep in mind that the fair-division problem is a very special case. Its template consists of many issues but each with only two resolutions. Furthermore, in this case we did not get involved with calculating Best Alternatives to a Negotiated Agreement and reservation values. In the next chapter we use the results of this special case to motivate the analysis of the more general case,

involving issues with more than two resolutions and with the complications of
BATNAs and RVs.

Inadequacies of other solutions.   The maximin approach has its drawbacks.
But not as many as some of the other commonly used methods.

Choosing sequentially. We have concocted some numbers to show how awful the result from a sequential choice procedure could be. Consider the data
displayed in Table 13.6. To be impeccably fair, let the first chooser be selected
by a toss of a coin. Suppose AAA chooses first. AAA would take item 1; BBB
would follow with item 2; AAA next takes item 3; and so on as shown. The chosen items are asterisked. AAA ends up with 91 and BBB with 44. Not very equitable. What went wrong?

When AAA takes number 1, BBB is out 25 points. With item number 1, the
ratio of scores of AAA over BBB is close to 1. When BBB selects item number 2,
AAA gives up only 2 points. In this case the ratio of BBB's score to AAA's score is
27/2, which doesn't hurt AAA much. When AAA continues and chooses number
3, once again BBB loses out on a whopping 17 points. It's pretty obvious what's
happening when the numbers are crisp and side by side; with a little ambiguity
thrown in, the parties may not be aware of just how inequitable the sequential
procedure could be.

There are other things wrong with sequential choice. This procedure always
results in both parties getting the same number of items—or the first chooser
gets one more item if the number of items is odd. That may not be very equitable. You can imagine cases where it makes sense for one party to get far fewer
items than the other party—but the resulting values to each will be more or less
equivalent.

Sequential choice can also lead to inefficiencies if the parties act strategically. AAA, for example, may not choose his best alternative, because he thinks
there's no danger that BBB will select that item when BBB selects next. If both
parties choose strategically and if there are misperceptions, then the final outcome may be inefficient as well as inequitable.

Divide and choose. Now let's look at the weaknesses of the divide-and-
choose procedure. Would you rather be the divider or the chooser? A good answer is—it depends.

Let's start from the observation that if AAA divides, BBB must get at least 50
points. AAA can guarantee getting 50 points for himself by dividing the items
into two equally desirable piles. Because of lumpiness, he may not be able to do
this exactly. He could manage to do so if he could assign one of the items using
randomization. But let's ignore this wrinkle. Splitting the value 50–50, or nearly
50–50, can be done in several ways. While AAA may be roughly indifferent between these splits, BBB may have strong preferences between them. There is no
guarantee that when BBB selects, the resulting joint payoff will be efficient.

Table 13.6   Choosing sequentially: AAA selects first

Item	AAA	BBB
1	27*	25
2	2	27*
3	18*	17
4	3	13*
5	17*	12
6	1	3*
7	15*	2
8	3	1*
9	14*	0
10	0	0
Total asterisked	91*	44*

Note: Asterisked numbers indicate choices.

Using the data in Table 13.6, suppose AAA knows BBB's scores and divides the items as follows: {1, 2} and {3, 4, 5, 6, 7, 8, 9, 10}. BBB would then choose {1, 2}, yielding 52 points to herself, and AAA would end up with 71 points. Quite a bonanza for AAA. So it may be better to be the divider. But although this is a good deal for AAA, the joint evaluation is not even efficient.

Consider another case. Suppose AAA thinks he knows BBB's evaluations but he may have misperceptions. For example, suppose AAA divides the items into the following two piles {2, 3} and {1, 4, 5, 6, 7, 8, 9, 10}, thinking that BBB would choose {2, 3}. But she surprises him and chooses the other set of items. A fiasco! AAA ends up with 20 points and BBB with 56—far from being either efficient or equitable.

The lessons of experience.   Try the following thought experiment. You are one of two individuals who have to share twenty items. You have examined your preferences and you know something about the preferences of the other party. Which procedure would you prefer?

1. Toss a coin to see who goes first and choose sequentially.
2. Toss a coin to determine the divider or chooser.
3. If the other party agrees, you both confidentially give your scores to an analytical intervenor who then chooses an allocation that gives you equal amounts on the efficient frontier.

I (HR) am experienced at acting strategically, but I know, on an expected value basis, that I would be better off with option 3, using an analytic intervenor. Would I tell the intervenor my true scores? Would you? I can imagine circumstances where I know I would be tempted to lie—by giving lesser points to an item that I like when I suspect the other person will probably give very low values to that item. But she may also be acting in strange ways. After a lot of experience in playing these games in the laboratory, I most often choose to disclose my values truthfully. The more erratic I believe the other person is, the more inclined I am to be truthful. Think hard about this provocative statement, because it generalizes to many complex negotiations.

FOTE Analysis with Spreadsheets

Organizing the Spreadsheet

Our next task is to show how we can solve the fair-division problem using mathematical programming—linear, nonlinear, with and without integer constraints. Those readers who are unfamiliar with this type of operation are the lucky ones! You'll learn a lot with a minimum of fuss. Our interest in doing this is pedagogical: we already have shown how to analyze the fair-division problem, so there is no need to pull out a sophisticated mathematical technique. It's like killing a gnat with a cannon. But if you understand the mathematical programming analysis for this problem, it will be much easier for you to follow the logic for the more general problem, where the technique will prove to be indispensable.

Look at Table 13.7. Notice in the spreadsheet that the columns are labeled with letters, A, B, and so on, and rows are labeled with numbers. To illustrate the referencing system, in Table 13.7 the number in cell C14 is 30. The data of the problem is exhibited, boxed, in columns A to C—ignore rows 29 to 31 for the time being. For readers who want to actively learn how to program the computer, we suggest that, for starters, you type the numbers in the grid from A5 to C26 inclusive into an Excel worksheet.

Columns E and F, labeled "Allocations," for Janet and Marty need some explanation. Columns E and F exhibit an illustrative allocation. Look at entries 0 and 1 in E7 and F7, which indicate that for this allocation, Marty gets item 1. Similarly, Janet gets items 4, 5, 9, 11, 15, 16, 17. We could have put 0's and 1's in any pattern in columns E and F, but we must always make sure that the entries in each row sum to 1. Otherwise we might give the same item to both Janet and Marty (1,1) or give the item to neither (0,0). This means we have twenty constraints: the entries in E7 plus F7 must always sum to 1; so must E8 plus F8; and so on up to E26 plus F26. The entry G7 is the sum of E7 and F7. Similarly down the G column, labeled Sum. We require all sums in the G column to be 1, and in the language of Excel this is denoted by G7:G26=1.

We can and do relax the assumption of using only 0's and 1's in the array columns E and F in Table 13.7. For example, if we used the numbers .72 in E7 and .28 in F7, they would sum to 1 as required. This would indicate that Janet gets item 1 with probability .72 and Marty gets item 1 with probability .28. To re-

Table 13.7   Organizing the spreadsheet

	A	B	C	D	E	F	G
1							
2							
3		Values			Allocations		
4	Item	Janet	Marty		Janet	Marty	Sum
5							
6							
7	1	1	2		0	1	1
8	2	1.5	1		0	1	1
9	3	8	5		0	1	1
10	4	1.5	3		1	0	1
11	5	9	7		1	0	1
12	6	2	3		0	1	1
13	7	3	8		0	1	1
14	8	14	30		0	1	1
15	9	0.5	1		1	0	1
16	10	0	1		0	1	1
17	11	7	4		1	0	1
18	12	0.5	0.5		0	1	1
19	13	25	18		0	1	1
20	14	0.5	1		0	1	1
21	15	10	5		1	0	1
22	16	4.5	3		1	0	1
23	17	3	0.5		1	0	1
24	18	8	4		0	1	1
25	19	0.5	1		0	1	1
26	20	0.5	2		0	1	1
27		100	100				
28							
29		Values					
30		Janet	Marty		Sum	Min	Product
31		35.5	76.5		112	35.5	2715.8

iterate, the numbers, row by row in columns E and F, must sum to 1. The numbers also have to be between 0 and 1. It suffices to require the numbers to be non-negative and sum (row by row) to 1. To summarize, in Excel: G7:G26 = 1; E7:F26 >= 0; E7:F26 <= 1. With these constraints, randomization is possible. If we want to, we can stipulate that randomization is not possible, by requiring E7:F26 = integer. All of these constraints can be put into SOLVER, which will "solve" our mathematical programming problem.

Once an allocation of items is legitimately specified in E7:F26 (by legitimate we mean that E7:F26>=0 and G7:G26=1), the computer is then programmed to record the payoff to Janet in cell B31 and to Marty in cell C31. In the table, this illustrative allocation nets Janet 35.5 and Marty 76.5. Nice for

Marty but not so nice for Janet. It's now like an algebra problem—almost. Given the data in A7:C26, we seek a set of numbers to put into E7:F26 subject to the constraints of legitimacy to get a pair of "desirable payoffs" in cells B31 and C31. The unknowns in the problem, the cells we instruct SOLVER to change, are the entries E7:F26. That defines the problem. For the resolution of this problem, we must decide what it is we want to optimize—or what constitutes a "desirable pay-off."

Desirable Payoffs

Cell E31 of Table 13.7 gives the sum of the two payoffs to Janet and Marty (in Excel: +B31+C31). For this illustrative case the *sum* is 35.5 + 76.5 or 112. Cell F31 presents the *minimum* of the two payoffs (in Excel: +MIN(B31:C31)) and cell G31 the *product* of the two payoffs (in Excel: +B31*C31). To answer our question—"what should we optimize?"—we'll investigate results from three ob-jective functions: (1) maximize the sum of the two payoffs; (2) maximize the minimum of the two payoffs; (3) maximize the product of the two payoffs.

Maximizing the sum.   We first set up the spreadsheet using the illustrative al-location displayed in Table 13.8 Then we call up SOLVER, the special mathe-matical program that works within Excel. SOLVER does most of the work for us. You can find SOLVER in the Tools menu.

Once the SOLVER dialogue box appears on the screen, it asks us what we want to maximize or minimize. We respond by selecting cell E31 in Table 13.7 and by choosing to maximize the value it contains. E31 gives us the sum of the payoffs. The next step is to identify the range of our variables, the cells whose values we want to change. These are the unknowns, the future alloca-tions. We input any set of legitimate numbers in E7:F26. After this SOLVER needs to know what the constraints are. We respond that: (1) E7:F26>=0, and (2) G7:G26=1. This completes the task, so click "SOLVE." Complete this exer-cise yourself. The formulation and solution of the problem that maximize the sum of payoffs to Janet and Marty are given in Table 13.8.

The answer pops up in a matter of seconds, because this is a linear problem and SOLVER finds it terribly easy. When I was a graduate student this problem would have taken hours to solve. The maximum sum attainable is 128.5 (you might want to check this), and notice that the optimal allocation for this objec-tive function does not employ any randomized strategies—all the allocations are 0's and 1's. Of course we could have solved this problem using our earlier man-ual techniques, but SOLVER reduces the workload, and we'll be in desperate need of the enhanced computational power when we confront more complex problems later on.

We can and do relax the assumption of using only 0's and 1's in the array

Table 13.8   Maximizing the sum

	A	B	C	D	E	F	G
1							
2							
3			Values			Allocations	
4	Item	Janet	Marty		Janet	Marty	Sum
5							
6							
7	1	1	2		0	1	1
8	2	1.5	1		1	0	1
9	3	8	5		1	0	1
10	4	1.5	3		0	1	1
11	5	9	7		1	0	1
12	6	2	3		0	1	1
13	7	3	8		0	1	1
14	8	14	30		0	1	1
15	9	0.5	1		0	1	1
16	10	0	1		0	1	1
17	11	7	4		1	0	1
18	12	0.5	0.5		0	1	1
19	13	25	18		1	0	1
20	14	0.5	1		0	1	1
21	15	10	5		1	0	1
22	16	4.5	3		1	0	1
23	17	3	0.5		1	0	1
24	18	8	4		1	0	1
25	19	0.5	1		0	1	1
26	20	0.5	2		0	1	1
27		100	100				
28							
29			Values				
30		Janet	Marty		Sum	Min	Product
31		76	52.5		128.5	52.5	3990

Notes:

Cell	Input
G7	+E7+F7
G8:G26	Drag G7
B31	+sumproduct(B7:B26, E7:E26)
C31	+sumproduct(C7:C26, F7:F26)
	(or drag B31)
E31	+B31+C31
F31	+min(B31,C31)
G31	+prod(B31,C31)

columns E and F in Table 13.8. For example, if we used the number .72 in E7 and .28 in F7, they would sum to 1 as required. This would be interpreted as Janet gets item 1 with probability .72 and Marty gets item 1 with probability .28. It suffices to require the numbers to be non-negative and sum (row by row) to 1. To summarize, in Excel language: G7:G26=1; E7:F26>=0; E7:F26<=1. With these constraints randomization is possible. If we want to, we can stipulate that randomization is not possible by requiring E7:F26 = integer. All of these constraints can be put into SOLVER, which will solve our mathematical programming problem.

Maximizing the minimum.   We now change the objective function from maximizing the sum of the payoffs to Janet and Marty to maximizing the lesser of the two payoffs. In other words, maximizing the minimum or, in game theory parlance, finding the "maximin" solution.

Before maximizing the minimum let's look at the previous solution: maximizing the sum gave us payoffs (76,52.5) and the minimum of the two is therefore 52.5. Using our new objective function, we want to exceed 52.5, but that will mean reducing the score of 76. When our maximin solution pops up, we expect that the two payoffs will be equal. Knowing this in advance means we could simply have added another constraint in the SOLVER dialogue box and imposed the additional constraint that the two payoffs be equal and then maximized Janet's (or Marty's) payoff. Doing this has a distinct computational benefit, because it changes a nonlinear operation (that is, getting the minimum of two numbers) into a linear one, and solving linear problems is now a piece of cake. But because this would not generalize so easily to the case of more than two parties, we stick with the more generalizable approach. The analysis is done using the setup in Table 13.7 with the solution shown in Table 13.9.

Mixed strategy. Let's return to the details of our maximin problem: given the data in A7:C26, find the variables in E7:F26 subject to the constraints that the variables are non-negative (E7:F26>=0) and each of the row sums equals 1 (G7:G26=1) with the aim of maximizing the minimum payoff (maximize F31 in Table 13.7). With this data entered, SOLVER searches through the possible allocations of items to Janet and Marty. This problem is stated in nonlinear terms, so SOLVER takes slightly longer to produce its answer—all of a minute or so. The maximin solution is given in Table 13.9. Janet and Marty each achieve 62.6395 of their total maximum satisfaction of 100. Notice that the solution uses a randomized allocation for item 13, giving Janet a .83 chance at that and Marty a .17 chance at it.

Pure strategy. If we don't want to use randomized strategies, we can impose a further restriction on the variables by requiring them to be integers (Excel: E7:F26=int). This yields a solution in pure strategies that gives Janet and Marty each a payoff of 62.5 each. This solution is not an extreme-efficient contract—it falls ever so slightly in the valley along the ridge. To come up with this solution

Table 13.9 Maximizing the minimum of the payoffs

	A	B	C	D	E	F	G
1							
2							
3			Values			Allocations	
4	Item	Janet	Marty		Janet	Marty	Sum
5							
6							
7	1	1	2		0	1	1
8	2	1.5	1		1	0	1
9	3	8	5		1	0	1
10	4	1.5	3		0	1	1
11	5	9	7		0	1	1
12	6	2	3		0	1	1
13	7	3	8		0	1	1
14	8	14	30		0	1	1
15	9	0.5	1		0	1	1
16	10	0	1		0	1	1
17	11	7	4		1	0	1
18	12	0.5	0.5		0	1	1
19	13	25	18		0.83	0.17	1
20	14	0.5	1		0	1	1
21	15	10	5		1	0	1
22	16	4.5	3		1	0	1
23	17	3	0.5		1	0	1
24	18	8	4		1	0	1
25	19	0.5	1		0	1	1
26	20	0.5	2		0	1	1
27		100	100				
28							
29		Values					
30		Janet	Marty		Sum	Min	Product
31		62.64	62.64		125.28	62.64	3923.71

SOLVER had to work to the limits of its power to solve an integer, nonlinear programming problem. So it toiled for a couple of minutes.

Maximizing the product. Instead of maximizing the minimum payoff, we can maximize the product of the two payoffs. Why maximize the product? This question goes to the heart of the notion of fairness, and we discuss it at length in Chapter 19, on arbitration. But note that this solution was proposed by John Nash, who received a Nobel Prize in economics in part for contributing the "Nash solution." Although we do not include all the relevant tables here—the details are just too repetitive—we do exhibit the results in a table comparing the different objective criteria.

Table 13.10   Comparison of procedures

Maximize	Janet	Marty	Sum	Min	Prod
Sum					
M	76	52.5	128.5	52.5	3,990
P	76	52.5	128.5	52.5	3,990
Min					
M	62.64	62.64	125.3	62.64	3,924
P	62.5	62.5	125	62.5	3,906
Prod					
M	71.75	55.81	127.6	55.81	4,004
P	76	52.5	128.5	52.5	3,990

We can repeat the exercise of maximizing the product of the payoffs, but now add in the constraint that all variables are integers (E7:F26=int) in Table 13.7. With this additional restriction Janet gets less and Marty gets a little more (see Table 13.10).

Comparison of Procedures

Table 13.10 summarizes the results of our optimization procedures using different objective functions. Compare the solutions given by maximizing the sum, the minimum, and the product with mixed (M) or pure (P) strategies. Which one do you prefer? Maximizing the sum is not attractive to many because of the inequality that usually results. It might be adopted by negotiators who place a very high value on each other's satisfaction, or by a pair who will negotiate again and again and expect that it will all even out eventually, or by those who value efficiency above all. There is a real debate over whether to maximize the minimum or maximize the product. We can't grapple with the pros and cons of the two unless we discuss the objective criteria of fairness that the two proposals are based on. We will have a lot more to say about these principles later on, especially in Chapter 19.

Comparing the different optimization procedures produces results that are not starkly different. The differences will be more dramatic when we increase the number of parties, as we will do in the last part of the book. But still, let's compare the payoffs when maximizing the minimum (or equating the payoffs of Marty and Janet) and maximizing the product. And let's make the comparison using mixed strategies, which means using a randomization procedure for only one of the twenty items. The common maximin value for Janet and Marty is 62.64. Now if the procedure changes to maximizing the product, then Janet gets a proportional increase of 14.5 percent and Marty gets a proportional decrement of 11.5 percent. We'll see later on that equating proportional increases and decreases is a feature of the product proposal. One can argue that in a state of ignorance, not knowing whether the change from "maxing the min" to "maxing the prod" will benefit Janet or Marty, they might a priori opt for the "prod." But many subjects take comfort in the equality of the two payoffs, and Marty would have a hard time going from "min" to "prod" if the results were not covered by a veil of ignorance.

Core Concepts

This chapter has given joint advice to two parties who negotiate using Full, Open, Truthful Exchange. We examined the case in which each possible negotiated contract is scored separately by each; their joint evaluations can be plotted as a point.

- A contract is *dominated* if there exists another contract that is preferred by both—or, stated alternatively, a contract is dominated if its contract point evaluation lies to the left and below (that is, southwesterly) of another contract point.
- The efficient boundary consists of the complete set of nondominated points—or the northeasterly ridge of contract points. Since the number of possible contracts may be huge, it is important to be able to find and plot the efficient frontier.
- If we journey along the efficient frontier from northwest to southeast we shall traverse some indentations, or valleys. If we rest a ruler on the frontier, the supporting points of the ruler are *extreme-efficient* points; and each extreme-efficient point is a supporting point for the ruler for some slope. Mathematically an extreme-efficient point is a maximizer of A's value plus C times B's value of the contract, and changing the constant, C, is tantamount to changing the slope of the ruler. We'll see in Chapter 14 how this bit of geometry will enable us to find all the extreme-efficient contracts for a jointly scored template involving many issues, each with many resolutions.

The chapter presented a thorough examination of the fair-division problem for two parties negotiating in a FOTE manner. The two parties must allocate to themselves n indivisible items—no sharing of the items over time. The template for this problem is especially simple, since each issue corresponds to an item to be allocated and this can be done in only two ways. Furthermore in this problem the problem of BATNAs and RVs do not present any analytical difficulties. Hence the analysis of this problem, interesting in its own right, served as a illustration of what we would like to replicate for more complicated templates (that is, with issues involving more than two resolutions) and with the complications of BATNAs and RVs. We developed the case in which the parties assign 100 points to the items to be divided. We showed that if we then arrange the ordering of the items according to the ratios of their scores for the two parties, any extreme-efficient contract arises by giving one player the top portion of the sorted items. We also demonstrated how inequitable some reasonable, commonly applied procedures (like choose sequentially or divide and choose) turn out to be.

We also posed the allocation problem as a mathematical programming problem that we analyzed using SOLVER, a special program of Excel. Although this may have been overkill for the needs of this chapter, the methodology will be useful when the problem gets more complicated. We also previewed equitable outcomes. In choosing an equitable outcome among the extreme-efficient points, we suggested three possibilities: (a) maximizing the sum of payoffs; (b) maximizing the product of payoffs; (c) maximizing the minimum of the payoffs (which means for the two-party case equalizing the two payoffs). Detailed arguments for these alternatives are deferred to Chapter 19.

14

Template Analysis (II)

From negotiations over issues that can have only two resolutions, we now move to the more general case where some issues may have more than two resolutions and where BATNAs and RVs play an important role. We assume that the two negotiating parties have already engaged alone and together in preparing a jointly held template; each has scored the template confidentially; and each has determined its own BATNA and RV confidentially. The parties, after due reflection, decide to cooperate jointly in a FOTE manner and simultaneously disclose to each other their scored template and RVs. Furthermore, each believes the other, so that the scored template with RVs is common knowledge. Later, in this and succeeding chapters, we'll disengage from these extreme assumptions. We hasten to add, however, that the ideal case is far from unrealistic. There are many cases where the parties choose to negotiate in this joint FOTE way.

Case Study: Nelson vs. Amstore

The Scored Template

Nelson is a contractor in a negotiation with a large retail firm, Amstore, for the construction of a new store.[1] In this artificial example, there are only three issues to be decided: *price*, with five possible resolutions; *design*, with two resolutions; and *completion date*, with seven resolutions. Hence the number of potential contracts is $5 \times 2 \times 7 = 70$. You see in Table 14.1 the full template of the problem, fully scored by both players.

Keeping in mind the ranges involved on each issue, Nelson has given 60,

---

1. This case has been used by Gordon Kaufman in his course at the Sloan School, MIT; it was based on an earlier case of mine.

Table 14.1   Nelson vs. Amstore: Scored template

| Issue | Level | Payoffs | |
		Nelson	Amstore
Price	10	0	70*
($000,000s)	10.5	25	60
	11	40	45
	11.5	55	25
	12	60*	0
Design	Enhanced	0	10*
	Basic	20*	0
Completion	20	0	20*
(months)	21	8	19
	22	12	18
	23	15	16
	24	17	12
	25	19	7
	26	20*	0

Note: Asterisked numbers indicate importance weights designated by each party.

20, and 20 points to the three issues of price, design, and time, respectively. Amstore designates importance weights of 70, 10, and 20, respectively. What's new in this example are the partial scores for resolutions in between the worst and the best. For example, Nelson feels that going from a price of $10 million to $11 million is more important than going from $11 million to $12 million. Think of a teacher who is grading three questions and, marking on a basis of 100, gives 60 points to a perfect answer to Question 1, 20 points to Question 2, and 20 points to Question 3. She now has to decide how to give partial credit.

We'll try to use a standard (canonical) format that orients the resolutions for each issue so that the first party (in this case Nelson) will increase her scores as we go downward in the table from 0 to the maximum for that issue. The opposite orientation holds for the second party (in this case Amstore).

For this simple example, in Table 14.2, we first exhibit and plot all the possible contracts.

There are a lot of numbers in Table 14.2, but the punch line is that we had to enter only those numbers that are starred and the computer did the rest. A few words of explanation. The listing of the seventy possible contracts is shown in columns A, B, C. (Note: 70 = 5 × 2 × 7.) The power of the spreadsheet is that it can copy an array of numbers and paste them anywhere on the spreadsheet. You should note the systematic format used in displaying these contracts and how different arrays get repeated. In columns D through I the scores are given — once again notice how different arrays get repeated. Columns J and K exhibit the accumulated scores for the contracts.

Table 14.2   Systematic listing of all contracts

A	B	C	D	E	F	G	H	I	J	K	L
			Price		Design		Completion		Totals		Contract
Price	Design	Completion	N	A	N	A	N	A	N	A	no.
10*	**Enhanced**	20*	0*	70*	0*	10*	0*	20*	0*	100*	1
10	Enhanced	21*	0	70	0	10	8*	19*	8	99	2
10	Enhanced	22*	0	70	0	10	12*	18*	12	98	3
10	Enhanced	23*	0	70	0	10	15*	16*	15	96	4
10	Enhanced	24*	0	70	0	10	17*	12*	17	92	5
10	Enhanced	25*	0	70	0	10	19*	7*	19	87	6
10	Enhanced	26*	0	70	0	10	20*	0*	20	80	7
10	**Basic**	20	0	70	20*	0*	0	20	20	90	8
10	Basic	21	0	70	20	0	8	19	28	89	9
10	Basic	22	0	70	20	0	12	18	32	88	10
10	Basic	23	0	70	20	0	15	16	35	86	11
10	Basic	24	0	70	20	0	17	12	37	82	12
10	Basic	25	0	70	20	0	19	7	39	77	13
10	Basic	26	0	70	20	0	20	0	40	70	14
10.5*	Enhanced	20	25*	60*	0	10	0	20	25	90	15
10.5	Enhanced	21	25	60	0	10	8	19	33	89	16
10.5	Enhanced	20	25	60	0	10	12	18	37	88	17
10.5	Enhanced	21	25	60	0	10	15	16	40	86	18
10.5	Enhanced	22	25	60	0	10	17	12	42	82	19
10.5	Enhanced	23	25	60	0	10	19	7	44	77	20
10.5	Enhanced	24	25	60	0	10	20	0	45	70	21
10.5	Enhanced	25	25	60	20	0	0	20	45	80	22
10.5	Enhanced	26	25	60	20	0	8	19	53	79	23
10.5	Basic	20	25	60	20	0	12	18	57	78	24
10.5	Basic	21	25	60	20	0	15	16	60	76	25
10.5	Basic	22	25	60	20	0	17	12	62	72	26
10.5	Basic	23	25	60	20	0	19	7	64	67	27
10.5	Basic	24	25	60	20	0	20	0	65	60	28
10.5	Basic	25	25	60	0	10	0	20	25	90	29
10.5	Basic	26	25	60	0	10	8	19	33	89	30
11*	Enhanced	20	40*	45*	0	10	12	18	52	73	31
11	Enhanced	21	40	45	0	10	15	16	55	71	32
11	Enhanced	24	40	45	0	10	17	12	57	67	33
11	Enhanced	25	40	45	0	10	19	7	59	62	34
11	Enhanced	26	40	45	0	10	20	0	60	55	35
11	Basic	20	40	45	20	0	0	20	60	65	36
11	Basic	21	40	45	20	0	8	19	68	64	37
11	Basic	22	40	45	20	0	12	18	72	63	38
11	Basic	23	40	45	20	0	15	16	75	61	39
11	Basic	24	40	45	20	0	17	12	77	57	40
11	Basic	25	40	45	20	0	19	7	79	52	41
11	Basic	26	40	45	20	0	20	0	80	45	42
11.5*	Enhanced	20	55*	25*	0	10	0	20	55	55	43
11.5	Enhanced	21	55	25	0	10	8	19	63	54	44
11.5	Enhanced	22	55	25	0	10	12	18	67	53	45

Table 14.2　(continued)

A	B	C	D	E	F	G	H	I	J	K	L
			Price		Design		Completion		Totals		Contract
Price	Design	Completion	N	A	N	A	N	A	N	A	no.
11.5	Enhanced	23	55	25	0	10	15	16	70	51	46
11.5	Enhanced	24	55	25	0	10	17	12	72	47	47
11.5	Enhanced	25	55	25	0	10	19	7	74	42	48
11.5	Enhanced	26	55	25	0	10	20	0	75	35	49
11.5	Basic	20	55	25	20	0	0	20	75	45	50
11.5	Basic	21	55	25	20	0	8	19	83	44	51
11.5	Basic	22	55	25	20	0	12	18	87	43	52
11.5	Basic	23	55	25	20	0	15	16	90	41	53
11.5	Basic	24	55	25	20	0	17	12	92	37	54
11.5	Basic	25	55	25	20	0	19	7	94	32	55
11.5	Basic	26	55	25	20	0	20	0	95	25	56
12*	Enhanced	20	60*	0*	0	10	0	20	60	30	57
12	Enhanced	21	60	0	0	10	8	19	68	29	58
12	Enhanced	22	60	0	0	10	12	18	72	28	59
12	Enhanced	23	60	0	0	10	15	16	75	26	60
12	Enhanced	24	60	0	0	10	17	12	77	22	61
12	Enhanced	25	60	0	0	10	19	7	79	17	62
12	Enhanced	26	60	0	0	10	20	0	80	10	63
12	Basic	20	60	0	20	0	0	20	80	20	64
12	Basic	21	60	0	20	0	8	19	88	19	65
12	Basic	22	60	0	20	0	12	18	92	18	66
12	Basic	23	60	0	20	0	15	16	95	16	67
12	Basic	24	60	0	20	0	17	12	97	12	68
12	Basic	25	60	0	20	0	19	7	99	7	69
12	Basic	26	60	0	20	0	20	0	100	0	70

Note: N = Nelson; A = Amstore. Asterisks indicate direct entry.

Generation and Evaluation of All Contracts

Graph of all joint evaluations. In Figure 14.1 we plot the joint evaluations (columns J and K of Table 14.2) of the seventy possible contracts. From the graph we can identify those contracts on the efficiency frontier and identify those that are extreme-efficient contracts.

It is also possible to find the extreme points by sorting columns J, K, and L in Table 14.2. This is shown in Table 14.3 and plotted in Figure 14.2. The efficient contracts are starred in Table 14.3. They are shown separately in Table 14.4 and are plotted in Figure 14.2.

Extreme-efficient contracts. Besides plotting the entire efficient frontier we can, by visual inspection, identify those points that are extreme efficient.

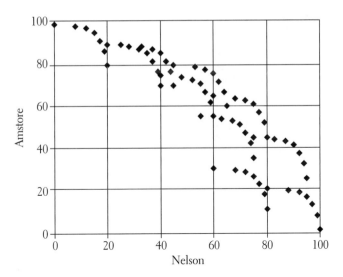

Figure 14.1.   Plot of all contracts.

Table 14.3   Graph of all contracts sorted (asterisk indicates efficient contract)

Contract number	Totals sorted	
	Nelson	Amstore
1	0*	100*
2	8*	99*
3	12*	98*
4	15*	96*
5	17*	92*
6	19	87
7	20	80
8	20	90
15	25*	90*
9	28	89
10	32	88
16	33*	89*
11	35	86
12	37	82
17	37*	88*
13	39	77
14	40	70
29	40	75
18	40*	86*
19	42*	82*
20	44	77
21	45	70
22	45*	80*
30	48	74

Table 14.3 (continued)

Contract number	Totals sorted	
	Nelson	Amstore
31	52	73
23	53*	79*
43	55	55
32	55	71
33	57	67
24	57*	78*
34	59	62
57	60	30
35	60	55
36	60	65
25	60*	76*
26	62*	72*
44	63	54
27	64*	67*
28	65	60
45	67	53
58	68	29
37	68*	64*
46	70	51
59	72	28
47	72	47
38	72*	63*
48	74	42
60	75	26
49	75	35
50	75	45
39	75*	61*
61	77	22
40	77*	57*
62	79	17
41	79*	52*
63	80	10
64	80	20
42	80*	45*
51	83*	44*
52	87*	43*
65	88	19
53	90*	41*
66	92	18
54	92*	37*
55	94*	32*
67	95	16
56	95*	25*
68	97*	12*
69	99*	7*
70	100*	0*

Table 14.4   Listing of all efficient contracts

Contract number	Nelson	Amstore
1	0	100
2	8	99
3	12	98
4	15	96
5	17	92
15	25	90
16	33	89
17	37	88
18	40	86
19	42	82
22	45	80
23	53	79
24	57	78
25	60	76
27	64	67
37	68	64
38	72	63
39	75	61
40	77	57
41	79	52
42	80	45
51	83	44
52	87	43
53	90	41
54	92	37
55	94	32
56	95	25
68	97	12
69	99	7
70	100	0

Contract 25, with coordinates (60, 76), is an extreme-efficient point, whereas contract 22, with coordinates (45, 80), is not, since it falls in the valley along the ridge. Remember that if we had millions of contracts, it would not be possible to exhibit all of them. But there is a way to find the extreme-efficient points analytically.

Decomposition of Large Problems

We would like to make an important aside here. Take the case in which there are many issues and many resolutions so that there are millions of contracts. Analysis often proceeds by decomposing the problem and working simulta-

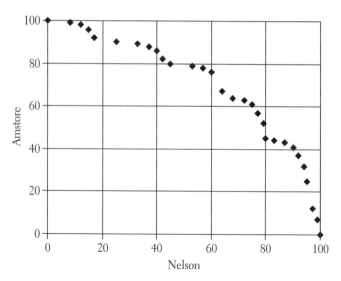

Figure 14.2.   Plot of all efficient contracts.

neously on the separate parts without making a decision on any one part. Thus, for example, if there were twenty issues, including a grouping of the three issues that had the structure of the Nelson versus Amstore problem, a subcommittee might make an analysis of the seventy contracts for those three issues and identify those ten that were efficient. The report might also single out three or four of these that span the concerns of both parties. This activity places a lesser premium on claiming. Collaborators from both sides engage in simplifying the vast number of tradeoffs that will ultimately have to be made. Incidentally, when the problem is decomposed in this fashion, it is not necessary to do the full scoring initially. It suffices to first score each decomposed part separately. Indeed the parties may be willing to work under FOTE conditions for some of the decomposed subparts of the problem, but not at the last stages, when the final compromises are made.

Finding Extreme-Efficient Contracts

Use of C-weights.   In the previous chapter we showed how to obtain extreme-efficient points by finding the contract that maximized AAA's score plus a constant times BBB's score. We imagined a ruler sliding up and down until it rested on an extreme point. You manipulate the slope of the ruler by changing the constant. We now show how this is done in Table 14.5.

Columns A to D in Table 14.5 present the scored template for the Nelson versus Amstore negotiation. We use the constant C = 2.35 in this illustration. (We could have used any non-negative number.) Column E displays Nelson's values plus the constant, 2.35, times Amstore's values, which we symbolically la-

Table 14.5   Use of C-weights to find efficient frontier

A	B	C	D	E	F
				C = 2.35	
		Payoffs			
Issue	Level	Nelson	Amstore	N + C*A	Contract
Price	10	0	70	164.5	0
($000,000s)	10.5	25	60	166	1
	11	40	45	145.75	0
	11.5	55	25	113.75	0
	12	60	0	60	0
Design	Enhanced	0	10	23.5	1
	Basic	20	0	20	0
Completion	20	0	20	47	0
(months)	21	8	19	52.65	0
	22	12	18	54.3	1
	23	15	16	52.6	0
	24	17	12	45.2	0
	25	19	7	35.45	0
	26	20	0	20	0
Efficient joint value		37	88		

> These entries use SOLVER'S SUMPRODUCT command. It multiplies col. C × col. F and col. D × col. F.

bel N + C*A. The second entry in this column is 166, which equals 25 + 2.35 × 60. Of the first five entries in column E, the maximum is 166, and to maximize the sum of N + C*A, for the *price* issue, we choose level $10.5 million and put a 1 next to that resolution level and 0's for the other 4 levels. Similarly the optimal level under *design* is enhanced, and under *completion date* is 22 months; and accordingly those levels are given the entry of 1 in column F. Hence column F depicts the optimal contract for C = 2.35. As before, we now assert that the contract exhibited in column F yields an extreme-efficient point. To find how this contract scores, we get Nelson's score of 37 by taking the sumproduct of columns C and F. Since the entries in column F are 0's and 1's, this boils down to adding up Nelson's scores corresponding to the placement of 1's in column F—for example, 37 = 25 + 0 + 12. A similar story holds for Amstore's accumulated score of 88 for the contract. You can check the figures in Table 14.4 to see that (37, 88) is indeed an extreme-efficient point.

The beauty of the spreadsheet is that once the sheet is set up, we merely have to type in a new number for the constant C—say, change it from 2.35 to 1.83, and the numbers change almost instantaneously in columns E and F, giving us a new extreme-efficient point. This, as we said before, is akin to changing the slope of the ruler. Of course, if the slope of the ruler is changed slightly—as, for example, from 2.35 to 2.38—the numbers in column E will change but the maxima will occur at the same levels and the placement of the 1's in column F will remain the same.

We have now shown how to find extreme-efficient points by using C-weights, but in the analysis of the fair-division problem we had an essentially simple way for finding all the extreme-efficient points by sorting and dealing from the top. The next section is designed to show how this procedure generalizes to the present case, where some issues have more than two levels.

Incremental Analysis

A little algebra and a few symbols will go a long way. Consider the case where AAA and BBB must jointly determine which of n levels of resolution they should adopt for issue XYZ. For level i, denoted by L_i, party AAA's and BBB's scores will be denoted by A_i and B_i respectively. Let's assume the scores increase for AAA and decrease for BBB as we proceed down the list.

Suppose the parties wish to find the level that maximizes AAA's score plus C times BBB's score. We know why this is of prime concern. In standard incremental analysis, they should move from L_i to L_{i+1} if and only if

$$A_{i+1} + C \times B_{i+1} \geq A_i + C \times B_i$$

or if

$$\frac{A_{i+1} - A_i}{B_i - B_{i+1}} \geq C. \tag{14.1}$$

The left-hand side of expression (14.1) is called the "critical ratio," going from level L_i to L_{i+1}. The numerator is the *increment* to AAA in going from L_i to L_{i+1}; the denominator is the *decrement* to BBB for going from L_i to L_{i+1}. The ratio can be interpreted as the increase to AAA per unit of decrease to BBB. If the critical ratio is larger than C, then we get an improvement (given our objective of maximizing AAA's score plus C times BBB's) by going from L_i to L_{i+1}.

Numerical illustration. Table 14.6 and Figure 14.3 show how we put this algebra to use with concrete numbers. Columns E and F in the table display the increment to AAA and decrement to BBB as we move down the list. The critical ratio in column G is the ratio of the increment to the decrement. When we plot the joint scores in the figure, we can see that the critical ratios show up as the slopes of the line segments connecting these extreme-efficient points.

Table 14.6   Maximize VAL(A) + C* VAL(B)

A	B	C	D	E	F	G	H	I
							Range	
				Increment	Decrement	Critical		
Issue	Resolution	AAA	BBB	to AAA	to BBB	ratio	From	To
XYZ	L1	0	30				2	Infinity
	L2	10	25	10	5	2	0.714	2
	L3	15	18	5	7	0.714	0.375	0.714
	L4	18	10	3	8	0.375	0.2	0.375
	L5	20	0	2	10	0.2	0	0.2

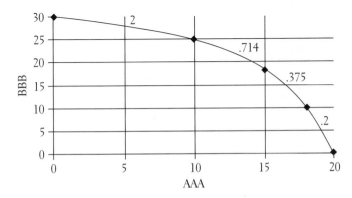

Figure 14.3.   Plot of results from Maximize VAL(A) + C*VAL(B).

Suppose now that the constant C is greater than 2. We shouldn't move from L_1 to L_2, since the critical ratio for this move is not greater than C. So for any C-value between 2 and infinity, level 1 is optimal; for C-values between .714 and 2, level 2 is optimal; and so on. In this numerical example, the exhibited points were all chosen to be on the extreme-efficient frontier. When this is the case, the critical ratios monotonically decrease as we go down the list.

Now we're ready to show how we can generalize the "dealing from the top" procedure that we developed for the fair-division problem—that is, for the special case where there were exactly two resolutions for each issue. To show how this works, we turn back to the Nelson versus Amstore problem.

Analysis by Critical Ratios

Tables 14.7a and 14.7b show how we can get the entire extreme-efficient frontier by "dealing off the top." The data of the scored template are given in columns A to D of Table 14.7a.

Table 14.7a   Nelson Contracting: Analysis by critical ratios

A	B	C	D	E	F	G
Issue	Level	Nelson	Amstore	Increment Nelson	Decrement Amstore	Critical ratio
Price	10	0	70			
Price	10.5	25	60	25	10	2.5
Price	11	40	45	15	15	1
Price	11.5	55	25	15	20	0.75
Price	12	60	0	5	25	0.2
Design	Enhanced	0	10			
Design	Basic	20	0	20	10	2
Completion	20	0	20			
Completion	21	8	19	8	1	8
Completion	22	12	18	4	1	4
Completion	23	15	16	3	2	1.5
Completion	24	17	12	2	4	0.5
Completion	25	19	7	2	5	0.4
Completion	26	20	0	1	7	0.14

Column E in Table 14.7a shows the increments to Nelson as we move down the levels for each issue; column F gives the decrements for Amstore; column G gives the critical ratios—the ratios of increments and decrements.

Table 14.7a is copied into the first 7 columns of Table 14.7b, but now the rows are rearranged in descending order of the critical ratios. This will be the order for dealing off the top. The way this is done is shown in columns H to L.

We start off by giving everything to Amstore, Nelson getting 0 and Amstore getting 100. See cells H3 and I3. This contract involves a price of $10 million, enhanced design, and a completion date of 20 months. We proceed by changing the completion date from 20 to 21 months—yielding Nelson an increment of 8 and Amstore a decrement of 1. We started this way because this gives the highest critical ratio of 8. Next we give more to Nelson; Nelson gets another month, going from 21 to 22, yielding Nelson an increment of 4 and Amstore a decrement of 1 for a critical ratio of 4. You should now study the boxed entries in columns H to L.

That completes the story—a satisfying one. It should now be self-evident how this analysis extends to any number of issues and any number of resolutions. A word of caution, however. The analysis depends on first purifying each issue by removing all nonextreme levels of resolution for that issue and ordering the levels for each issue so that the critical ratios decrease for each issue. Once this is done we interleave the orderings using all the critical ratios. Ties are a bit bothersome, but not enough to warrant any special attention.

Figure 14.4 plots the extreme-efficient points of columns H and I of Table 14.7b. This completes what we have to say about two-party efficiency. Onward to BATNA and equity analysis.

Table 14.7b Extreme-efficient contracts: Nelson Contracting—analysis by critical ratios

	A	B	C	D	E	F	G
1					Increment	Decrement	Critical
2	Issue	Level	Nelson	Amstore	Nelson	Amstore	ratio
3							
4	Completion	21	8	19	8	1	8
5	Completion	22	12	18	4	1	4
6	Price	10.5	25	60	25	10	2.5
7	Design	Basic	20	0	20	10	2
8	Completion	23	15	16	3	2	1.5
9	Price	11	40	45	15	15	1
10	Price	11.5	55	25	15	20	0.75
11	Completion	24	17	12	2	4	0.5
12	Completion	25	19	7	2	5	0.4
13	Price	12	60	0	5	25	0.2
14	Completion	26	20	0	1	7	0.14
15	Price	10	0	70			
16	Design	Enhanced	0	10			
17	Completion	20	0	20			

H	I	J	K	L
Extreme-eff. values		Extreme-eff. contracts		
Nelson	Amstore	Price	Design	Compl.
0	100	10	Enhanced	20
8	99			21
12	98	10.5	Basic	22
37	88			
57	78			23
60	76			
75	61	11		
90	41	11.5		
92	37			24
94	32			25
99	7	12		
100	0			26

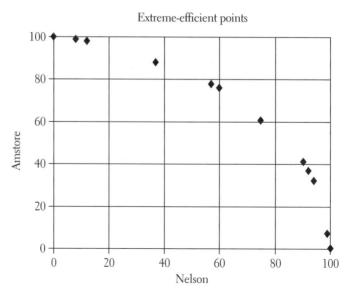

Figure 14.4.   Plot of extreme-efficient points.

BATNAs and Reservation Values

If negotiators AAA and BBB have each scored their commonly agreed-upon template, then it is easy, in principle, to divide contracts into those that are acceptable and those that are not. In the currency of the scoring system, each protagonist must decide on a reservation value associated with its BATNA. Not an easy task, but not impossible.

In Figure 14.5 we assume values for AAA's and BBB's reservation values and designate the point of the joint reservation value. Any contract to the northeast of the joint reservation value will be said to be *feasible*. In the diagram, we illustrate the case in which feasible contracts exist. If the joint reservation value fell to the northeast of the efficient frontier, the feasible region would be void—or in the parlance of mathematics, it would be an empty set. In actual negotiations the parties may never discover a point in the feasible region, even though this region exists; other times the region may not exist, and each negotiator may mistakenly think it the fault of the other that prevented the successful search for a solution.

Maximum feasible values.   A few observations and definitions: first, the efficient frontier can be explored without identifying reservation values; second, the parties may choose to do what we call efficiency analysis, using a FOTE approach, but draw the line at being trustworthy when it comes to identification of reservation values. Suppose the negotiators choose to do full FOTE analysis (including reservation values) with the assistance of an analytical facilitator, who gathers confidential information from both. The facilitator can then draw a di-

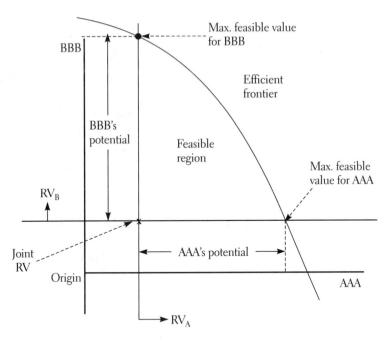

Figure 14.5. Feasibility and efficiency.

agram as shown in Figure 14.5. In the figure, we draw your attention to the point on the lower right labeled *maximum feasible value for AAA*. That's the point that maximizes AAA's score, subject to the constraint that it remains feasible to BBB. The value that is the maximum feasible for AAA can be ascertained by linear programming using SOLVER.

Analogously, SOLVER can find the maximum feasible value (MFV) for BBB. The interval from the reservation value for AAA to the maximum feasible value for AAA will be called the *potential* for AAA. A similar definition holds for BBB. The potentials will be used in the latter part of this chapter dealing with equity.

If the template is not scored, life becomes more complicated. It is no longer possible to talk about a numerical reservation value, and the diagram in Figure 14.5 becomes only suggestive rather than operational. In practice, without a template the parties never clearly identify what is feasible for them. They usually negotiate, perhaps keeping their BATNAs in mind, and when confronted with a possible compromise solution, they intuitively judge whether the contract is acceptable. Without a scored template, it is terribly difficult to systematically explore the feasible region, except, perhaps, on a point-by-point basis using holistic, intuitive judgments. The power of a solution algorithm such as SOLVER is lost. Consequently feasible points may not be found when they exist; and feasible solutions, when found, may not be efficient or equitable. Still to come is the incorporation into our discussion of AAA's and BBB's potentials.

Monetary Scoring of the Template with Side Payments

The analysis is much simplified when the template is scored by each side in monetary units and where monetary transfers are possible. In this case the parties should jointly maximize the sum of their scores (now in monetary units) and then use a monetary transfer to achieve equity. Furthermore, in calculating the overall sum of their scores, they need only maximize the sum of their scores for each issue in the template. That's as simple as it can get.

In this special case, once the template is scored, the analysis breaks into two parts: first, create the biggest pie by maximizing the sum of monetary payoffs, and second, seek equity by a final transfer payment. In monetary units the efficient frontier of Table 14.5 becomes just one straight line (see Figure 14.6). In comparing points X, Y, and Z, the point that maximizes the joint payoff is at X, and an efficient, equitable point such as W is achieved by a transfer payment from BBB to AAA.

Analysis with SOLVER

(*Note:* The reader who wants to get on with our story without the full mathematical development can skip the remainder of this chapter. If you want to read the gory details, however, which are really quite elegant, you will hobble yourself if you haven't previously struggled with the latter part of the last chapter.)

The Programming Problem

We now return to the negotiation between Nelson and Amstore, assuming that Full, Open, Truthful Exchange has taken place. Let's assume that each party has

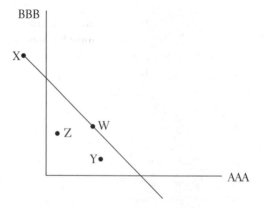

Figure 14.6.   Efficient frontier with monetary transfers.

done its BATNA analysis and has scored reservation values in the currency of the scored template. Nelson's reservation value is 60 and Amstore's is 20.

The scored template is shown in columns A, B, C and D in Table 14.8. Ignore for the time being the entries in the lower left (C19 and D19). Now turn your attention to entries E4:E17, which consists of 0's and 1's. This is an illustrative contract that is interpreted as using resolution \$10.5 million for price, basic resolution for design, and 24 months for completion time. The entries E4:E17 are the variables that SOLVER will manipulate subject to the constraints we're about to impose in order to maximize an objective function that is designed to capture our conception of fairness.

We require that the entries in E4:E17 are non-negative. Furthermore, the sum of the five entries E4 to E8, exhibited in F8, must be 1; similarly the sum of the two entries E9 and E10, exhibited in F10, must be 1; and finally the sum of the 7 entries from E11 to E17, exhibited in F17, must also sum to 1. We could— but for the time being we don't—require entries in E4:E17 to be integers. So at this point it is possible, as far as SOLVER is concerned, that some of the entries in E4:E17 may be real numbers between 0 and 1. Thus, if E5, say, were .72, then this would mean that the contract would select resolution \$10.5 million for price with probability .72. Any contract with just 0's and 1's will be termed a *pure* contract in contrast to contracts that use randomization, which will be termed *mixed* contracts.

Legitimacy and feasibility. A set of entries in E4:E17 that satisfy our constraints will be called a *legitimate contract*. Our aim is to get SOLVER to find a legitimate contract that will do something else for us. So let's push on. For any set of *legitimate* contracts in E4:E17, we get the spreadsheet to exhibit the values of that contract to Nelson and Amstore in cells C19 and D19 respectively. C19 is the sumproduct of Column C and E; and D19 the sumproduct of Column D and E. So for the exhibited contract, Nelson's score is 62 and Amstore's 72.

Now look at the group of entries G3:J12. Entries I5 and J5 just repeat the scores for the exhibited contract. In the next row, the reservation values 60 and 20 are exhibited and the excesses calculated in entries I7 and J7. For a legitimate contract to be *feasible*, these excesses must be non-negative.

Maximum feasible contracts. Now we come to the entries labeled "maximum feasible." As shown, the maximum value that Nelson can achieve that meets Amstore's reservation value is 96.3. How did we arrive at that number? We asked SOLVER to manipulate the entries E4:E17 subject to the

legitimacy constraints
    E4:E17>=0; +sum(E4:E8)=1; +sum(E9:E10)=1; +sum(E11:E17)=1,
and the *feasibility constraint*
    I7:J7>=0

Table 14.8   Analysis with SOLVER

		Payoffs			
Issue	Level	Nelson	Amstore	Contract	Sum
				Contract	
Price	10	0	70	0	0
($000s)	10.5	25	60	1	
	11	40	45	0	
	11.5	55	25	0	
	12	60	0	0	1
Design	Enhanced	0	10	0	
	Basic	20	0	1	1
Completion	20	0	20	0	
	21	8	19	0	
	22	12	18	0	
	23	15	16	0	
	24	17	12	1	
	25	19	7	0	
	26	20	0	0	1
Value of contract		62	72		

		Nelson	Amstore	Min. POP
Contract values		62	72	
Reservation values		60	20	
Excess values		2	52	
Maximum feasible		96.3	75.9	
Prop. of potential		0.06	0.93	
Sum		134.00		
Product		104.00		
Min. POP		0.06		
Max		Nelson	Amstore	
Sum	M	67	68.9	0.196
Min	M	80.7	51.9	0.571
Min	P	79	52	0.52
Prod	M	82.9	50.4	0.544

that maximize the entry in cell C19. SOLVER recognizes this as a linear programming problem, and in a matter of seconds it gives us the solution, indicating that Nelson can get a maximum of 96.3 points. Analogously, SOLVER tells us, when suitably queried, that Amstore could achieve a maximum of 75.9.

The next line (I9:J9) gives us the *proportion of the potential (POP)* values for Nelson and Amstore for the contract displayed. Nelson's POP of .06 is calculated from his excess of value of 2 with this contract, divided by the potential of 36.3, which is the maximum feasible (96.3) minus the reservation value (60). Making the same calculation, we find that Amstore's POP turns out to be .93. The minimum of these two POPs is .06 and shown in cell I12.

Using SOLVER to Find Equitable Contracts

Using Table 14.8, we are now ready to request SOLVER's help in finding the contract that would maximize one of the following three criteria: (1) *maximize the sum* of Nelson's and Amstore's scores—the entry in cell I10; or (2) *maximize the product* of Nelson's and Amstore's excesses—the entry in cell I11, or; (3) *maximize the minimum of Nelson's and Amstore's POPs*—the entry in cell I12.

Let's illustrate finding the maximin POP solution, that is, maximizing cell I12. We call up SOLVER and instruct it as follows: choose the variables in E4:E17 subject to the legitimacy constraints (that is, the variables must be non-negative and the sum of entries for each issue must be 1) and subject to the feasibility constraints (that is, the two excesses must be non-negative) that maximize the entry in I12 (that is, maximize the minimum POP). It turns out that the solution is a mixed strategy, labeled M, that yields a common POP of .571, which attests to the modestly bowed nature of the efficient frontier. If, in Figure 14.5, the efficient frontier were a straight line connecting the two maximum feasible points, then the contract that would maximize the minimum of the POPs would yield equal POP values of .50. With a bowed efficient frontier, each party can get more than half of its potential. We next ask SOLVER to solve the same problem using a pure (P) strategy by adding integer constraints—E4:E17=integer. Now in pure strategies the maximum of the minimum POP is .544, and in this case the POP values are not equal.

To find the Nash solution we ask for the contract that is legitimate and feasible that maximizes the entry in cell I11, which we programmed to be the product of the excesses. A comparison of these solutions is exhibited in Table 14.8 in entries G14:K19.

This completes the solution of the problem under the assumption of FOTE negotiations with an analytic helper. To fully understand how SOLVER works, you'll have to get your hands a bit dirty and struggle some with the program. The procedure is quite general and can be easily employed in negotiation tasks with more issues and resolutions. The scored template does *not* have to be processed so as to be in so-called canonical form—rather, for any issue the levels of resolution can be in arbitrary order. In some sense the mathematical programming analysis takes less care than the analysis exhibited in Table 14.7b. But the real power of the programming solution, of course, comes from its generalizability to more than two negotiating parties, and we demonstrate this later on.

Core Concepts

This chapter has considered negotiations involving a jointly scorable template (with many issues, some with more than two resolutions) with BATNAs and RVs to consider. We continued examining our polar case of FOTE-like negotiations and showed how the analysis of the special fair-division problem of the preceding chapter generalizes. Once again we characterized and showed how to find and exhibit all the extreme-efficient points, and once again we used SOLVER to help us—but now to more advantage.

A contract is said to be *feasible* if it is acceptable to each party—that is, its joint evaluation lies northeasterly of the joint reservation value. For any contract, its *excess* value to either party is the difference between its score and its RV. Thus feasibility requires non-negative excesses.

We showed how SOLVER can be used to find the *maximum feasible value (MFV)* for each party, and then we defined each party's *potential* as the difference between its MFV and its RV. For each party, and for each feasible contract, we defined its *proportion of potential (POP)* as its excess divided by its potential. For equity analysis, we continued to focus on the three standard alternatives: (a) maximizing the sum of excesses; (b) maximizing the product of excesses; and (c) maximizing the minimum proportion of potentials (or equivalently, maximizing the common POP).

Whereas this chapter has had a jointly normative flavor, the next one will be mostly jointly descriptive and the following one asymmetrically prescriptive/descriptive.

15

Behavioral Realities

Time for a change of pace. Instead of theorizing about how a couple of idealists might negotiate in a FOTE manner, let's examine how real negotiators behave. We'll put "real" in quotes, since we are talking about professional students, primarily MBAs, early in a course in negotiations playing simulated laboratory exercises.

One might suspect that the results would be unrepresentative. After all, the cases were not part of the students' lives, and so perhaps they were not as invested in the outcomes. And the atmosphere of a school may have encouraged more cooperative behavior than the actual marketplace. So shouldn't the outcomes be quite rosy? Yet the results were not so encouraging, and there were reasons to think that they were not so unrealistic after all. The students were real people, who were raised by the same sorts of families and grew up in the same cultures as negotiators out in the real world. These MBA candidates had incentives that paralleled real life—some material benefit (grades were based partly on results) and mostly the same emotional and psychological motives to earn others' (and indeed, their own) respect. Moreover, the culture of the business school is one that prizes displays of toughness and cleverness—students are motivated to protect their social position at least as well by being tough as by being nice—which may also correspond to the real world.

We think that the results from MBA students would be replicated in a survey of how business people and other professionals behave—if a researcher could persuade them to allow their behavior to be fully observed. Thus the student results will provide us with a useful body of data and an abundance of anecdotes substantiating our main conclusions.

Most negotiators do not prepare in anywhere near the detail we proposed in Chapters 11 and 12. How much preparation is undertaken depends on the magnitude of an agreement's consequences, on negotiators' experience and expertise, on cultural traits, and so forth. But as a general rule, negotiators do not prepare adequately, alone or together. They do not do their homework. They do not examine their fundamental interests or seriously explore their alternatives to negotiation. They don't brainstorm together, and they often start with positional bargaining, initiating claiming tactics far too early on and locking out possible joint gains. Rarely do they jointly decide on the issues they need to address and on the shape of a template to guide further negotiations. Most negotiators are not clear about their tradeoffs, their BATNAs, or their reservation values.

By the same token, real negotiations do not follow the FOTE approach set out in Chapters 13 and 14. Negotiators dissemble and spend more time on cutting up the pie than on baking a larger pie to share. Moves to claim value often chase out moves to create value. Without realizing it, negotiators leave potential gains on the table and fall into innumerable psychological traps. After completing a negotiation, most negotiators think that they have performed well and blame the other side for difficulties encountered along the way.

Laboratory and Classroom Simulations

Scorable Negotiations

Numerous simulated negotiations are conducted in classrooms and academic laboratories using role-playing exercises. Most instructors who teach negotiations, including ourselves, use "scorable games"—not exclusively but often enough to provide us with a good deal of data. In scorable games, student-subjects play roles in specially designed negotiations. The case of AMPO versus City, described briefly in Chapter 11, is one such game.

The stage is set by supplying all the subjects with common information about the problem to be resolved by negotiations. In most such simulations, it is assumed that the type of preparation we outlined in Chapters 10 and 11 has been done. The subjects are presented with a template listing the issues to be decided and possible resolution levels for each issue. Each role player is given confidential material that unambiguously and in quantitative terms scores the template and specifies reservation values for that subject's role. It is presumed that each role player enters the negotiation after a good deal of preparation. Each role player thus acts as an agent to represent his or her side in final negotiations with complete, unambiguous instructions about payoffs and reservation values. In addition, some rather vague information is given to each side about the other side's interests.

The rules of conduct of the game specify that the written confidential information cannot be shown to the other side. A player might choose to verbally paraphrase his or her instructions, but questions about the completeness or veracity of such communication are always a factor. Subjects are encouraged to play the game as if it were for real. By and large, these simulations are instructive and enjoyable.

The subjects are graded on each negotiation simulation by how well they do in their role in comparison with all the others playing a similar role. This is akin to duplicate bridge scoring. Each student, on each game, converts his or her raw score into a z-score that records the number of standard deviations the raw score falls from the mean. Student-subjects keep a cumulative tally on the average of their individual z-scores and normalize this again into a so-called z of z's.

An aside: in the early 1980s my students talked me (HR) into basing a substantial part (one-third) of their grade on how well they did on the totality of simulated games they played. It was shocking just how desperately they cared about their results. One student, let me call him X, playing against Y, excused himself to go to the bathroom and left behind his written confidential instructions. Y couldn't resist noticing X's confidential RV and took advantage of X. Or so Y thought. Y did not know that X doctored up his RV on his instructions and put in a false number. Y lost mightily and X was triumphant. Y complained to the instructor of the unfair practices of X. The instructor docked points from both.

So let's imagine that there is a class of fifty students playing a simulated two-party negotiation between AAA and BBB. Each student is assigned an AAA or BBB role, and each is given common general information and the appropriate confidential information. The subjects are given time to assimilate the written material and are encouraged to think about the tactics they will use in the up-coming negotiations. For pedagogical purposes, the subjects are told to keep to the limitations of the template—they are not to propose new issues or resolution levels. Obviously that's not something we would want to encourage in real negotiations.

Debriefing

So let's suppose that twenty-five negotiation pairs have completed their exercise and filled out forms describing what happened and some of their reactions. The class congregates for a postnegotiation debriefing.

Let the negotiating pairs be labeled 1, 2, . . . , 25. The members of Pair-17 volunteer the information that they successfully completed the negotiations and came to an agreement that was acceptable to each side. AAA-17 exclaims: "I was instructed to get at least 40 points and ended up with 52, so I'm pleased with the result." And BBB-17 adds: "I ended up with 46 points, but I was only required to get 35, so I did pretty well too." They are eager to tell the class how they accomplished this feat. They feel that if this were a real-life situation they would happily report the outcome to their principals and could rightfully expect a bonus for their impressive achievement.

The instructor calls on other pairs and eventually posts the collected results of the class. Table 15.1 lists a set of typical results, and Figure 15.1 plots these outcome scores. For example, Pair-17, with a joint score of (52, 46), is shown in the two-dimensional graphical display of Figure 15.1.

Wide dispersal of results.   The thing that amazes us about this type of exercise is how varied the results are. Outcomes are scattered from no agreement, to grossly inequitable and inefficient agreements, to points on the efficient frontier. When we began accumulating data using scorable games, we naively thought that since the scoring instructions and reservation values were explicitly and numerically defined, the empirical contract solutions would be tightly distributed. This certainly was not the case. There is an important message here: final outcomes depend critically on the negotiation skills of the parties. It is not true that once the template is carved in stone and explicit quantitative scores and reservation values are provided to the negotiators that the ballgame is essentially over.

Table 15.1   Class results

Pair	AAA	BBB
1	44	58
2	51	40
3	53	56
4	40	35
5	56	57
6	42	60
7	60	36
8	71	33
9	45	48
10	40	35
11	50	60
12	60	43
13	70	38
14	62	55
15	43	46
16	48	52
17*	52*	46*
18	55	40
19	40	35
20	70	38
21	67	39
22	45	40
23	61	53
24	64	50
25	50	60

Note: Asterisk indicates pair shown in Figure 15.1.

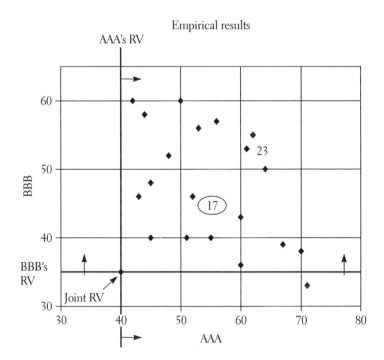

Figure 15.1.   Plot of class results.

Quite the contrary, it has only just begun! The sheer diversity of outcomes is one of the most surprising results.

Gains left on the table.   Before seeing the results of other pairs, the members of Pair-17 were happy with their joint outcome of (52, 46). When those in some other pair, say 23, announced their joint score of (61, 53), Pair-17 was initially incredulous. How did Pair-23 come up with those numbers? There must be a calculation error. No, Pair-23 agreed on quite a different contract from that of Pair-17. Gradually it dawned on the duo in Pair-17 that despite their initial enthusiasm and confidence, their contract had left significant potential joint gains on the table.

  The simple truth was that the members of Pair-23 dovetailed their differences in a manner to bake a bigger pie than did those in Pair-17. Pair-17's naive overconfidence had been revealed. They no longer felt that they deserved bonuses from their principals. They had managed to reach a *feasible solution* (that is, a solution above each side's reservation value), but it was not an *efficient solution* in the sense that there were other contracts that would have been preferred by both parties. But let's not overlook a crucial point here: in the real world no one would know that Pair-17 had left joint gains on the table. In the real world, negotiators are evaluated primarily on their ability to achieve feasible agreements, and no one ever knows if potential gains are left on the table.

Let's look at the results of our simulated negotiation from another angle. Occasionally pairs in the laboratory come up with contracts that do not meet the feasibility test. Take, for example, Pair-8, with a joint payoff of (71, 33). "But you were instructed to get at least 35 points! What happened?" someone admonishes BBB-8. The unfortunate BBB-8 has three possible retorts: "What? Where does it say that?" And a little later: "Oh yes, you're right, I didn't see that." A more belligerent BBB-8 might answer, "I saw that I had to get 35 points, but it wasn't possible—the instructions were wrong." Or else, "We worked so hard in trying to find a solution and I felt sorry for my partner so I agreed to settle on 33." Imagine BBB-8's reaction when he or she finds out that AAA-8 had been dissimulating and exaggerating to claim the biggest possible piece of the pie. BBB-8 was taken for a ride. Let's look at this in a real-world context. BBB-8 will be in hot water with his or her principal for agreeing to a contract below the party's reservation value. AAA-8 would have been congratulated for surpassing his or her reservation value. This much is clear. What is less clear is whether the parties would fully realize how successful AAA-8 had been at claiming a large piece of the pie.

Notice that pairs 4, 10, and 19 did not come to an agreement, and therefore they are each scored at the joint reservation value of (40, 35). Pair-20's outcome was highly inequitable. AAA-20 soared above her 40-point hurdle, whereas the BBB-20 party barely squeaked by his. In such cases it may be a surprise to AAA that he or she did so well.

Negotiating an Agreement

Since outcomes vary so widely in the classroom or laboratory simulation, a typical debriefing now centers on how different pairs negotiated. What seems to work to both parties' advantage? Which tactics should be avoided? Without going into the nitty-gritty of detailed offer and counteroffer exchanges, it is instructive to examine the broad characteristics of how subjects behave. In doing so we will contrast two styles of negotiation tactics: (a) the dance of packages (in which complete contract proposals are placed on the table); and (b) joint construction of a compromise contract.

The dance of packages.   We don't often see the style of negotiation we are about to describe in the laboratory, but it occurs frequently in real-world negotiations. We call it the *dance of packages*. For convenience in the use of pronouns, we'll imagine that AAA (a woman) is negotiating with BBB (a man). In preparing for the final phase of give-and-take negotiations, AAA prepares a complete package that she finds appealing and hopes BBB will accept. AAA presents her proposal to BBB, accompanied by a short speech: "I thought hard about what a final agreement should be like. I thought of your interests as well as my own and what would be fair for both us. Here is my suggestion." The package is a contract with proposed resolution levels tentatively selected for each issue. If BBB is smart, he should not get anchored by AAA's initial offer and should put it to one side. The worst thing that BBB can do, in fact, is to go into the details of AAA's package and propose amendments to it. Such a response would be distinctly unwise.

Typically BBB will answer with a complete package of his own, accompanied by a preliminary injunction of, "I think this is a fairer suggestion that takes your interests into account." Now there are two packages on the table, and each side describes the merits of its own offer and criticizes the other. If there are many issues to be resolved, this is not the ideal way to seek creative compromises. In a slight variation of this procedure, AAA and BBB might not offer packages in sequence, and instead both of them might simultaneously put offers on the table.

In the laboratory setting in which both sides are given confidential scoring systems, we can plot, as is done in Figure 15.2, the joint evaluative scores of each proposed package. As one would expect, AAA's initial package might be evaluated at point A_1, which yields a wonderful score for AAA and an unacceptable score for BBB. The counteroffer from BBB might be scored at a point like B_1. Now let's suppose that points A_1 and B_1 fall outside the feasible range—the zone of possible agreement. After some discussion, AAA and BBB make counteroffers that are evaluated as points A_2 and B_2. (Remember that we have assumed the unrealistic perspective of an omniscient analyst. We have knowledge of the contents of Figure 15.2, but our two protagonists are groping around in the dark.)

Suppose the movement from A_1 to A_2 pierces the reservation value of BBB, as shown in the figure. If she queries, "Does my new offer reach the minimum that you can sign up to?" BBB would be utterly naive if he responded, "Indeed! That's more like it. I want you to make some more concessions but I can just about accept that last package." If he were so innocent and AAA wanted to claim as much as possible, he might end up with little more than his score at A_2, an inequitable result from our perspective. BBB would be better advised to reply to A_2 by saying, "Going from your first to second package is an improvement, but I'm hoping for more than that. How about this package instead?" and then putting forward a new offer, one that we-in-the-know score as B_2.

And so the dance of packages proceeds. Much of the time the dance may bounce back and forth within the feasibility domain. In other cases it may take several iterations to enter the feasibility domain, and the parties, desperately seeking a solution, might satisfice: sign up to the first feasible package in the zone of possible agreement and quit the search for more individual or joint gains. The dance may at times be disorderly: AAA may do all the work, proposing a succession of offers without any intervening offers from BBB. Or the path followed by AAA's packages (A_1, A_2, A_3) may seem to bounce around in a disorderly fashion, not always moving northwesterly. Let's not forget that without accurate information about BBB's tradeoffs and preferences, there is no guarantee

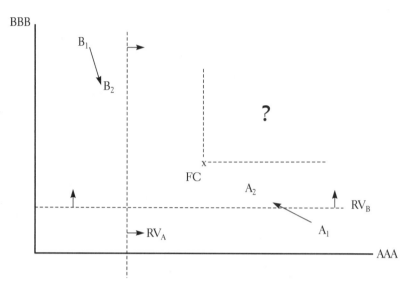

Figure 15.2.   The dance of packages.

that AAA's concessions—her moves from one resolution level to another that decrease the attractiveness of the package to her—will actually benefit BBB.

Now let's assume the dance produces a final accepted agreement at the point labeled FC for "final contract" in Figure 15.2 and that this point is in the feasible set of potential agreements. It could very well happen that this FC may be highly inequitable (from our omniscient vantage point), giving, for example, far more to one party than to the other. But AAA and BBB may remain blissfully ignorant of this fact. There may be points to the northeast of point FC, implying that the final agreement is inefficient and that different contracts would have better satisfied the interests of AAA and BBB. In this scenario, joint gains are left on the table. Once again, in the real world, the parties may be totally unaware that they have divided a small pie. What can be done about this? We'll offer some advice for exploiting unrealized joint gains in a later chapter when we talk about postsettlement settlements and contract embellishment.

Joint construction of a compromise contract.   Another common approach to reaching an agreement is to build up a package with a succession of compromises on individual issues or subgroups of issues.

Getting started. In order to get started, a pair of negotiators may select some issue and seek a compromise on that issue in isolation. More often than not they may have difficulty doing this, because on that issue alone preferences may be strictly opposing. When you think about it, using a temporary joint choice on a single issue as a way to begin is conceptually hard to do. How will our tentative agreement on issue-13, say, affect how we'll settle other issues?

But the negotiators may feel compelled to try something; so let's say they latch on to issue-7 and temporarily agree on some psychologically prominent focal point. For instance, if issue-7 can be resolved at three levels, low, medium, and high, the negotiating pairs feel a natural pull to agree on "medium." Furthermore it may be clear that one party will never settle on "low" and the other will never settle on "high," so in desperation, they may agree on "medium"—but only temporarily, to get started. *Nothing is settled until everything is settled* is the operative mantra. The temporary settlement may prompt one party to say, "Look, I'm tentatively willing to go along with our tentative agreement, but you should know that I think you owe me one later on."

Some astute pairs try to get started by discussing two issues for joint compromise: one issue is resolved in favor of each negotiator. When building up a contract, most subjects proceed tentatively: agreements made early on are not treated as irrevocable, but are reviewed later as the package evolves and gains in complexity. At times a negotiator may believe he or she is being swindled by the other side, but, as in real negotiations, there is a premium on maintaining momentum and getting results.

Some pairs give up trying to get definitive resolutions on single issues and opt for achieving compromises in the form of ranges. "On salary, can we agree tentatively—always tentatively—that the average salary level will be between 32 and 35 thou? If you agree, let's turn to vacation days for officers with 5 to 10 years of seniority."

Making trades. Another approach, which may be used at various points in a negotiation, is to make trades across issues. We can imagine a conversation between two hypothetical negotiators: Mr. A (representing AMPO) and Ms. C (representing City).

Table 15.2 shows an abbreviated scoring schedule depicting possible beneficial tradeoffs. Let's suppose that Mr. A and Ms. C. have already built up a compromise contract and are now looking for jointly beneficial improvements. In building up the contract, issue by issue, let's say they have tentatively agreed to a $600 increase in starting salary, to a five-day increase in vacation for officers with less than five years' service, and to a greater than 20 percent increase in two-man patrols (those entries are boxed in Table 15.2). Keep in mind that the information in Table 15.2 is not common knowledge and that most pairs in the laboratory—and certainly in real negotiations—would not be willing to disclose their confidential evaluations. Indeed, in the real world they may not have their own scorable preferences. Or perhaps in the laboratory the instructions to the negotiating agent may be purposefully imprecise.

After a few suggestions for joint improvement have been offered and rejected, Ms. C says, "How would you feel about increasing starting salaries from $600 to $700 and simultaneously lowering our agreed increase of vacation days from five days to three? This would be roughly equivalent for me [not quite

Table 15.2   Abbreviated scoring schedule for Mr. A and Ms. C

Issue	Level	City	Ampo
Increase in			
Starting salary	$\boxed{600}$	−4	700
	700	−8	850
	800	−16	1,000
Vacation	3	−6	60
	4	−10	85
	$\boxed{5}$	−15	110
Two-man patrols	Status quo	+15	−25
	$\boxed{\text{Greater than 20\%}}$	−5	25

Note: Box indicates tentative agreement.

true], but if it would offer advantages to you, I would go along with it, provided that you would then try to help me out." Actually, Ms. C gains a modest 5 points for this simultaneous change, whereas Mr. A gains 100 points. (Be aware that their scoring systems are not comparable.) "That's something of an improvement for me," he responds cautiously, "but how about going from $700 to $800 in starting salaries? Then I could possibly go back to the status quo on the two-man patrol cars."

Ms. C at first sounds very dubious, "Well, that's an awfully high starting salary. On the other hand, I do hate to see so many two-man patrols in these financially hard-up times. City Hall might not like this, but—all right, it's a deal." So she nets an increase of 12 points, and he an increase of 75 points. And the migration continues northeasterly.

Closing the contract. As the contract begins to take shape and many issues have been temporarily resolved, the negotiators begin to become increasingly aware of the need to clear the hurdle of their RVs. The time for claiming has arrived! The negotiations tend to become increasingly competitive and take on zero-sum overtones. The endgame can become particularly nasty if the parties begin to threaten, "If I don't get this, I'll walk." We have seen many negotiations conducted in a FOTE-like atmosphere turn sour at the endgame and slip from FOTE to POTE (Partial, Open, Truthful Exchange—the truth but not all the truth) to what we dub NOTE (No Open Truthful Exchange). It's particularly galling when one party keeps on playing FOTE and the other is convinced that his or her fellow negotiator is lying. Trust evaporates.

Behavioral Realities in Negotiations: Biases, Anomalies, and Errors

In Chapter 3 we showed that when real people make decisions, they do not always act in accord with the dictates of normative decision theory. Individual choices are affected by anchoring, the status quo trap, the sunk-cost bias, framing effects, overconfidence, and many other so-called deviations from rationality. We now broaden the scope of this laundry list to include biases and anomalies that occur when two or more individuals are involved, not just one.

We start by considering some behavioral realities in preparing for negotiations.[1]

The Zero-Sum Bias

There are zero-sum situations in which what you gain the other side loses and vice versa. In such situations minimizing your adversary's return is tantamount to maximizing your own. Negotiating the price of a commodity at a bazaar is one such illustration. The trouble is that often, in reality, the game is not strictly zero-sum and the players do not realize it. There are elements of potential cooperation and collaboration that go unrealized because the parties are concentrating exclusively on the competitive aspects of the problem. Players almost always destroy value or damage their relationship when they waste time and energy in protracted wrangling.

In a classroom setting, I assigned one hundred students into fifty A and B pairs and had them play a many-issue negotiation already processed for them into template form. Each side had its own confidential qualitative scoring system, and qualitative information about its BATNA. From the fifty A players, I then selected twenty-five at random and gave them supplemental information: quantitative information about tradeoffs, quantitative information about RVs, and more information about the value of B's preferences. So there were twenty-five A/B control pairs and twenty-five experimental pairs with A having privileged information. In the debriefing, before analyzing the data resulting from the negotiations, I asked the class what they would expect to see. There was general consensus that the A's with added information would do better. "And how about the B's?" I asked. "Those playing against privileged A's would do worse on average," came the answer. That was the consensus, even though in the previous class we had discussed zero-sum fallacies. The truth of the matter is that the privileged A's were in a better position to squeeze out joint gains, and their B partners did better, not worse, than the control group B's. This observation deserves reflection.

---

1. For readers with an appetite for more about biases, we suggest Max Bazerman and Margaret Neale's excellent book, *Negotiating Rationally* (1992). Much of the ground-breaking research into biases was done by Daniel Kahneman and Amos Tversky, although we warn you that their work is intended for specialists.

Social Utility

Individual negotiators evaluate the acceptability of their own outcomes relative to the outcomes obtained by the other negotiators. This socially influenced preference structure is called "social utility." Negotiators concerned with social utility will reject offers that represent significant gains for themselves on the grounds that the gains for the other side are disparately and unreasonably much larger. In a less-than-rational world, negotiators would rather have nothing at all than be subjected to unfair treatment and see a selfish negotiator profit excessively. "Darn it, I might be hurting myself, but I'll rot before I let him take advantage of me." This behavioral attitude feeds the zero-sum view of the world of negotiations.

But there is a flip side to all of this antisocial behavior: it's the affect of wanting to be fair, of empathizing with others, of incorporating concerns about others in your own utility calculations. As one of your basic objectives, you might not want to benefit tangibly at the expense of others.

In a classroom setting, I again experimented with one hundred student-subjects divided into fifty A and B pairs, playing this time a scorable game with quantitative RVs as well. Once again we chose twenty-five pairs as the control group, and for the other twenty-five pairs we modified their instructions as follows: your final score will be the sum of two components, (a) your "hedonistic" score (or your usual score from adding up the additive contributions of your scored template), and (b) your "empathetic" score (which is 10 percent of your adversary's hedonistic score). The empathetic component was introduced to build in a more collaborative atmosphere among the experimental subjects.

After computing the data, I calculated the following averages:

- $\bar{v}_A'$: The mean of the hedonistic component for the control group.
- $\bar{v}_A''$: The mean of the hedonistic component for the experimental group.

To the class's surprise, it turned out that

$$\bar{v}_A'' > \bar{v}_A';$$

that is, the experimental group, on average, did better than the control group, even counting only the hedonistic component of their payoff function. Why? The experimental group was more empathetic and less combative. Its members probably wasted less time and effort on confrontation. They may well have evoked a reciprocal goodwill. The message: even if you are solely interested in your own hedonistic outcome, you may be better off if you act in an empathetic, socially desirable fashion.

If V_A and V_B, respectively, represent A and B's value (hedonistic) functions (with comparable scales), and V_A^* and V_B^* their "social" value functions, then let

$$V_A^* = V_A + k_A V_B \quad \text{and} \quad V_B^* = V_B + k_B V_A,$$

where k_A and k_B can be manipulated to capture varied nuances of cooperation and competitive interpersonal attitudes. Each is instructed to maximize social value not knowing the other's k-values. Now consider cases such as:

a. $k_A = .1$, $k_B = .1$ (both cooperative)
b. $k_A = .1$, $k_B = 0$ (A cooperative and B neutral)
c. $k_A = .1$, $k_B = -.1$ (A cooperative and B competitive)

Reactive Devaluation

Proposals and offers suggested by the other side often tend to be devalued simply because they have been proposed by the other side. This is aptly called *reactive devaluation* by Lee Ross, who extensively examined this phenomenon. "If the other guy thinks this is a good idea, I'm suspicious. There must be some hidden jokers involved. I don't know what they are, but they are lurking somewhere."

As director of the International Institute for Applied Systems Analysis (IIASA), I was very conscious of the problem of reactive devaluation. In some domains, my actions as director of the institute needed a positive endorsement by the institute council, which convened about four times a year. The chairman of the council, Jerman Gvishiani, was a prominent figure in the Soviet hierarchy. At some council meetings he would adopt a very noncollaborative stance. This especially occurred when there was some flare-up of antagonistic relations between the United States and the Soviets (such as an incident at the Berlin Wall, or some botched spy attempt having nothing at all to do with the operations of IIASA). In the midst of such episodes I felt that I was (or would be) a victim of reactive devaluation. I therefore either refrained from making new proposals, waiting for a more propitious time, or engaged in a bit of creative manipulation: I would meet separately with a council member and in private discussions lead him to generate a variant of the proposal I wanted introduced. Coming from the council member of country XYZ, the proposal did not evoke the knee-jerk reactive devaluation that I wished to avoid. I'm sure that at times the Soviet chairman of the council engaged in similar tactics to avoid reactive devaluations of what he wanted the council to adopt. I hasten to add that at other times the spirit of true collaboration prevailed.

Negative Attribution

Alice and Jim are going nowhere in their vitriolic negotiations. Alice is convinced that Jim has taken certain noncooperative stances because he's just like that: mean, ornery. It's true that she has not acted in an exemplary fashion, either, but after all, she was forced to act that way by Jim.

There is a common tendency for negotiators to overemphasize *dispositional factors* (stable personal traits) when explaining and interpreting the behavior of the other side, while stressing *situational factors* to account for their own negative behavior. In a negotiation setting, this double standard in the judgment of behavior creates a perception that the conflictual elements of the negotiation are caused primarily by the other side's purposeful intransigence, to a lesser extent by situational factors affecting both sides, and least of all by the negotiator's own behavior. Still worse, when the opposing side's behavior is ambiguous, negotiators are all too willing to attribute the other's actions to malevolent ulterior motives.

In general, negotiators behave as if the other side is a fairly inactive participant in the process, without interests, concerns, incentives, desires, emotions, or expectations of their own. By ignoring the cognitions of the other negotiator, a vast swath of essential information is not incorporated into the negotiator's thinking. This lack of sophistication feeds the notion that there is no reason for the other party's behavior other than pure malice.

Because of their superficial analysis of the other side, negotiators often make faulty inferences about the motivation behind the other side's actions. These inferences have the effect of exaggerating the perception of conflict inherent in the negotiation. These biases interact with and reinforce each other. The zero-sum bias feeds fallacies about negative attributions, which, in turn, corroborate the party's zero-sum convictions. Poor inferences are also partly based on an erroneous belief that all parties have access to the same information, whereas most negotiations involve asymmetric information.

Not only do biases interact with and reinforce each other within the mind of one party, but the biases and anomalies reverberate back and forth from one party to the other.

In the early 1950s, following in the footsteps of Merrill Flood, then with the RAND Corporation, I had student-subjects play what is now known as the repeated prisoner's dilemma game, and after each trial, I asked them to keep a log of what they believed the other player was thinking about. The logs were instructive. Lots of negative attributions. It's always the other guy who is to blame, and we seem always to be able to rationalize our noncooperative behavior as being not aggressive but realistically defensive and not typical of us. Occasionally there would be a complete mismatch in the logs of competing players. For example, player A initially might not quite understand the rules of the game and is just thrashing around. B misinterprets A's actions, seeing them as devious, well-thought-out attacks, and endeavors to teach A a lesson; but then it dawns on A that B is really mean and so A lashes back. It could evolve that A gets positive satisfaction from hurting B. It's as if A's value function is of the form discussed above:

$$V_A^* = V_A + k_A V_B,$$

where k_A becomes increasingly negative with negative feedback.

Negative images of the other side often remain unchanged during a negotiation despite that side's conciliatory behavior. This social cognitive bias exists because positive actions are more ambiguous than negative ones; the effect of selective perception and selective memory is to create a vicious circle reconfirming negative images. In addition, attributional biases contribute to our view of behavior consistent with our (negative) view of the other side as being dispositional, and our interpretation of behavior discrepant with our negative view of the other side as being situational. In other words, when a negotiator has a persistent negative image of the other side, the opponents are perceived to make concessions only when forced to do so by circumstance, not by active choice. And when each is privy to the same bias, they solidify each other's faulty analyses.

I am purposefully painting a gloomy picture. This is how bad it can get. But the reverse can also be true: A makes an error that hurts himself and helps B. B mistakenly thinks that A is being nice and reciprocates the action. A cycle develops that benefits each, provided that each takes satisfaction in the well-being of the other. Here each is not acting in accord with his or her original value function but with the modified social function

$$V_A^* = V_A + k_A V_B \quad \text{and} \quad V_B^* = V_B + k_B V_A$$

where the k's are positive.

Sometimes people cooperate because they consciously think it is better for all to do so. Sometimes the cooperative behavior is sustained because the players repeatedly take actions that are good for the whole, even though they are not optimum for themselves. Sometimes they do not know that they are doing it; we all occasionally act without much analysis or hard thinking. So-called anomalies, errors, and biases do occasionally work in favor of society rather than against it.

Cultural Differences in Negotiations

Mathematicians talk regularly about things they refuse to or cannot define. In geometry one talks about points and lines but considers points as undefinables. We would like to talk about culture without stepping up to the plate and defining culture or what a cultural grouping might be. Our goal is not to talk about the comparative behavior of different cultural groups in general but, in this chapter, to describe how cultural differences might affect two-party negotiations and, in the next, to prescribe what our client might do about it. We do believe that cultural differences affect the way we negotiate.

By "culture" we *roughly* mean a set of norms and practices that are commonly held by the members of a group and perpetuated over time. The group itself might be a nation, a religion, a profession, a gender, or even a family. While cultures are usually associated with different practices around food, social rituals, and the like, what concerns us here is their effect on negotiation.

Differences Within and Between

For any culture you might name it would be easy to think of individuals who don't conform to the profile associated with that culture. But it is probably even easier to think of individuals who do. Take, for example, gender differences between males and females. Let us look at a trait like assertiveness—we could for that matter look at height or weight or even at the g-factor (for intelligence). Psychometricians have developed measurement scales that purport to measure individual scores on innumerable traits. The scores are designed so that they are normally distributed, and if one looks at the distribution of scores on some trait such as assertiveness, there will be differences between males and females. Let's say for the population as a whole, the mean of the individual scores is arbitrarily put at 1,000 with a standard deviation of 100. Then the differences of the population means between males and females may be on the order of 50 points—the male and female populations having means of 1,025 and 975, respectively. These numbers are only suggestive. So yes, there is a gender difference, but still the chance that a randomly chosen female is more assertive than a randomly chosen male may be on the order of .35. And if one looks not at the whole population but at students in MBA programs, say, then the distributions of scores within the subpopulations become even more overlapping. Similarly, the difference between Japanese and Americans on a host of factors may be tiny when examined in the diplomatic community. In summary, *within* differences often swamp *between* differences—but still there may be differences.

Joint Membership

Keep in mind that each person does not belong to a single culture—rather, we are all members of many cultures simultaneously. You might think, "I don't have anything in common with someone from the U.K." But the same person—actually David Metcalfe—might consider himself a European, a Briton, a Welshman, a Methodist, a social democrat, an Oxford man, an economist, a wine lover, a supporter of Leeds Football Club, and a middle child, all at once.

Differences in Interests

The refrain goes, "This is one tough negotiation. The other guys come from such different cultural backgrounds!" Let's see why this might and might not be the case.

The constant immersion of membership in a culture seeps into a subject's interest profile, and two negotiating parties from radically different cultures may have radically different interests, objectives, and aspirations. But differences can also exist among members of a single culture. In some cultures land is considered the most desirable commodity, in others leisure time or a retinue of servants might be considered the mark of the good life. But do differences of interests make it harder to negotiate? Without even raising the question of culture, differences of interests play a central part in negotiation theory. We have argued that the more people differ, the larger the potential for joint gains. If Albans want X and don't give a damn about Y, and if Batians, coming from a different culture, want Y and couldn't care less about X, then it may be easy to strike a deal with lots of joint gains. So, in theory at least, if differences in culture manifest themselves as differences in interests, then cultural differences make it easier rather than harder to get favorable negotiated outcomes. But differing interests are just one part of the story. Parties from differing cultures might differ radically in thinking about the role of negotiations in settling disputes, about how such negotiations if they occur at all, should be conducted, and about how one behaves during negotiations.

Differences in Negotiation Style

Negotiators from different cultures may have different negotiating styles: they may have radically different norms

- about disclosing information about their interests
- about the veracity of their disclosures
- about their use of threats and other hardball claiming tactics
- about the speed and timing of concessions
- about the number of concessions that are expected before agreement
- about the standards by which they judge fairness
- about their willingness to accept outside help from mediators or arbitrators

The Us/Them Bias

In a previous section we talked about behavioral realities: about a negotiator's anomalies, biases, and errors. Cultural differences between negotiators tend to exacerbate these behavioral anomalies.

One characteristic common to people of all cultures is the tendency to prefer to deal with those who come from the same background. People constantly divide the world into "us" and "them," and culture is one of the lines along which those divisions are made. Some cosmopolitan cultures may show less of this bias, and some xenophobic cultures may show it more strongly, but social psychologists say that the phenomenon is robust enough to be considered universal.

An us-versus-them orientation makes it easier for the negotiators to fall into the traps

- of zero-sum thinking
- of engaging in reactive devaluations
- of naive realism
- of misinterpreting conciliatory gestures
- of simplifying uncertainties by emphasizing worst-case analysis
- of causing self-fulfilling negative prophecies
- of getting trapped into destructive escalatory behavior and not being able to extricate oneself

When distrust is fueled by differing cultures with memories of the other's past untrustworthy behavior, it may be naive to pursue empathetic trust. It still may be possible, however, to generate a delicate form of working trust.

All this may also happen with two disputants from the same cultural grouping, of course, but matters become much worse with the complications of interactions across cultures.

Core Concepts

While the preceding chapter had a *joint normative* flavor, this one has provided a *descriptive* account of actual behavior.

In discussing laboratory/classroom experiences with scorable negotiation simulations, we began by commenting on experiences of our own and many others in conducting what are now called "scorable negotiations." Student-subjects are paired, and each pair is given identical background information in the form of a confidentially scored template with specified RVs. Subjects are scored in terms of how well they do in comparison to others playing identical positions.

Some pairings that achieve a feasible outcome are completely surprised to find that they may have left substantial joint gains on the negotiation table. In the real world there is just one pair negotiating a specific deal or dispute, however, and shortfalls from efficiency go unnoticed.

It was a surprise how variable the outcomes of the simulation exercises were despite the fact that instructions to each pair were identical and quantitatively unambiguous and crisp.

The second part of the chapter gathered from the vast literature about real-world negotiations some common anomalies, errors, and biases of negotiators. Anomalies, biases, and errors made in an interactive negotiation setting can be especially pernicious because A's biases feed B's, which in turn feed A's and so on in a cyclic spiral.

In many negotiations there are elements of potential cooperation and collaboration that go unrealized because the parties are concentrating exclusively on the strictly competitive aspects of the problem. They develop a *zero-sum* mentality that views the world through adversarial spectacles.

They engage in *reactive devaluation*: proposals and offers suggested by the other side tend to be devalued simply because they have been proposed by the other side.

They make *negative attributions* of the other's behavior. Their analysis of the motivations of the other side may be so simplistic that they attribute no reason for the other party's behavior other than pure malice. We tend to see the worst in others (overemphasizing dispositional factors) and rationalize our own aggressive behavior as being not intrinsic but situationally induced because of peculiar circumstances.

Misperceptions beget misperceptions in an escalatory dynamic, which may end up altering each player's basic objective, to do right by himself, to doing right by himself plus doing harm to the other. All this is exacerbated when there are cultural differences present.

Two negotiating parties from radically different cultures may have radically different interests, objectives, and aspirations in part because of their cultures. But at least in theory, if differences in culture manifest themselves as differences in interests, then cultural differences make it easier rather than harder to achieve joint gains. But differences of interests are just one part of the story. Parties from differing cultures may think radically differently about the role of negotiations in settling disputes, about how such negotiations should be conducted, if at all, and about how they themselves behave during negotiations.

16

Noncooperative Others

We are about to give partisan, coaching advice to a wonderful client: not only is he or she—let's settle on "he"—analytical, but he would like to engage in a collaborative negotiation with a like-minded partner.

In Chapters 11–14 we assumed that both sides are idealistic collaborators and would embrace the FOTE approach. They engage in joint template design and so on. But here we step aside from the normative to employ an asymmetric prescriptive/descriptive approach. It's time for a dash of reality.

As we have said before, it takes two to tango, and now suppose the negotiator across the table is not of similar mind or character or disposition as our protagonist. How do we advise our client? It depends on what type of person the other individual is. There are a myriad of possible types, depending on a negotiator's willingness to cooperate, and to brainstorm; her analytical inclinations; her willingness to tell the truth; her aggressiveness in claiming tactics; her willingness to seek joint gains; her manipulative antics; her belligerence and need for posturing. We could go on and on and complicate matters by referring to differences in culture. Instead we simplify. We'll confine our remarks to three types:

1. The simple cooperative
2. The competitive, vigorous claimer (who doesn't measure up to a POTE standard)
3. The disputatious, aggressive adversary

Although ideally we would like to work with analytic cooperators on the other side, we quickly find that most negotiators do not share these characteristics. We have written this book in part to stimulate the spread of this jointly collaborative, analytic approach. We can imagine a ladder of types (it really is not a linear scale—but suppose it were) from the pits to the heavens. Part of our strategy might be to employ tactics to move our negotiating counterparts (the other side) up the ladder of cooperative behavior by selectively commending their exemplary behavior and/or by preaching to them—or even, perhaps, by trying to convince them of the merits of more cooperative behavior. But first let's take them as they are and try to act in a way that lets us do right by our client.

Negotiating against a Simple Cooperator

Characteristics

So let's suppose the other side views herself not as a combatant but as a "problem solver." She is prepared to share her interests with us (but doesn't quantify them) provided we do likewise, and she is eager to brainstorm. In short, she wants to behave like we want to behave, but she is distrustful of putting numbers on everything. Indeed she has a lurking suspicion that quantitative types are aggressive claimers and are not to be trusted. Imagine that! We type her as a *simple* cooperator in contrast to an *analytical* cooperator. We use the word "simple" not in any derogatory sense, but to distinguish this perspective from the quantitative (and mathematical) intricacies of the analytic approach. Many of simple cooperation's leading lights and practitioners are lawyers and other professionals who have a great facility with words, but who are perhaps not as comfortable with numbers.

Most simple cooperators are a friendly sort. You may find them quite amenable to trying new methods. At the same time, they tend to be people who have thought a good deal about their approach to negotiation, and may be loath to let go of their favorite methods. Most are suspicious of quantification. They view themselves as being philosophically opposed to competitive negotiators who "focus on the numbers" instead of more humane values. It can be easy for them to see an analytical type as one of the bad guys. So beware of flaunting your proclivity to be quantitative lest you push them *down* the ladder of cooperative types. We're uncomfortable giving this advice, though, because we're supposed to be giving FOTE advice—and here we're starting off being devious.

Creating Value with a Simple Cooperator

Here's what our client can do:

- Prepare alone à la Chapter 11.
- Share interests; brainstorm with the simple cooperator; and jointly design a template as we discussed in Chapter 11.
- Privately evaluate the template (à la Chapter 12) both qualitatively and quantitatively (surreptitiously) and do BATNA and RV analyses.

Let us think of our side as the AAAs and the other as the BBBs. To be somewhat concrete, consider an abstract negotiation with an evaluated template as shown in Table 16.1. For convenience, we list the importance weights for the issues in Table 16.2. We have also done our BATNA and RV analysis.

The principle of tentative agreement.   We agree to negotiate according to the *principle of tentative agreement*—nothing is agreed until everything is agreed. Both sides know, however, that hours of preparation and meetings create a commitment to the process of negotiation that spills over into a vague, implicit commitment to a desire for agreement. To bind BBB into the process, we try to give them the feeling of joint authorship and ownership of the process and growing consensus. They are undoubtedly trying to do the same for us.

Eliminating and rearranging resolutions: Within-issue, qualitative, efficiency analysis.   We suggest that we would be willing to exhibit our *qualitative* rankings of the resolutions for each issue if BBB would do likewise (simultaneous revelation). So be it. When the rankings are not simple reversals of each other, we eliminate dominated entries. So we bring the template into a so-called canonical form. No big deal so far.

Sharing information about the relative importance of issues.   We resume the conversation by suggesting the selective, simultaneous disclosure of information. We propose: "Here's a suggestion. Not all issues are equally important to us. We would be willing to tell you truthfully the issues that are of high, medium, and low importance to us, if you would simultaneously do the same." After thinking about this for a while, those on the other side concur with the merits of our suggestion, and after simultaneous disclosure we come up with the results shown in Table 16.3.

We write the table on a flip chart so that we both can see at a glance our comparative qualitative evaluations. It is heartening for us to see the potential for joint gains. In many situations, each party may already know which issues are of high, medium, and low importance to the other side, and not much new information may be conveyed by the simultaneous disclosure of this table. It is still useful, however, because the presentation itself is designed as a decision aid for the next step.

There is admittedly room here for deception. We, AAA, for example, may feel mildly about issue 1 but strategically try to mislead BBB by saying that it has high importance to ourselves. But at this stage we have no intention of departing from our POTE orientation. Remember, the more we can give them without substantially hurting ourselves, the easier it will be to offer them something better than their BATNA.

Table 16.1   Scored Template for AAA

Issue	Resolution	Value AAA
1	*a*	0
	b	3
	c	6
	d	7
	e	8*
2	*a*	0
	b	2
	c	4
	d	5*
3	*a*	0
	b	2
	c	3
	d	8
	e	10*
4	*a*	0
	b	18
	c	30*
5	*a*	0
	b	1*
6	*a*	0
	b	5
	c	15
	d	18
	e	20*
7	*a*	0
	b	2
	c	7
	d	10
	e	13
	f	15*
8	*a*	0
	b	1
	c	2*
9	*a*	0
	b	1
	c	2
	d	3*
10	*a*	0
	b	2
	c	4
	d	4.5
	e	5
	f	5.5
	g	6*

Note: Asterisk indicates the importance weight of the issue.

Table 16.2   Importance weights for AAA

Issue	1	2	3	4	5	6	7	8	9	10
Imp. wt.	8	5	10	30	1	20	15	2	3	6

Table 16.3   Joint display of the importance of issues in broad categories

Issue	1	2	3	4	5	6	7	8	9	10
AAA	M	L	M	H	L	H	H	L	L	M
BBB	H	L	L	M	L	H	L	M	L	H

Dealing with imbalances in evaluations of importance.   Now let's take a cue from what we developed in fair-division theory under FOTE conditions. The efficient frontier was obtained by looking at the ratio of values for the two parties (Janet and Marty) and then "dealing from the top." Keeping that in mind, we find that the obvious thing to do is to tentatively—and we emphasize tentatively—give us (AAA) the benefit when we feel some issue is of greater importance to us than to them; and to give them the benefit for those issues they feel more strongly about. Because we have the same valences for issues 2, 5, 6, and 9, we leave them aside for the time being.

- Issue 1: They are high and we are medium; we give it to them. There are 5 levels of resolution for issue 1 and they should get a low letter; they push for *a*, we suggest *c*, and we tentatively agree on *b*.
- Issue 3: We are M and they are L. Since there are 5 levels, we would like *e* but tentatively agree on *d*.
- Issue 4: The importance levels are H and M, so we should get the benefit of the doubt. Since there are only three levels we opt for *c* and they resist, so we agree on *b*; but we register our unhappiness.
- Issue 7: We meet no resistance in getting our best on this issue.
- Issue 8: We give them their best, *a*.
- Issue 10: We dicker and end at *c*.

On a flip chart we exhibit what we have tentatively agreed on (see Table 16.4).

AAA's points, shown in the table, are not shared with BBB. From these tentative agreements, we net a total of 48 points. Is this any good? It depends on our RV.

Claiming Value with a Simple Cooperator

When we evaluated the template, our side did a BATNA analysis and determined that our RV was rather good, a sizable 52. Given that we still have issues

Table 16.4   AAA's partial score

Issue	1	2	3	4	5	6	7	8	9	10	*Total*
Tentative allocation	b	—	d	b	—	—	f	a	—	c	
AAA's pts. assigned	3	—	8	18	—	—	15	0	—	4	48
AAA's pts. unassigned	—	5	—	—	1	20	—	—	3	—	29

2, 5, 6, and 9 to assign, which are valued at 29 points, we are feeling rather smug, and decide to engage in a bit of theatrics. When the other side expresses some concern over the allocations made thus far, we don't quite believe their protestations and dissemble by saying that we are hoping for some consideration on the remaining issues. Not exactly FOTE behavior, but perhaps in the realm of POTE.

RVs without a scorable template.   In most negotiations where our side has a scorable template, we strongly advocate choosing an RV that is a composite for the entire contract. The RV can be thought of as a summative requirement, balancing all the issues. Whenever a negotiator does not have a clear scorable template to work with, which means most of the time, the negotiator may not even have a clear idea of his or her BATNA, and cannot think of a summative RV index. Near the end of negotiations, the negotiator can examine suggested contracts and subjectively make judgments about their acceptability. In the early stages of building up a contract, however, the negotiator without a scored template and without a summative RV has little guidance in knowing what is acceptable. So that negotiator has a tendency to act prudently by imposing separate RVs on each of the important issues. There is little flexibility to create options. In a rather perverse way, doing this may give an advantage to the ill-prepared party, who must protect his or her interests by being cautious.

Thinking about our aspiration level.   In Chapters 6 and 7 on distributive bargaining (the pure claiming case) we saw the importance of keeping in mind an *aspiration level* as well as an RV. There's a tendency with just an RV in mind to settle too modestly, to satisfice. How does this get translated into the integrative case?

We fully score the template, but we know the other side hasn't. We don't know how bowed the final frontier would be if the other were to score the template and if we were to do a full FOTE analysis. However, we have some idea how bowed the frontier might be from the information already gleaned. It depends mostly on how differently we and they appraise the importance of the issues. We know our reservation value, but since we don't know theirs—they probably don't know theirs, either—we can't easily judge what maximum feasible value (MFV) we could achieve and what our side's *potential* might be. (See Figure 14.5.)

But somehow we glean some insight into what these values might be. We think they have a mediocre RV (despite their protestations), and given that we believe that the efficient frontier is reasonably bowed, we make the heroic assumption that our maximum feasible value is about 75. With an MFV of 75, our potential would be $75 - 52$ or 23. We start off with some aspiration level to achieve, like our reservation value plus 60 percent of our guesstimate of our potential. This yields an aspiration value of $52 + .6 \times 23$ or 65.8. We're being the tough claimers now. If we are to finalize the tentative agreements on issues 1, 3, 4, 7, 8, and 10, we would have to get 17.8 of the remaining 29 points to be allocated. We shouldn't be locked into this, though, because it depends on a lot of assumptions.

Positional bargaining with the aid of a template.   BBB's negotiators suggest that we jointly examine each of the remaining unassigned issues. We tell them that we are willing to tell them about our strengths of preferences for intermediate levels of resolution if they would be forthcoming with similar disclosures. They don't understand what we have in mind, so we take the case of issue 6, which gets high valences from both AAA and BBB. We show the other side our evaluations (see Table 16.5).

We explain to them that we scored our worst level, *a*, at 0 points and our best at 100 points. And on the basis of 100 points, we scored *b* at 25, *c* at 75, and *d* at 90. "In other words, we're telling you that getting *c* is more than halfway for us." They don't quite catch on at first, but when they volunteer that for them *b* is almost as good as *a* and that the big gaps for them were from *e* to *d*, we start playing around with numbers for them and finally end with Table 16.6, which fills in the question marks. We plot these evaluations in the top part of Figure 16.1.

We disclose our preferences truthfully and have every reason to believe that BBB will also. Because most of these preferences are obvious to the other side, dissembling at this stage would be counterproductive. There are extreme examples in which lying at this stage might gain one side a temporary advantage. Let's assume that is not the case in our carefully constructed example.

It is pretty clear to them, and we don't resist, that the preferred compromise for issue 6 is level *c* (yielding for us 75 out of 100, or $.75 \times 20$, or 15 points). We move on. After a little posturing on our side, we give them resolution *a* on issue 5, and this leads us to issues 2 and 9. We follow the same procedure for these as we did for issue 6. We agree to level *b* on issue 2 (yielding for us 40 out of 100, or $.4 \times 5$, or 2 points); and to level *c* on issue 9 (yielding for us 80 out of 100, or $.8 \times 3$, or 2.4 points). At this juncture we make an evaluation of where we stand (see Table 16.7).

Table 16.5   Joint analysis of issue 6 with blanks

Resolution	AAA	BBB
a	0	100
b	25	?
c	75	?
d	90	?
e	100	0

Table 16.6   Joint analysis of issue 6

Resolution	AAA	BBB
a	0	100
b	25	95
c	75	70
d	90	30
e	100	0

With this allocation, we end up with a total of 67.4 points. They seem happy and we are delighted. But still, can more be squeezed out? We try to embellish the contract.

We suggest:

- How about going from *b* to *c* on issue 1 (yielding us a gain of 3), and from *b* to *a* on issue 2 (yielding us a loss of 2)? Fine by them. So agreed.
- How about from *b* to *c* on issue 4 (yielding a gain of 12 to us), for a switch of *f* to *d* on 7 and *c* to *b* on 10 (yielding a combined loss of 7)? No deal. How about if we also throw in a shift from *c* to *a* on 9 (for a loss of 2)? They agree. After these agreed-upon embellishments we end up with our final allocation (see Table 16.8).

After all these embellishments, we finish with a total of 71 points. They think they have done well also! An agreement is signed.

Handling Deadlock

That was relatively easy. But sometimes things are harder, even when the parties are well intentioned. Those on the other side might be now claiming harder — perhaps they have an internal constituency that demands more. Perhaps they have a better BATNA, leaving a smaller zone of possible agreement. Let's go back and revise Table 16.4 to make things more difficult.

Joint Analysis of Issue 6

Resolution	AAA	BBB
a	0	100
b	25	95
c	75	70
d	90	30
e	100	0

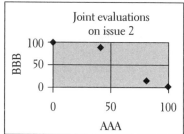

Joint Analysis of Issue 2

Resolution	AAA	BBB
a	0	100
b	40	88
c	80	20
d	100	0

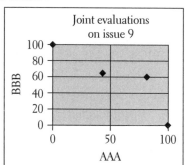

Joint Analysis of Issue 9

Resolution	AAA	BBB
a	0	100
b	40	65
c	80	60
d	100	0

Figure 16.1.   Within-issue joint conditional evaluations.

After perusing Table 16.3, BBB suggests a tentative partial allocation, shown in Table 16.9. The last two lines of 16.9 are not shared with the other side. There's no way that we can agree to Table 16.9 and hope to meet our RV of 52! So we announce to BBB that we can't live with those allocations. They do not know whether they can believe us, and so they toughen up in their claiming behavior. We're getting entrapped, and trust is deteriorating. A deal is turning into a dispute. The reality might be that there is no zone of agreement. Claiming is souring our creative efforts.

Table 16.7 The tentative contract and AAA's evaluation

Issue	1	2	3	4	5	6	7	8	9	10
Resolution	b	b	d	b	a	c	f	a	c	c
AAA's pts.	3	2	8	18	0	15	15	0	2.4	4

Table 16.8 Final allocation

Issue	1	2	3	4	5	6	7	8	9	10
Resolution	c	a	d	c	a	c	d	a	a	b
AAA's pts.	6	0	8	30	0	15	10	0	0	2

Table 16.9 AAA's partial score against a tougher claimer

Issue	1	2	3	4	5	6	7	8	9	10	Total
Tentative allocation	b	—	c	a	—	—	e	a	—	b	
AAA's pts. assigned	3	—	3	0	—	—	13	0	—	3	22
AAA's pts. unassigned	—	5	—	—	1	20	—	—	3	—	29

Suggest a process deemphasizing claiming.   Let's almost start over. Both sides have seen that claiming can easily get out of hand. So we propose the following. Let's not worry for a while about generating a single final contract, but let us jointly explore the identification of a couple dozen or so lesser contracts that are efficient or nearly efficient and range quite widely over the efficient frontier so that some are tilted toward AAA, some toward BBB. Instead of BBB saying, "I want such and such a contract," and AAA saying, "No way," let AAA respond: "Fine, let that be one of the contracts under review." Remember the whole idea so far is to finesse the tactics of claiming. The process of jointly identifying a set of contracts that are somewhat efficient and range over the efficient frontier deemphasizes claiming activities, and interactions take on a more cooperative tone. After a sufficient number of contracts are jointly identified, the trick now is to accept one of them. Here is one way we can proceed.

The double-auction method.   Both AAA and BBB proceed to rank order the selected contracts and reveal their orderings simultaneously. If the orderings are not reversals of one another, jointly dominated contracts are eliminated. Now comes the hard part. Each negotiator is asked to divide the list into two parts: acceptable contracts (contracts that beat that party's BATNA) and unacceptable ones. For our client, who has scored the template and established an RV, this task is easy. It's tougher for BBB. This procedure, however, encourages BBB to think about a summative RV instead of specifying reservation values for each issue. This is an important feature in favor of this procedure.

If AAA's and BBB's sets of acceptable contracts have contracts in common (that is, if their intersection is nonempty) a zone of possible agreement exists. If not, the no-agreement state is adopted. Each party has some motivation to tell the truth, because, as we saw in Chapter 7, two exaggerators can eliminate all jointly feasible contracts. If the intersection is not empty, then the remaining task is to select a compromise contract. The focal choice is the midpoint in the range.

As an illustration, suppose the negotiators identify twenty contracts that are well distributed along the efficient frontier. The parties simultaneously reveal their rankings. Suppose that they are not quite reversals of each other; after some inspection, we eliminate two contracts that turn out to be dominated. We now take the remaining eighteen contracts and number them from 1 to 18, with AAA preferring higher numbers and BBB lower ones. AAA's RV is such that AAA finds contracts 10 to 18 acceptable. BBB, which doesn't have a crisp RV at its disposal, has a tougher time deciding, and finally agrees that contracts on or below 12 beat its BATNA. If BBB has disclosed truthfully, then contracts 10, 11, and 12 are feasible and contract 11 is the obvious compromise. If AAA announced, "12 or higher is acceptable," and BBB announced, "10 or lower is acceptable," both slight exaggerations, then the double-auction procedure would falsely make it appear that no agreement was possible. The moral is that my exaggerations might help me a little, but they might hurt us a lot. It can be worthwhile to explicitly mention the incentive to be honest.

Negotiating against an Aggressive Claimer

We climb down the rungs of the ladder of types until we come to the manipulative, strategic, aggressive-claimer type. Such a negotiation partner deserves to be referred to as our "adversary" across the table. This type has a competitive win-lose orientation, engages in premature positional bargaining, refuses to share information, exaggerates his BATNA, adds phony issues for negotiation in an effort to get leverage on more relevant issues, engages in brinkmanship, and tries to worsen the opposing side's BATNA. The aggressive, unprincipled claimer assumes a zero-summish attitude—not only does he want to maximize his gain, but he wants to beat his adversary; he sticks to an uncompromising position, and does not engage in reciprocal exchange of sensitive information. The lesson is simple: Full, Open, Truthful Exchange of information by one party is highly inadvisable when the other party uses aggressive, unprincipled claiming behavior. If we naively implement a cooperative approach and imagine we can play FOTE on a nonreciprocal basis, we will be taken to the cleaners. But if both parties behave strategically, the outcome is likely to be inefficient. The dilemma is clear. Can we deal with strategic behavior and continue to pursue mutual gains? In operational terms, can we adapt the collaborative approach to negotiate securely and confidently with a party who plays according to different rules? Why negotiate with someone like that? Perhaps our client simply has to.

Preparation

We would want to start by getting ready for a different kind of interaction. Let our client be Ada A. Altman (AAA) and our adversary be Barney B. Beatty (BBB).

Phase I preparation: Anticipating a change from FOTE to POTE.   As we begin, our client, Ada, does not yet realize how unscrupulous Barney really is. She starts Phase I of her preparation hoping to enter into interest-based negotiations seeking joint gains, but learns from a trusted outsider that Barney Beatty has a history of aggressive negotiation behavior. We suggest that Ada's minions collect as much information as possible about the other party. What are his interests? What are his alternatives? What is his social and professional background? What happened in his previous deals and what can we learn from them? How do we expect him to behave? How should she respond? We set up a simulation of the prenegotiation, with one of our team playing the role of BBB and Ada playing herself. This gives her a better feel for the situation and a jot more confidence.

The reality is that our side will have to move away from a FOTE stance. Can we, however, maintain a POTE stance—telling the truth but not necessarily all the truth? That's one of our key interests. Let's try. There are a few tricks to be learned about how to do this.

Phase II preparation: Prenegotiations.   In a meeting with the other side, Ada starts off: "In preparing for our first meeting, I have considered my interests, thought about issues and tradeoffs, imagined what I want Allied Alarm Associates to look like in ten years' time, and generated some options for negotiation. Have you done the same? If so, shall we reciprocally share this information?" BBB's eyes light up. AAA thinks that he will reciprocate. "What a good idea, tell me about your interests!" Barney suggests. Ada begins to run through the fundamental interests that she wants to satisfy, some of her tradeoffs, and favored options for agreement. Forty minutes later BBB has an excellent idea of what AAA's interests are, which issues are important for her, how he can exploit her, how he can reduce his exposure to risk on the investment and increase his expected return at AAA's expense. Before Ada put all her cards on the table, she should have tested whether Barney would reciprocate in a forthcoming manner. She should have sought feedback early on before naively completing her list of disclosures.

He begins, "We value your company at $3 million now, and with a $1 million injection from us, we expect 65 percent of the equity." "What?" cries AAA, "That's outrageous. Our company is worth far more than $3 million and you know it!" "That may be so," retorts the aggressive claimer, "but we have you over a barrel. We know you need the money desperately and can't get it elsewhere. We expect 65 percent of your equity or no deal."

Ada tried to be open and selectively truthful, but it's clear that BBB is starting off by playing hardball and putting forth an unreasonable initial demand. So much for joint brainstorming a template for negotiation. She may have to engage in a dance of proposals (see Chapters 6 and 7). But first things first. We don't want to get anchored by exclusively examining the only proposal on the table and gradually backing off from that. Ada might respond: "You can demand all you want, but that's not what you're going to get. No way! I expected that you might jump in with a proposal so I have prepared a counteroffer." She spells this out. Then she adds: "Can't we back up? I told you about some of our interests. This potential deal between the two of us involves a lot more than money and share of equity. What are some of your interests?" The idea is to offer Barney an alternative to beginning an unpalatable battle of ultimatums. If he takes the easy way out, then things are on the right track.

But let's say Barney ignores the suggestion and offers a slight modification of his original offer. Somewhere along the line, he asks, "What is the minimum you would accept? I'll see if we can improve on it." Answering truthfully would be a negotiating disaster. In anticipation of this, we coached Ada how she could maintain a POTE stance. Here are two ploys:

- *Deflect the question.* "Gee, I would much prefer to hear what is the maximum you would be willing to offer."
- *Ignore the question.* There is no rule that says that you have to answer all questions posed to you. Change the subject rather than lie. This helps in maintaining a POTE stance.

Reluctantly, Ada is drawn into a dance of proposed contracts. She didn't want to plunge directly into the give-and-take of hardball negotiations before exploring, in a collegial manner, possible sources of joint gains. Dialoguing and brainstorming have been skipped.

Fostering Collaboration: Moving Up the Ladder

As we remarked earlier, part of our strategy might be to employ tactics to move our negotiating counterparts up the ladder of cooperative behavior by selectively commending their exemplary behavior and/or by preaching to them—or even by trying to convince them of the merits of more cooperative behavior. So far we have taken them as they are; now it's time to try to foster better cooperation by pushing them and ourselves up the ladder of types. Let's suppose the guy on the other side of the table appears to be a principled vigorous claimer, and we want to push him upward.

We open a prenegotiation dialogue (Phase II of preparation) by saying something along these lines: "As I see it, there are two ways we can proceed. We can share information about the strength of our interests and try to arrange a mutually beneficial deal, and find a composite contract by trying to figure out what's fair. Or we can play hardball—try to trick each other, refuse to budge until we exact a price from the other, and that sort of thing. I think we'll both be better off if we work together. If we play hardball, I think we'll be here a long time, leave a lot of value on the table, and probably hurt our chances of making more deals in the future. I know how to play both of those games. I assume you do as well. I'd prefer the first, but I am willing to go along with you on the one you choose. By choosing a strategy, you will also be choosing the one I'm going to use when I negotiate with you—so think carefully about your choice."

Compliance with agreements like that are pretty high in experiments. We see no reason to think that they would not be high in real negotiations as well. Of course, it is possible that your counterpart will agree to cooperate, and then clobber you by taking competitive unilateral action. How can we lessen the allure of that course?

Emphasize future interactions.   In the laboratory, iterated prisoner's dilemmas have far higher collaboration rates than single-turn games. The threat of retaliation in the future discourages players from angering the future partner by defecting. If negotiators can reframe their interaction as an iterated game, they should be able to reduce the incentive to defect. The happy reality is that an iterated prisoner's dilemma is a better model for negotiations than a one-round affair.

Most negotiations, especially most important negotiations, involve repeat players. A venture capitalist and an entrepreneur will have many future interactions. Two lawyers settling a case may not have future dealings on the same matter, but if they practice before the same courts in the same part of the country they are likely to run into each other again. In addition to their chance of meeting again, there is the chance that one of them will face a partner or associate of the other's firm in the future. In most firms, lawyers send around memos asking for information about opposing counsel: "Plaintiffs in this case are being represented by Louise Smith—has anyone had any prior dealings with her?" It must be acknowledged, however, that sometimes one-shot interactions are just that, and some negotiators believe that their behavior in one negotiation has no ramifications for future negotiations. The homeowner moving to another city or the insurance company dealing with a claimant may rightly conclude that the chance of seeing this party again is vanishingly small.

Trust for the untrustworthy.   As we said above, it's easier to maintain some semblance of cooperative stability if the game is seen to be repetitive without an imminent ending. In dynamic situations where AAA and BBB have to repeatedly interact with each other despite the fact that they hate each other, they may agree to a level of tacit cooperation for fear that any deviation will be punished, with dire consequences for both. If BBB myopically hurts AAA in trial X, then AAA in a fit of rage will in turn try to hurt BBB in ensuing trials. It may take longer than you expect to get back on the track of cooperation. In the repeated prisoner's dilemma game, the advice offered in Chapter 4 was:

- Try to be *cooperative at first* and don't be the first to defect;
- Be *provocable*: if the other defects, be sure to punish him in ensuing trials. How many? Enough for him to realize that he gains no advantage in defecting in the first place.
- Do not be *vindictive*. If you punish him too much, he might go berserk and try to demolish you even at the expense of ruining himself in the process.
- Be *simple* and *transparent*. Don't signal him in ways that he might not comprehend.

You may have to be ingenious in modifying a seemingly one-shot situation into a dynamic repeat-play situation, but that's something to think hard about.

Negotiating against a Disputatious Adversary

The Essence of Disputes

Disputatious behavior in disputes differs markedly from aggressive claiming behavior in deal making.

- A dispute is often triggered by a crisis situation.
- The participants often view themselves as antagonists.
- The mentality of the negotiators is often zero-sum: there will be a winner and a loser.
- Interactions are emotionally laden.
- It is in the interest of one side to try to hurt the other.
- It is in the interest of each not to try to improve relations with the other side.

- There is little trust between the contending parties.
- Contracts are not secure because of this lack of trust.
- Each side thinks the other side will shade the truth, act strategically, exaggerate, and lie, and therefore, in defense, each must do the same.
- Hurtful actions against the other can escalate, and as in the escalation game, the parties may inflict damage to each other that far exceeds the tangible value of the prizes at stake.
- Threats (for example, to lower the other's BATNA) are made and sometimes executed.
- Negotiations are viewed as taking place in the negative domain: the negotiators think their task is to minimize their losses rather than to maximize their gains.
- When one side acts in a manner that confirms the other side's worst suspicions, these acts are considered purposeful.
- When the acts disconfirm negative expectations, they are construed as being acts forced on the other side by circumstance.

Case Study: A Business Dispute

To illustrate how to deal with disputatious behavior, we have concocted a dispute between two business partners. We'll imagine that Bill and Tony have together built up a wine brokerage business. Eighteen months ago they made a strategic decision to significantly expand their warehousing capacity. The plan has backfired, however, and profits have taken a nosedive. Bill has blown his top and blames Tony for the fiasco, but they need to jointly devise a route out of the quagmire. Their relations have deteriorated to the point that they can't talk civilly to each other, and the list of grievances is growing exponentially. It's apparent that they must dissolve their partnership, but how? There are so many interlocking parts. Tony has come to us for advice. How do you think we should advise him to deal with his disputatious partner?

Creative preparation alone.   To begin with, we advise Tony to conduct Phase I preparations. We help him to elaborate his interests, define his alternatives, set out visions for the future, generate some options, identify uncertainties, and come up with objective criteria. This exercise is certainly useful, but the preparatory blueprint of the collaborative approach does not cover the raft of process problems posed by Bill's disputatious behavior.

Focus on process.   When the other side behaves disputatiously, it puts a premium on our side to think more deeply and creatively about the design of a legitimate and effective negotiation process. We might invite third-party help;

formally separate inventing options from commitment (by delegation to a sub-committee); seek to test and understand the other side's interests; and so on.

Anticipating disputatious behavior.   To ready Tony for his meeting with Bill, we brainstorm with him to come up with a more concrete idea of what he can expect at the first meeting. More likely than not, Bill will frame the problem as an adversarial win-lose situation, fundamentally believing his only goal (get rid of the additional warehouse space) is utterly incompatible with Tony's only goal (keep the extra space). A disputatious negotiator like Bill searches for information that confirms his partisan beliefs, ignores information that contradicts these beliefs, and interprets (and distorts) information through selective perception. We also suggest to Tony that Bill will blame him for the "bad" decision, perceive him to be representing a hostile group rather than participating as an individual, mistrust his behavior, devalue genuine concessions, reject proposals that are a priori favorable to Bill, and may define success as imposing losses on Tony. Bill may attribute negative characteristics to Tony—untrustworthiness, selfishness, lack of moral virtue, professional incompetency, and congenital hostility.

Threat and counterthreat.   Having briefed Tony on how he will be perceived, we now consider the actions that Bill is likely to take. Typically, disputatious behavior involves threats. Bill may threaten Tony that if he doesn't make concession X, then Bill will irrevocably commit himself to implementing the costly and painful action Y. Such threats are very difficult to deal with. How should Tony respond? Uppermost in our minds is the need to avoid escalating the conflict. Given imputations of malign intentions, responding with a threat is unlikely to stabilize or deescalate the conflict. We strongly advise Tony not to retaliate with a counterthreat—however angry he may be. He should neither yield to Bill's pressure nor break off negotiations. Both of these responses can also trigger escalation.

Irreversible commitments.   What other actions can we anticipate? Bill may make an irrevocable commitment to a given course of action. This differs from a threat, because it is not contingent on Tony's acting or failing to act. Bill can unilaterally commit to a certain action whatever Tony does, in the hope of coercing Tony into making concessions. For instance, Bill may commit himself to telling major clients that Tony is incompetent. Tony will have to think about how to react to irreversible commitments. The committing party often makes the commitment public to demonstrate its resolve and increase the costs to itself of backing down. Because of their irreversibility, commitments entail high risks—for both parties.

Not only should Tony be aware of the contentious actions that Bill may take, but he also needs to remember that disputatious behavior is self-reinforcing. Partisan perceptions and poor decision making can easily lead both parties to overcommit and entrap themselves in an escalatory spiral with its own dynamic. Our advice is designed to help Tony confront the realities of the other party's disputatious behavior and navigate between two natural reactions: playing the adversary's game and escalating the conflict; or yielding to the coercive behavior. In conflicts, it is common for negotiators to lose sight of their interests, reservation values, aspiration levels, and tradeoffs. We impress on our client the need to keep his entitlements firmly in mind.

Changing the Geometry

How do our preparations need to change to cope with emotional confrontation? How can we avoid becoming disputatious ourselves? How can we create and claim value in a dispute? What psychological traps associated with conflicts do we need to avoid? Should we continue to pursue mutual gains? How can we use coercive tactics to get what we want?

Read Bill Ury's eminently readable and instructive book *Getting Past No.* It contains a lot of practical advice that's worth your deliberation if you have to negotiate with such reprehensible characters.

Also keep in mind three additional alternatives.

Do you want to play?   Before entering into a highly contentious negotiation with a highly disputatious adversary, take stock of what may be likely outcomes and decide whether you want to play in that game. There may be other possibilities elsewhere. This is a decision problem under uncertainty, and much of what we have developed in Chapters 2, 8, and 9 has methodological relevance to this tough decision problem.

Negotiate through agents.   Consider the case in which your client (the reasonable one) learns that the other side is obstreperous, and the negotiation is fueling the fires of the other side's discontent. Our side has tried the usual catalogue of calming ploys suggested quite succinctly by Bill Ury. But none of these ploys can get the other side to settle down enough for us to pursue joint gains. Perhaps every gain to us is interpreted as a calamity to him. It may make sense to say, "Look, you can't work with me, nor I with you, so let's both appoint professional agents to negotiate for us. Perhaps our agents, who are less emotionally involved, can find some compromise that we would both prefer to a no-final-agreement state that could ruin each of us."

Use of a third-party helper.   Or we may say, "Look, we're either getting nowhere or going somewhere neither you nor I want to go. It's stupid for us to holler at each other when there may be joint compromises that will benefit each of us. I'm not sure that we can become any more rational working together, so how about bringing in a mediator who can keep us on track and who will admonish us not to just whack at each other?"

Core Concepts

In this chapter we have given partisan advice to an analytic type who would prefer to engage in collaborative negotiations in a FOTE atmosphere, but whose opponent may be on a different wavelength. We imagine a ladder of negotiating types going from a simple cooperative (who disdains the use of quantitative analysis) to an aggressive claimer to an obstreperous disputant.

Most simple cooperators share many of our ideals but are suspicious of those who "focus on numbers." They do not resist in jointly designing a template for negotiations, but they would not themselves dream of scoring the template. And while they would establish a BATNA, they would not, or could not, establish its associated RV. Much of the idealistic behavior suggested in Chapters 11 to 14 is relevant to negotiations with a simple cooperator.

We have discussed how our client could share qualitative information in the creation process, but the claiming process gets a bit more complicated. The other side, not having the security of a crisp RV, tends to establish cutoff levels for several of the issues separately.

When the other side is an aggressive claimer it forces us to adopt a POTE stance (rather than FOTE), and we must be wary of our adversary's exaggerations. Several techniques we suggested for dealing with the simple cooperator can be adapted, with caution, for the aggressive claimer; in addition, some of the material in Part II on distributive bargaining is of relevance here. We have offered a few hints on how we can maintain our desire not to lie without succumbing to the relentless attacks of the other side.

We then examined the essence of heated disputes, and suggested ways of negotiating with obstreperous adversaries. Some of the techniques introduced for responding to the aggressive claimer can be adapted for the disputatious opponent, but the temptation may be great for us to counter threat by counterthreat. We must be aware of getting entrapped in jointly devastating escalation games. Lessons learned from the repeated prisoner's dilemma game of Chapter 4 provide ways of securing insecure contracts. We closed by offering some final thoughts: Do we really want to play in such games? Can we change the rules of the negotiating game? How about negotiating through agents? How about using third-party help (facilitators, mediators, or arbitrators)?

External Help

In some negotiations—mostly disputes rather than deals—the parties, by themselves, may not be able to arrive at a compromise agreement without some outside assistance. They may lack the necessary skills; or they might rub each other the wrong way and be unable to talk civilly to each other; or, left by themselves, they might find the experience too traumatic; or they might fear that the interchanges between them will escalate into verbal warfare. So they mutually agree to invite in an outsider—a facilitator, mediator, or arbitrator—to help them overcome their emotional conflict. In my own university, personnel disputes go through a round of mediation and then arbitration before the case goes to court. There are also cases where external helpers are not invitees but inviters. The outsider may be a deal maker or matchmaker who invites two parties to the negotiating table because he believes there is a synergy to be gained by the parties in combining their talents and comparative capabilities. Then there is the case—rare, to be sure—where the parties would like to conduct their negotiations in a FOTE manner, but they can't take the risk of confiding fully to the other side. They are prepared to tell all truthfully, however, to a reputable analyst who has a track record in helping negotiating parties achieve efficient and equitable outcomes. This is the subject matter of Part IV.

The first chapter, Chapter 17, starts out by considering conventional facilitation and mediation, reserving the next chapter for conventional arbitration. Rather than getting involved in disputes about nomenclature, we adopt the neutral language of "external helper" or "third-party intervenor." In our abstraction, we talk about the {A, B; H} dynamic, where A and B are the negotiating parties and H the helper. We start by drawing up lists of conventional roles that H might perform, saving those that have a more analytical flavor for later development. What role H plays is itself to be negotiated, and depends on how H gets involved (as invitee or invitor or a mixture of these), on the context of the problem, and on the particular temperaments of the three actors involved.

We talk about the interests of H as another player—but a different type of player. We review the many reasons why external helpers are not used when perhaps they should be.

The chapter then considers more active roles for the mediator by involving H in generating proposals for the consideration of A and B. We discuss President Jimmy Carter's role at Camp David as a proactive mediator with clout in the negotiations between Anwar Sadat of Egypt and Menachem Begin of Israel.

Chapter 18 explores the world of arbitration. In practice, a negotiated impasse often triggers mandatory arbitration, where H, in an evaluative mode, is expected to impose a "solution" that is binding on the parties. There are all sorts of variations: arbitration may be not mandatory but a voluntary option on the part of the negotiators; the proposal of H may be not binding but suggestive; and in some circumstances H may be expected not only to be neutral initially but to be nonevaluative throughout.

In the standard distributive bargain where A wants a higher value and B a lower value, instead of asking H, after fact finding, to suggest his ideal compromise value, we ask A and B to make sealed final offers and then require H to select one of these offers. In the chapter's appendix we examine a game-theoretic treatment of this type of final-offer arbitration and, surprisingly, find it somewhat flawed from a normative perspective even though it seems to work in practice.

The latter part of the chapter deals with cases of complex integrative negotiations in which H, acting as a neutral joint analyst (NJA), helps the parties achieve an ideal collaborative compromise solution described at length in Chapters 11 to 14. Such an NJA acts as a special kind of nonevaluative arbiter. The reluctance of the parties to truthfully reveal their reservation values complicates the analysis, and a double-auction bidding system is introduced to help resolve this complication.

We also consider an intervention in which I helped divide an art collection between two brothers who knew ahead of time that they could not do it alone without jeopardizing their relationship.

After the completion of an unassisted negotiation, the parties, realizing that there may still be joint gains to be had, might invite in an NJA to try to embellish their agreement in a so-called postsettlement settlement. In this case the parties do not have to reveal their RVs to the NJA—the previously negotiated compromise acts as a pseudo–joint reservation value.

There remains the nagging question for the NJA: which point on the efficient frontier should be selected as the most equitable? What is fair? Chapter 19 discusses this question in some detail.

Chapter 20 examines the common nature of intractable disputes (mainly between feuding contiguous countries or national entities or cultural groupings), identifies a set of barriers that prevents constructive negotiations from taking place, and suggests how parallel negotiations could help.

The chapter documents a seemingly successful intervention in the Peru-Ecuador conflict. Influential, nonofficial surrogates from the belligerent countries were brought to Harvard by the Conflict Management Group to engage in a week-long activity called facilitated joint brainstorming (FJB). The facilitator, Roger Fisher (backed by a team of helpers) engaged the group in generating creative options to resolve the controversy, ruling out all claiming tactics. The surrogates were evidently inspired enough at week's end to return to their countries and to influence a change of heart and mindset among key official people. This is the way it was supposed to work, and in interviews with the subjects it seemed to have worked that way. Fisher was publicly commended by both sides for his positive contributions to peace in the region.

The FJB exercise involved a logistical feat: getting eleven distinguished surrogates around a table in a neutral spot. The chapter discusses a variant of the FJB. An academically minded researcher (perhaps a dissertation writer) shuttles back and forth interviewing key subjects on their perceptions of the controversy, on their interests, on their tradeoffs, trying to elicit creative options, and prepares a document we dub a premediation briefing report (PMBR), addressed to a later facilitator/mediator but read by key representatives on both sides of the controversy.

17

Mostly Facilitation and Mediation

We continue our discussion of two-party negotiations of both disputes and deals, and now we introduce the possibility of including an external party acting for the benefit of both contending parties. Typically these external helpers are called facilitators, mediators, or arbitrators, and there are endless disputes in the literature over whether some interventions should be classified one way or another. We reserve this chapter for those roles we consider mostly facilitation or mediation without strictly distinguishing between the two.

The next chapter, on arbitration, deals with an intervenor who takes a very proactive role in suggesting a final outcome for the negotiations. That suggestion may or may not be binding.

In the next chapter we shall also introduce an external helper who bends over backward not to be evaluative or judgmental about the negotiators, but nevertheless suggests a final contract that squeezes out joint gains (that is, is efficient) and is fair (equitable); this process too we deem arbitration. Similarly, another helper may design a double auction that will help the parties select a point on the frontier computed by the helper after a period of fact finding, and this individual is thought, again by us, to be doing more arbitration than mediation. Hence we shall postpone discussion of those procedures to Chapter 18.

In terms of the degree to which a method involves evaluation, the standard, but not universal, vernacular defines *facilitation* as the least evaluative, then *mediation*, and then *arbitration* as the most evaluative.[1] But as we shall see, there are analytical interventions that upset this apple cart of nomenclature. Since it will be difficult to catalogue some important but nonconventional roles as either facilitation, mediation, or arbitration, for the time being, let's finesse this contro-

---

1. *Evaluative arbitration:* The arbiter often determines who gets what on the basis of his determination of which party is right; which party has behaved more appropriately; which party's case would hold up better in a court of law; who deserves more sympathy. If H determines that A is more

versy and refer to any such outsider as an "external helper" or simply a "helper." To remain flexible in notation, we'll talk about a negotiation between protagonists A and B with helper H.

In two-party disputes we can refer to the external intervenor as a third-party helper. In disputes with more than two parties, it is also common to talk of the third-party helper, although using the "$(n+1)$th party helper" would be more accurate.

It's obvious that the mission of the external helper is to help. But how? We'll start our discussion by presenting a series of sparse scenarios that fall into the {A, B; H} paradigm we have in mind. As we go through this list of types of interventions, keep in mind the following questions:

- Is it a dispute that needs to be resolved or a deal that needs to be structured?
- Is it H's role to be evaluative (judging in favor of A or B) or to be neutral?
- If H is to be evaluative, does H have the authority to impose a final contract?
- Is H concerned only with helping with inter- and intrapersonal problems, or is H there to help the protagonists find a "good" solution to their problem?
- Is H being proactive (by suggesting options, say) or reactive (for example, by recording what A and B agree upon along the way)?

Types of Interventions

Consider the following broad classes of interventions.

Quarreling Siblings with Mom as the H

Al and Betty are bickering over who should get the new fire truck, and mom intervenes. Ideally mom wants not to be evaluative but to teach the children the value of sharing. She might be proactive in suggesting to the children that they could share the fire engine over time or in introducing the new toy bear into the

---

righteous than B, then H might tilt his chosen arbitrated solution in A's favor. These are cases of *evaluative* arbitration. The judge's role is to be evaluative.

Now for *nonevaluative arbitration*: H has helped A and B to determine the efficiency frontier and joint RV. H must now determine an appropriate contract. When H debates whether he should opt for the Nash solution or the equitil solution, H is bending over backwards not to be evaluative, not to take sides. (For the two players, the *equitil* solution maximizes the common, equal utility value of the two players. It is the same as the *maximin* solution.) It's not a question of who is right or wrong. Such deliberations will be considered *nonevaluative*.

picture and suggesting a time-sharing deal between the two toys. Notice that H must cope with interpersonal antagonisms before suggesting problem-solving options. Is this arbitration? Dad could play similar roles.

The Manager as Intervenor

A and B are vice presidents in charge of different divisions in a large firm. There has been an embarrassing mini-disaster, with A and B each declaring the other responsible for the problem. Monetary losses are involved, and corrections in procedures will have to be initiated to avoid a repetition of the fiasco. A and B are so furious with each other that they cannot agree on anything. Out trots the president, H, to help resolve the disaster. H has the mandate to be evaluative and impose a solution, but he would rather not. H starts his intervention, through his mere presence, by getting A and B to calm down enough to resolve their internal dispute. Given an impasse, H might have to shift his role to that of judge, imposing penalties and dictating a resolution, but he may wish to find an efficient solution without being judgmental about it. This same story can be retold with H at any level of management and with A and B as any two of his inferiors in the corporate hierarchy.

Alternative Dispute Resolution for Contract Disputes

A and B are two firms engaged in a contract dispute. Private negotiations have failed to provide a solution, and before going to court they are required to submit their case to the Alternative Dispute Resolution Society. A and B are corporate members of this group, comprising hundreds of members, all of whom pledge that any dispute among members will first be mediated using a roster of professionally trained Helpers. An arbitrated proposal is not binding: either A or B may choose to go to court without publicly divulging the details of the ADR intervention to the judge. At the beginning of the twenty-first century, the number of firms voluntarily committing themselves to such ADR societies is large and growing. A firm not belonging to an ADR group may be shunned by those who are members.

Collaborative Lawyering

Lawyers may choose to join a society each of whose members pledge to conduct negotiations in a FOTE or POTE style. This variation, pioneered in Minneapolis and Cincinnati and growing, is known as "collaborative lawyering." If negotiations break down between two collaborative lawyers, the principal disputants then have to hire new lawyers for court proceedings. Hence with the joint representation of collaborative lawyers, there are no adverse incentives for the lawyers to be unduly adversarial and to claim strenuously enough to push the dispute into court—which may be financially remunerative to the lawyering class.

Divorce or Family Disputes

These disputes can be highly contentious, and negotiations are often conducted by lawyers representing each side. The lawyer-agents conduct prenegotiation dialogues, negotiations, and then represent their clients in court if need be. There is an alternative. The bickering parties can employ a single divorce mediator (usually a lawyer as well) to play an H role in their dispute. The hope is that H can help them avoid paying monstrous lawyers' fees and prevent protracted deliberations that could escalate either in or out of court. H tries to keep the deliberations between A and B civil and may be proactive in suggesting options for joint gains. H might refrain from being evaluative unless asked by both A and B, "Well, what do you think is right?"

Court-Mandated, Nonbinding Arbitration

The courts are clogged, and in an attempt to expedite cases through the judicial system, various states (including California) mandate that plaintiff and defendant first submit their case to a court-appointed arbiter (often a retired judge), who gathers facts and in an evaluative fashion recommends a disposition of the case. The case is often heard in a much less formal style than found in courtrooms, and often principals are not represented by lawyers. Either side can reject the finding and insist on a jury trial, but often the loser in the informal procedure decides his or her case is not strong enough to warrant going to court. This intervention is also used in other countries, with the added incentive that if the loser in the arbitration also later loses in the formal court hearings, he or she must pay the court costs, which can be sizable.

The Judge as Mediator

Judges often act as third-party helpers. The Federal Rules of Civil Procedure, for example, empower judges to hold pretrial conferences for the purpose of "facilitating the settlement of the case" (F.R.C.P. 16 (a)(5)). Many judges make active use of this provision in an effort to clear as many cases as possible from their busy dockets.

There are a lot of judges, of course, and they have different approaches to settlement conferences. But the common view is that most judges do not really mediate—rather, they conduct shuttle bargaining, alternately pressuring each side to be more forthcoming in its settlement offers. They warn those on each side about the risks they face at trial, express skepticism about the weaker parts of their case, and cajole them to lower their expectations. As one judge in New York told a lawyer, "I bashed on them this morning and now I'm going to bash your case down." Judges usually don't quiz the parties at length about their interests, or encourage them to listen to each other's story about the case, generate a list of options—or any of the other techniques that mediators recognize.

Litigants fear being seen by the judge as an obstreperous party, with subsequent damage to their chances at trial. Typically they do make some concessions, though often as few as they think they can get away with without provoking the judge's displeasure. Naturally enough, litigants and their attorneys don't find the process a pleasant one. Those who are familiar with mediation theory complain about these sessions. "I've been in mediations, and that was not a mediation" is the frequent refrain.

Judges don't approach settlement the way professional mediators would, because most judges lack the expertise. They aren't chosen on the basis of their mediation skills, and they aren't trained in them. Further, the stance or attitude appropriate to a judge is far different from the perspective that best serves a mediator. Most people are not flexible enough to completely switch professional attitudes from one moment to the next.

Aside from their lack of expertise, judges are not good mediators because they are not truly neutral. They have interests of their own at stake (clear dockets and speedy resolution) that may interfere with their disinterested efforts as mediators to help the parties make up their own minds. The fact that they will decide cases that don't settle also means that parties will be extremely guarded about revealing information that might prejudice their chances at trial. Most mediations are kept confidential to encourage candor—but few people will be truly candid before an official of the state who has the power to take harsh action against them. Even if the same judge does not preside over both the mediation and the trial, litigants will be suspicious that judges talk to each other in their chambers.

Proceedings Initiated by H

The Vatican intervenes in a territorial dispute.   Argentina and Chile had mobilized and were threatening to go to war over three small islands at the tip of South America. "Stop," interjected the Vatican, "give us a chance to mediate this controversy." So a distinguished cardinal elected himself to the role of H and met with the belligerents to seek an accommodation. But first came a preliminary round of protocol, and another, and another. Who should sit where, and speak first, and come through which door? I (HR) followed the details of this debacle and had misgivings about the cardinal's ability to reach an accommodation. Time passed. The controversy, which occupied a good portion of the front pages of the papers with tempers at the boiling point and with both sides mobilizing like mad, became relegated to small blurbs on page 10, and gradually people became bored. "What, go to war over those islands? Where are they?" The whole affair simmered down. I'm not sure whether the cardinal was aware of his brilliance. I only know that if I had been the arbiter at that time, I would have made the egregious mistake of delving into the substance of the case.

The Camp David interventions.   It was President Jimmy Carter who appointed himself to the rank of H and invited A (Sadat) and B (Begin) to the conference table at Camp David. This case involves some interesting analytical interventions and will be discussed at length later in this chapter. But let us point out here that A and B refused to be in the same room together and H had to shuttle back and forth. H was not evaluative, but H was proactive (with the help of some U.S. analysts in the backrooms) and offered some compromises that were duly rejected until rich man H threw in a few enticements for A and B—enticements that A and B could not afford to refuse—and a peace deal was struck that has been stable ever since. Bribery? Sure, why not. H can be thought of as a mediator with clout who has lubricating powers.

The Valletta Accord.   In 1991 the countries of the Conference on Security and Cooperation in Europe (CSCE) adopted a mechanism for the peaceful resolution of disputes between members of the conference. After direct negotiations between antagonists A and B, both members of the CSCE, have proved fruitless, one side can call for the intervention of an H, provided by the CSCE. Bringing in H does not require both A and B to get on board, but it requires a ripening of the dispute before the CSCE can intervene. Is this maturing or ripening of the dispute a good requirement?

Track II Negotiations

For political reasons, countries A and B cannot invite H to help them. But perhaps they can quietly suggest that surrogates A' and B' work with H in clandestine dialogues—only dialogues, mind you, and not negotiations (or in negotiations with small letter n and not Negotiations). Or better yet, let's keep A and B out of this altogether and have A' and B' work with H on their own initiative, without the knowledge of A and B, so that A and B have maximum downside protection. Or maybe H is astute enough, as in the Oslo accords that we'll consider later, to extend invitations to A' and B' to have cocktails with him and his wife where a certain topic might be considered in passing.

The Psychiatrist as Mediator

There is a parallel between the roles played by the facilitator, mediator, or arbiter in standard two-party interventions and those played by the psychological counselor in dealing with controversies within a single but divided mind. The counselor tries to be nonevaluative and, depending on the situation, responds reactively or intervenes proactively in trying to achieve compromise solutions that yield joint gains. The counselor's role is to be nonevaluative, but there are exceptions.

Types of Help

Helpers may play many different roles. Before we examine them, we note that these roles are frequently played out in the context of two classes of problems, which should be kept in mind.

1. *Disputes with heavy interpersonal conflicts.* In some cases, A and B have to take some form of joint action, yet they can't work together because of past, present, and future animosities. Such concrete, baggage-laden problems arise in divorce proceedings, family disputes, civil suits arising from accidents, ordinary cases exacerbated by vitriolic cultural differences, festering disputes between antagonistic national neighbors, and so on.
2. *Deals with noncollaborative protagonists.* There are also cases where A and B need to work together to solve a problem, but they don't know how, are suspicious of each other, lack trust, and there is tension about sharing information. Many such cases arise in labor-management disputes, in turf problems within an organization, in mergers and joint ventures, in contracting, between contending political factions within a government agency, and so forth. In other cases the protagonists are collaborative in spirit, but they are not experienced, capable problem solvers and they need consulting help.

Then there are gradations between these two and all kinds of mixtures.

Roles of an External Helper

We start by listing in Table 17.1 some roles that have to do with convening and moderating meetings, then functioning as a human relations counselor. We continue this listing in Table 17.2 by considering roles played by an analytically minded helper who would like to squeeze out joint gains and see that these are distributed with a concern for fairness.

Table 17.1   Roles of a nonevaluative, external helper

The external helper can help by
- Convening meetings
- Providing logistical support
- Bringing parties together who should be negotiating but aren't (a brokerage function)
- Chairing or moderating meetings: maintaining rules of civilized debate
- Acting as neutral, nonsubstantive discussion leader
- Setting the agenda
- Helping reticent parties have their say
- Preparing neutral minutes
- Preparing public relations documents
- Attesting that good faith negotiations took place
- Articulating consensus as it develops

The external helper can act as a human relations counselor by
- Assisting with personal, dysfunctional problems
- Controlling emotions
- Shuttling back and forth while angry parties are kept apart
- Correcting miscommunications
- Mitigating the effects of unintended, cultural faux pas
- Enhancing relationships
- Establishing a constructive ambience for negotiation

Before the parties can seek joint gains, they must first learn how to work constructively together, so that the roles in Table 17.1 must precede those in Table 17.2. The analytically minded intervenor tries to get the protagonists to adopt a POTE-like stance and, if successful, to convert this joint behavior into a FOTE-like stance by doing a bit of preaching and helping with the analysis.

The analytical intervenor may be invited to make proposals to the negotiators for their consideration, since the intervenor may have special problem-solving skills and be privy to confidential information gathered from both sides. In this capacity the intervenor acts like a voluntary, nonbinding arbitrator. Short of this extreme, the intervenor can draft what is called a single negotiating text (SNT) and make successive modifications of it after receiving feedback from the negotiators. This technique is complicated enough, and used extensively enough, to warrant a special discussion later on in this chapter.

The analytical intervenor may not be encouraged during negotiations to elicit confidential information about the preferences or about the BATNAs or RVs of the protagonists, but instead to help marginally by making some non-substantive process suggestions. Then, after the parties achieve a final contract, the intervenor might shift gears, become much more active, and be challenged to seek an improvement over the final contract that will be advantageous to both parties—either side can veto the intervenor's suggestions. For the intervenor to be effective in this role, he or she will have to become privy to more confidential

Table 17.2   Roles of an analytically minded, external helper

The analytical helper can help by
  • Helping each party privately to structure its own interests (needs, wants, fears, objectives), sorting out means from ends, and identifying a few evaluative interests (by asking the "why, what, and how" questions)
  • Helping each party to articulate its separate vision
  • Helping each party to grapple with its uncertainties
  • Discouraging positional bargaining and premature claiming tactics
  • Helping the parties jointly to brainstorm together to identify options for consideration
  • Helping them construct a template for negotiation listing the issues that need to be addressed and their possible resolutions

The analytical helper can help each party to privately evaluate the template by
  • Providing needed expertise (about the law, economics, etc.)
  • Conducting qualitative within-issue and between-issue analyses
  • Conducting quantitative within-issue and between-issue analyses
  • Helping the parties jointly to refine their template by discarding dominated resolutions and bringing the template into canonical form
  • Helping each party separately and confidentially to find its Best Alternative to a Negotiated Agreement (BATNA) and to convert that to a reservation value if the template is scored, or, if not, to develop a way of determining the feasibility of any contract that is a candidate for agreement
  • Helping each party to check the reasonableness of its alternatives—reality testing
  • Helping the parties to jointly share information and to collaboratively seek joint gains
  • Helping them to remember what has been said by the use of charts and decision aids
  • Helping them jointly to learn more about their uncertainties by modeling, sampling, and experimenting
  • Helping them to seek an efficient, equitable outcome—or at least a feasible outcome—perhaps by using a FOTE or POTE interchange; helping with closure and document agreements
  • Helping with implementation and verification

information from each side. Will they give this information truthfully? We'll also discuss this procedure later under the heading of Postsettlement Settlements in the next chapter.

There are still other roles that an intervenor can play, but you should get the picture: there is a broad range of options.

Negotiating the Mandate for the External Helper

Who decides what the intervenor should do depends on the context of the dispute or deal. It makes a big difference if the intervenor is an inviter or an invitee. By and large, in mediation, in contrast to arbitration, the roles of an intervenor

Table 17.3   Why external helpers (H) are not used

No one thinks of it.
H not perceived to be neutral.
H perceived to have interests of his or her own.
H may not understand the subtleties of the situation, be uninformed.
The parties fear losing control—being no longer the master of their own affairs.
An incompetent H may escalate the conflict.
H may add issues better left untouched.
H may inappropriately wish to explore linked problems.
Parties accepting H may feel they lose face and react against it.
Confidential information may be leaked.
Best solution may be obvious and easily arrived at.
Flexibility may be lost; H might complicate discussions.
Parties may fear H will be too legalistic.
H may be perceived as being too judgmental or evaluative.
H may not take into account power imbalances.
H may increase pressure to accept compromises worse than the status quo.
H may secretly not want to resolve the conflict.
H may want to keep the talks going indefinitely.
Involvement of H will increase uncertainty of the outcome.
There may be disagreement over who selects H and who pays.
H may have a bias to agreement when no agreement should be reached.

are themselves subject to negotiation. In this negotiation about mandates be-tween the external helper, H, and the main protagonists, A and B, what are H's interests and goals? Here is one such list.

Interests of an External Helper
1. To do good for the negotiators in terms of obtaining a feasible, reason-ably efficient, and equitable outcome for A and B. Also to *appear* to do good.
2. To enhance his or her own reputation. The outcome and the process should both be perceived to be neutral and fair.
3. To further his or her own substantive interests. These interests may in-volve: the "world good"; protecting the weak or unrepresented stake-holder (for example, children of divorces, deceased benefactors, con-sumers).
4. There may also be personal goals: compensation; time to completion of intervention—there are perverse cases where intervenors are accused of wanting to keep the negotiations going for as long as possible; recom-mendations for follow-up business.

Just what H will be asked to do in a given intervention will depend on the context of the problem and his or her reputation. If A and B have doubts about the quality of H or about H's neutrality, they may want to restrict the mandate for intervention. One possibility is to increase the role of the helper adaptively as A and B's confidence in H increases or as they become more and more desperate in needing outside help.

Just another negotiator?   How is the external helper different from A and B? In most cases A and B have veto power over H. Either one can fire the helper — unless of course it's a case of mandatory, binding arbitration. Usually H has only the power bestowed on him or her by A and B. H's performance rating also depends on how well A and B do. H is trying, for the most part, to achieve good outcomes for A and B. Their perception of H is critical to H.

Why external helpers are not used.   Intervention is a growth industry, but all around us we see parties floundering in discordant negotiations that could profit from some sort of intervention. Why is this the case? Table 17.3 lists some of the reasons.

Case Study: The Camp David Negotiations

The historic Camp David negotiations will be used here to illustrate the role of a third-party intervenor with clout, and as a basis for discussing a technique that employs what is known as the "single negotiating text."

Historical Background

After meeting with his senior policy advisers in the summer of 1978, President Carter decided that without his presidential intervention the Egyptian-Israeli peace process would collapse, and that given Secretary of State Cyrus Vance's report about glimmers of flexibility, a three-nation summit would be a reasonable gamble.

On August 4, Vance flew to the Middle East in an effort to break the impasse that had developed between Egypt and Israel during the previous few months. His trip, however, had a more specific purpose than the American public was led to believe. In an attempt to revive the momentum toward peace that had been created by the visit to Jerusalem by Egypt's president, Anwar Sadat, Vance carried with him personal invitations from Carter to Sadat and to Israel's prime minister, Menachem Begin, to join him at Camp David, Maryland. On August 8, the White House issued the following statement: "The President is pleased to announce that President Sadat and Prime Minister Begin have accepted an invitation to come to Camp David on September 5 for a meeting with the President to seek a framework for peace in the Middle East. . . . Each of the three leaders will be accompanied by a small number of their principal advisors and no specific time has been set for the duration of the meeting."

To prepare for the upcoming U.S. mediating effort, Carter set up a task force that included Zbigniew Brzezinski and William Quandt of the National Security Council and, from the State Department, Harold H. Saunders and Alfred L. Atherton, Jr. (both assistant secretaries for Near Eastern and South Asian Affairs), as well as Vance. The task force was to derive methods or tools of mediation to be used by the president, to "invent" solutions, and to identify compromise language acceptable to both Egypt and Israel.

What were the United States' interests in the upcoming summit discussions? In 1975 a report entitled *Toward Peace in the Middle East,* prepared by a Brookings Institution group that included Brzezinski and Quandt, had presaged the Carter administration's comprehensive approach to the settlement of the conflict in that region. The report had reached the following conclusions. First, the United States had a strong moral, political, and economic interest in the resolution of the Middle East conflict. Second, unless the core issues of the Arab-Israeli dispute (such as the Palestinian issue) were addressed soon, the risk of another war would increase. Third, future negotiations should make use of informal multilateral meetings or a reconvened Geneva Conference. Fourth, the United States, "because it [enjoyed] a measure of confidence on both sides and [had] the means to assist them economically and militarily," should remain actively involved in the settlement.

Preparing the Mediator: The U.S. Team's Approach

The members of the team advising Carter were not new to the Egyptian-Israeli situation. They had already thought deeply about their preferred solutions. They knew what issues had to be debated at Camp David and they knew how the Military Committee and the Political Committee had already structured the issues dividing the two sides. In addition, the members of the U.S. team were familiar with the Israeli proposal of December 31, 1977, called the "twenty-six point self-rule plan." They knew a lot about both sides; they could have assessed—but evidently did not assess—a multi-attribute value function for each side and even one for the United States, as well as reservation values on packages and on individual issues. Note that the set of negotiators from each side did not have a monolithic position—to say nothing about the contending factions back home —and that many concerned parties hovered on the fringes: the Arab states, the PLO, the Soviet Union, and a number of oil-starved developed and developing nations. Crisp formalization would have been almost impossible.

Carter and his team decided that progress could not be made in a fishbowl atmosphere: privacy during the negotiations was vital. Carter also tried desperately (futilely, as it turned out) to create a cordial ambience for negotiations and to get the contending parties to approach the problem as a joint problem-solving exercise. In addition, it was critical for the world, and especially for the political forces within Israel and Egypt, to know that three very important world leaders were isolating themselves from all other duties in order to devise a compromise accord—an accord that could only be acceptable to Egypt and Israel if it did not come easily. Any quick agreement was destined to meet trouble at home.

The U.S. mediators did not want both sides to come to the negotiating table with fixed packages. A dance of packages had already been tried, and the gaps were formidable. The mediators tried initially to get the principals to build up a package issue by issue, but they expected that this strategy would not work—there would be too much claiming along the way that would interfere with creating joint value and too much posturing to appease the hard-liners on the negotiating teams. The mediators were right. By day two, Begin and Sadat would not talk to each other. What could be done?

The conflict was mediated through the use of a single negotiating text, a device suggested by Roger Fisher of Harvard Law School, who knew some of the key U.S. players (including Atherton, Quandt, and Brzezinski). Some sort of SNT is often employed in international negotiations, especially with multiparty negotiations. The U.S. team devised and proposed an entire package for the consideration of the two protagonists. They made it clear that the United States was not trying to push this first proposal, but that it was meant to serve as an initial, single negotiating text—a text to be criticized by both sides and then modified and remodified in an iterative manner. These modifications would be made by the U.S. team, on the basis of the criticisms of the two sides. The SNT was to be used as a means of concentrating the attention of both sides on the same composite text.

Neither side formalized its value tradeoffs; but if they had, then the United States might have generated a set of feasible joint evaluations and an efficient frontier (as shown in Figure 17.1). Assume that the ranges on each of the issues have been specified in advance; that each side has scored the worst possible agreement for its side at zero and the best agreement as 100; and that both sides have monolithic preferences. It is not necessarily true that the agreement that is worst for Israel is best for Egypt, or vice versa.

The United States starts the ball rolling by offering its first single negotiating text (point SNT-1 in the figure). Both Begin and Sadat protest that the proposal is ridiculous, whereupon the mediators assure them that SNT-1 is not intended to be a final settlement, but as a document to be criticized and improved upon: why, they ask, is it so unacceptable? The mediators know very well why each side is so vehement in its rejection of SNT-1. This is part of the ritual. After some of the most egregious flaws have been pointed out by each side, the U.S. team comes up with SNT-2. Begin and Sadat, although they may agree that this text is marginally better than SNT-1, still claim that it's so far from being acceptable that they feel they're wasting their time. Sadat picks up his bags and gets ready to go home, but Carter persuades him to stay for a few more rounds.

After SNT-2, the United States offers a new SNT, but the Israelis feel that

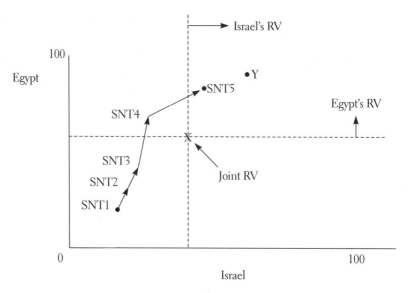

Figure 17.1. A hypothetical march of evaluations of a single negotiating text.

this "improved" text is marginally a step backward—and a step backward from a hopelessly unfair starting point. So the United States comes up with a revised SNT-3; then with SNT-4 and SNT-5. Now let's imagine that the improvement from SNT-3 to SNT-4 was a critical jump for the Egyptians, because the transition pierced their real reservation value—that is, Egypt truly preferred no agreement to SNT-3, but preferred SNT-4 to the no-agreement state. There still may be joint gains to be had, however, and if Egypt announces that SNT-4 is acceptable whereas Israel does not, then the ensuing gains are going to be tilted toward the Israeli side. This would not be a disaster for Egypt if this is the only way Israel can get over its reservation hurdle, but Sadat might think that the Israelis are already satisfied and are just trying to squeeze out more at Egypt's expense. So he still maintains that SNT-4 is unacceptable, but his protests are less vehement than before.

With the proposal of SNT-5, Israel's reservation value, too, is pierced. Will Begin announce this? Probably not, for the same reasons Sadat did not. But now it is no longer possible to squeeze out additional joint gains. If SNT-5 is modified to the advantage of one side, it is only at the expense of the other side. In Figure 17.1, SNT-5 is on the efficient frontier and no achievable joint evaluations are northeast of it.[2] Point X represents a composite reservation value: Egypt would rather have no agreement than any deal that yields an evaluation south of X; Israel would rather have no agreement than any deal that yields an evaluation to the west of X; both sides would prefer to have any point northeast of X rather

---

2. Our example is simplified—Carter's team reportedly prepared twenty-three drafts.

than the no-agreement state. But each, acting strategically, does not announce that SNT-5 is better than no agreement. Of course, if the composite reservation value were Y rather than X, then they would be acting sincerely in their rejections of SNT-5. We're dealing with idealizations here. The reservation values are vague, and a politically acceptable agreement is usually one that has been difficult to negotiate.

Assume that both sides claim that they cannot settle for SNT-5, and that it proves impossible for the mediating team to squeeze out further joint gains. What now? The mediators are very discouraged, since the United States, too, has a stake in the negotiations. It may now be propitious for President Carter to give up something. Perhaps Israel could accept SNT-5 if the United States funded the construction of new airfields in Israel to replace those of the Sinai. No? Well how about some oil guarantees as well? And might Egypt accept SNT-5 if the United States provided some financial aid for Egypt's ailing economy? So the president applies pressure and offers sweeteners, and a deal is struck.

Did Egypt and Israel expect the United States to sweeten the pie? Did they gamble by declining SNT-5 in anticipation of a U.S. contribution? Did the United States anticipate that it might have to offer inducements to each side in order to generate an agreement? In preparing for the meetings, did all three sides think hard about what the United States could offer, and about their trade-offs and reservation values with potential U.S. contributions? Did Egypt and Israel agree to come to Camp David because this would put pressure on the United States to get an agreement? Did the U.S. team know that Egypt and Israel were thinking this way? Did they know that the mediators knew that they knew? Did Egypt and Israel engage in tacit collusion to squeeze the United States?

Most of these questions probably have affirmative answers, at least to some degree. Is this morally wrong? Some might be tempted to say that it is, because they are not pleased with the outcome of the Camp David accords. But leaving that aside, would it be morally reprehensible, in principle, for statesmen to behave so strategically? Our general answer is that it is not morally reprehensible— that leaders who do not act strategically may not be behaving in the best interests of their constituencies.

Critique of the Single Negotiating Text Process

The SNT process worked at Camp David, and it has worked elsewhere. But the use of successive modifications of SNTs has weaknesses.

Where you end up depends on where you start. Consider Figure 17.2. Starting at SNT-1′, the parties might have ended up at X; starting at SNT-1″, the parties might have ended up at Y.

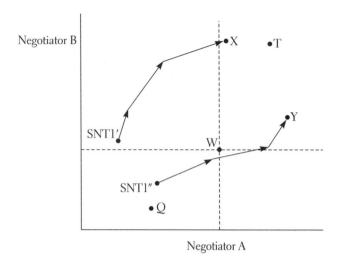

Figure 17.2.   The final outcome of a single negotiating text depends on the choice of starting point.

Now let's consider the effects of the paths leading to X and Y when viewed from the vantage points of different joint reservation values. If the joint RV were at Q, either X or Y would be pretty good outcomes: both would be efficient and tolerably equitable. If, however, the joint RV were at T, neither X nor Y would be a feasible outcome. Continuing, if the joint RV were at W, the path from SNT-1′ would start as acceptable to B and get better and better, whereas the path is brought into feasibility range for A only at the end; outcome X is jointly feasible, but highly inequitable for A since X is just barely acceptable for A and highly desirable for B. Starting at SNT-1″ and a joint RV at W would lead to a reasonably equitable outcome at Y. The point of all this is to show that revisions of SNTs can work, but can also lead to inequitable outcomes. The dynamics are complicated, and a lot depends on luck. In choosing a starting point for the SNT process, it's critical to have some idea where the joint RV lies.

One way to do better than the SNT process is to somehow convince the parties to formalize the template and to negotiate, with the help of the active intervenor, in a FOTE-like manner. Easier said than done, granted.

Core Concepts

Part IV discusses how external helpers can be of assistance in two-party negotiations. To set the stage for a discussion of the many ways a third party can help, we have mixed things up by looking at some nonstandard examples of interventions, such as a mother who acts sometimes as mediator and sometimes as arbiter in disputes among her children. The corporate world has seen a remarkable growth in firms who pledge to submit themselves to Alternative Dispute Resolution interventions before fighting it out in court. There is also an incipient trend among lawyers, called "collaborative lawyering," in which lawyers pledge themselves to negotiate only in a collaborative FOTE or POTE style.

We have listed possible roles and ways a third-party helper can be of assistance to two parties in need. These range from help with interpersonal frictions to help with problem-solving skills. There are many roles the helper H can perform, and which ones H takes on is often a matter of negotiation between the three parties.

The intervenor becomes a player in the drama, and it is instructive to examine H's interests (objectives) in and strategies for influencing the outcome of negotiations. It's important to reflect that while H is another player, H is a different kind of player. We have also commented on why external helpers are not used more frequently.

The Camp David case study provides an example in which the helper H (Jimmy Carter) was the inviter rather than the invitee. H's team played a very proactive but nonevaluative role in generating an acceptable compromise. Since the protagonists A (Sadat) and B (Begin) could not negotiate face to face despite the interpersonal interventions of the skilled H, H resorted to the device of having A and B critique a series of single negotiating texts until a compromise emerged. It helped when H lubricated the deal by offering enticements to A and B and offered them his special services in securing insecure contracts.

18

Arbitration: Conventional and Nonconventional

We continue with our abstraction {A, B; H}, which depicts negotiators A and B working with an external helper, H. In this chapter, H plays the role of arbiter or arbitrator. The first part of the chapter deals with evaluative interventions; the second examines nonevaluative interventions. This may seem strange: how can H decide an outcome in a dispute or deal without being evaluative? We mean "evaluative" in the sense of deciding what is fair—who is right, who is wrong, and declaring that such and such an outcome is *the* right one. The nonevaluative arbitrator supplies *a* solution. It may involve more general universal criteria without delving too deeply into the specifics of right and wrong in the case. We have already introduced such situations in Chapters 13 and 14, when we talked about resolving equity issues by using the Nash solution that maximizes the product of the players' excesses over their RVs or by invoking the principle that equalizes the proportions of their potentials (POPs). In the second part of this chapter we'll return to that discussion, this time with an external helper.

Evaluative Interventions

Conventional Arbitration

In the standard situation, disputants A and B can't resolve their dispute by themselves, and a third-party arbiter, H, after hearing the facts, unilaterally decides on an outcome that is binding on the parties. H sits in judgment and determines who is right or who has the more compelling argument or whose case is more backed by law or tradition.

In the usual case, the argument is about the determination of a single quantity: a wage rate, a penalty value for noncompliance, the amount of compensation A owes to B, the amount of time A must spend until his obligation to B is fulfilled. These are distributive negotiation problems and inherently zero-sum: what one party gains incrementally, the other party loses incrementally. But not all situations are of this kind. Some situations are integrative and require the arbiter to offer resolutions on several issues. Unlike adjudication in a court of law, there is usually no requirement that the arbiter be explicit about divulging his thought process for the record. The judge in a court proceeding is a special type of arbiter. Since parties can be dragged to court without their consent, and because the judge wields the coercive power of the state, she has a greater duty to guard against procedural unfairness. The judge not only has to resolve the dispute but to maintain consistency with past cases and articulate the reasons for her evaluation, which become part of the record.

So arbitration is not a true judicial proceeding. Neither is it a mediation or a facilitation. Still, a broad range of processes come under this rubric.

Other Types of Arbitration

For our purposes we can catalog arbitrations in the following categories. We will expand on some of the divisions; others are reasonably transparent.

- Is the process evaluative? If yes, how explicitly does it have to be documented?
- Is the intervention mandatory or voluntary?
- Is the final determination of the arbiter binding or suggestive?
- Is the dispute distributive or integrative or both?
- Does the arbiter take into account the BATNAs and RVs of the disputing parties?
- Is the arbitrated resolution restricted to a finite range created by the disputants?

These distinctions, of course, are not always so clear cut. As we saw in the last chapter, some arbitrations are undertaken informally, usually by a retired judge, who simply tells the parties how he would rule. The hope is that the loser will be dissuaded from wasting further resources on a losing case, but there is nothing stopping a party who insists on having a day in court. Similarly, the decisions of the World Court in The Hague are ostensibly binding on all countries, but the court has no way of enforcing its ruling on a nation that rejects the outcome.

Or the opposite can be true: an arbitrator may be so empowered that he essentially takes over from the parties. This role can be institutionalized. We know a company where A and B each own 49 percent of the voting shares and C owns the remaining 2 percent. A and B ran the company amicably, and C was called in to arbitrate any differences arising between A and B. It worked for twenty years, but now B claims C has secretly sold out to A and they are squeezing him out of his rightful ownership. Party C has an effective mandate for performing evaluative, mandatory, binding arbitration. Party C, by colluding selectively between A and B, can effectively run the company.

But now let's look at a few of these distinctions in more detail.

Bounded arbitration. Many different examples fit the following story line. Ms. A and Mr. B don't like each other, and instead of going to court to resolve their differences, they have been talked into submitting their dispute to binding arbitration. Both are anxious, wondering what that crazy arbiter will think appropriate and fair. The arbiter must come up with an amount x for Ms. A to pay Mr. B. To make life a bit simpler, suppose the value of x will be some dollar value between $0 and $100 million, and both A and B have confidential, subjective probability assessments for the x-value to be determined by the arbitrator. Both are risk averse, and their utility functions for x-values are shown in Figure 18.1.

Someone suggests to the disputants that they can, if they wish, bound the award (from A to B) of the arbiter by agreeing on a low (L) and a high (H) value. If x is below L, Ms. A agrees to pay Mr. B the amount L; if x is higher than H, Mr. B receives the value H; if x falls between L and H, that's the final value. The bounds are so chosen to give some reassurance to the parties that they will not suffer egregious losses. Ms. A wants to push H to the left to bound her losses and she is willing to push L to the right to make Mr. B happier. Ms. A and Mr. B now start negotiating about appropriate L and H values. If L and H turn out not to be too far apart, then the parties might agree to the suggestion of settling for $(L + H)/2$.

How should Ms. A decide on the merits of one (L, H) pair versus another? This is a standard individual decision problem. She starts by assessing her probability distribution for x—perhaps laying this out on a computer spreadsheet—and then determines her utility function for her payment x. For any (L, H) pair, there is then a risk profile that can be analyzed by computing A's expected utility for that bounded uncertain x value. With the computer, the spreadsheet can be structured in a way that enables Ms. A to explore different (L, H) pairs. Mr. B does likewise, and they now can explore efficient and equitable (L, H) pairs.

As a second-order consultant (consultant to a first-order consultant) I (HR) worked on a problem involving the breakup of a company into two divisions, A and B, where the arbitrated transfer payment x from A to B could possibly be between zero and one billion dollars—that's a big firm!—and I thought it would be helpful if the parties could find values like L = $200 million and H = $700 million to bound that risk. But the contending parties had such disparate probability distributions for x that they chose not to investigate that proposal.

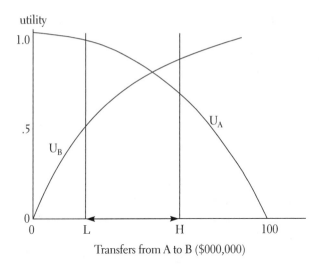

Figure 18.1. Bounded arbitration.

Selective arbitration. Parties A and B are engaged in integrative negotiations. They have adopted a POTE modality—the truth but not necessarily the whole truth will be told. They have jointly constructed a template listing the issues to be resolved and their possible resolutions. They have agreed on everything except issue 11, where they remain far apart. It's not worth giving up on an overall agreement because of their differences on issue 11, and jointly they agree to let some external neutral, H, choose one of a set of prespecified resolutions. It's a pragmatic approach for resolving their differences. If they argued it out, it might sour their relations. Note the relation to what we called bounded arbitration. The parties don't want to be surprised by H's choosing some outlandish alternative, and therefore they restrict H's choice. In this respect this procedure is like final-offer arbitration, but for a subset of issues. The arbiter's mandate for choice is limited. A and B think of H not as performing evaluatively but as playing a thoughtful role that beats using a random device.

Final-offer arbitration. The firemen of Podunk City are unhappy: they want more, just like everybody else (especially the city's policemen and sanitation workers), and they feel they deserve more. The city says it can't afford additional expenditures, especially for what it considers to be the outrageous demands of the firemen. The negotiators for the firemen and for the city have settled the fringe issues, and what remains to be settled is the "basic wage rate," which indexes salaries at all levels. Both sides have made known their ostensibly reasonable demands and are now playing a waiting game. Both sides are adamant, and time is passing. The city cannot lay off the firemen. The firemen, by law, cannot strike.

In such public-service disputes, compulsory, binding, interest arbitration—
"compulsory" as opposed to "voluntary," "binding" a opposed to "nonbinding,"
"interest" as opposed to "grievance"—has been mandated in many states. An arbitrator is appointed and, after determining the facts, must dictate the imposed
outcome.

Some of these states require what is known as *final-offer* or *last-offer* arbitration. With minor modifications, it works as follows. Negotiations are divided
into two phases. In Phase 1 the parties bargain directly, with or without the
aid of an intervenor (mediator). If the parties agree, there is no second phase.
If the parties disagree, the negotiations enter Phase 2, at which point the arbitrator—not the mediator wearing a different hat—enters the scene. In most
states the arbitrator does not obtain guidance or information from the mediator present in Phase 1. The arbitrator determines the facts and then demands
from each party a sealed final offer—or bid, if you will. These final offers are
submitted essentially simultaneously, and the arbitrator must then, by law, select one of these two final offers; no in-between compromises are permissible, and the selected final offer becomes binding on both sides. This final-offer
arrangement is often appealing to the arbitrator, who doesn't have to search
his or her soul to come up with an optimal resolution value: just choose between the two offers on the table. On the other hand, there are certainly cases
where the arbitrator is not sure which offer to select and would love to be able
to take the midpoint as a compromise; but this is not legitimate in final-offer
arbitration.

Baseball salary negotiations. Final-offer arbitration has been used in the resolution of salary disputes in major league baseball. In 1973 it was agreed that,
starting with 1974 contracts, final-offer arbitration could be invoked by either
players or by clubs in an impasse over salaries; once invoked, it would be binding on both sides. The guidelines for arbitrators were established in the 1973 basic agreement, which states:

> The criteria will be the quality of the Player's contribution to his Club during the
> past season (including but not limited to his overall performance, special qualities
> of leadership and public appeal), the length and consistency of his career contribu-
> tion, the record of the Player's past compensation, comparative baseball salaries,
> the existence of any physical or mental defects on the part of the Player, and the re-
> cent performance record of the Club including but not limited to its League stand-
> ing and attendance as an indication of public acceptance (subject to the exclusion
> stated in (a) below). Any evidence may be submitted which is relevant to the above
> criteria, and the arbitrator shall assign such weight to the evidence as shall to him
> appear appropriate under the circumstances. The following items, however, shall
> be excluded: (a) the financial position of the Player and the Club; (b) press com-
> ments, testimonials or similar material bearing on the performance of either the
> Player or the Club, except that recognized annual Player awards for playing excel-
> lence shall not be excluded; (c) offers made by either Player or Club prior to arbi-

tration; (d) the cost to the parties of their representatives, attorneys, etc.; (e) salaries in other sports or occupations. (Chelius and Dworkin 1980, p. 296)

Of special interest here is exclusion (c), which attempts to prevent concessions (or nonconcessions) made in Phase 1 negotiations from influencing the arbitrator in Phase 2 negotiations.

The rationale for final-offer arbitration. In practice, final-offer arbitration is quite effective in persuading parties to settle without an imposed, arbitrated solution. In the {A, B; H} abstraction, let A be a maximizing female and B a minimizing male. In order for A to increase her chance of winning (which means, in this case, having the arbiter prefer her sealed bid), she should be less demanding and lower her contemplated offer or bid. There is a similar logic for B. So the perception is that there is motivation for both sides to be reasonable in submitting their sealed offers. But it's more complicated. Party A should be interested not only in her probability of winning but in the amount she wins, if she wins. Increasing A's contemplated offer decreases the chance of her winning, but increases the potential size of her winnings. In the appendix to this chapter we develop this problem further from a game-theoretic perspective.

Nonevaluative Interventions

The Neutral Joint Analyst as Arbiter

My most rewarding experiences have been those rare occasions where I have been asked as an analyst to help both sides in complex negotiations. I have also sometimes been requested to give partisan advice to one party, and early on I asked whether my client would allow me to approach the other side as well with the idea of playing the role of neutral joint analyst. Mostly the responses have been negative, because of a zero-sum mentality: if you're going to help the other side, it must come at the expense of my side. That's a bit naive. Let me describe what I mean by *neutral joint analyst*, or NJA.

An NJA tries to get nonbelligerent negotiators to adopt a simple collaborative joint style for negotiating and then, as teacher and analyst, guides them through the analytic approach featured in Chapters 11 to 14. The diverse activities of the NJA are exhibited in Table 18.1.

Before disclosing the arbitrated final contract, the NJA should discuss whether the parties want this contract to be suggestive or binding. Note that the NJA's choice of an arbitrated contract would take into account the parties' reservation values, but it would not be evaluative in the sense of tilting in the direction of the more deserving party. The NJA must, however, choose some fair mechanism such as choosing the Nash outcome or equalizing the two POPs (proportions of potentials). This will be discussed at length in the next chapter.

Table 18.1   Roles of a neutral joint analyst (NJA)

The NJA can attempt to implement a full collaborative FOTE analysis by
- Discussing the merits of adopting a FOTE approach with the aim of achieving an efficient and equitable outcome
- Helping the parties implement collaborative negotiations as discussed in Chapters 11–14

In particular, the NJA may help with
Phase 1: Preparation
- Helping each party privately structure its own interests (needs, wants, fears, objectives), sorting out means from ends, and identifying a few evaluative interests (by asking the "why, what and how" questions)
- Helping each party articulate its separate vision
- Helping each party grapple with its uncertainties

Phase 2: Preparation (construction of the template)
- Encouraging the parties to share (adaptively) their interests and visions
- Addressing uncertainties
- Encouraging the parties to brainstorm together
- Getting them to work together (such as appointing joint subcommittees to explore uncertainties or to model future operations)
- Helping them jointly to design and adopt a template for further negotiations (listing issues and their possible resolutions)

Phase 3: Preparation (evaluation of the template)
The NJA can help each party privately to evaluate the template by
- Conducting qualitative within-issue and between-issue analyses
- Conducting quantitative within-issue and between-issue analyses
- Helping each party separately and confidentially to find its Best Alternative to a Negotiated Agreement (BATNA) and to convert that to a reservation value
- Helping the parties jointly to refine their template by discarding dominated resolutions and bringing the template into canonical form;

Actual negotiations (template analysis)
The NJA can obtain the confidential scorings of each party and privately construct (as described in Chapter 14) the efficient frontier. Then without showing this to the negotiating partners, the NJA can either
    a. guide the negotiators in their search to find some feasible, efficient contract, or
    b. select an equitable and efficient arbitrated final contract.

Also note that if the negotiators choose a FOTE style and are asked to reveal their RVs confidentially to the NJA, they should do that *before* the efficient frontier is revealed to them. It is far less tempting to slant one's selection of an RV strategically if there is a veil of ignorance about the efficient frontier. Not only does this help motivate each player to disclose his or her true value, but it provides more security to each that the other player is similarly motivated.

Combining POTE with a double-auction mechanism.   If the parties balk at disclosing their RVs, it will be difficult for the NJA to suggest a final agreement. Here is one mechanism that might be usefully employed.

Let the parties proceed in a POTE manner and submit to the NJA their confidential scoring of the template. Each party is then instructed to submit confidentially a *reservation bid* to the NJA that stipulates its minimum acceptable scoring value. The reservation bid can be the same as that party's reservation value, but it could also be something higher. The NJA then determines whether the two submitted bids are compatible—that is, whether there are feasible contracts that dominate the joint reservation bid. If so, the NJA would select a best outcome on the efficient frontier, treating the joint reservation bids as though they were the joint reservation values. If the joint reservation bids are incompatible, negotiations are terminated without a solution. No second thoughts. No rebidding. This rule is designed to force the parties not to exaggerate too much. Indeed, in the veil of ignorance regarding the exact whereabouts of the efficient frontier, the parties may select reservation bids that are close to their reservation values. This resolution mechanism is akin to the use of a double-auction procedure in the distributive bargaining problem. (See Chapter 8.) But the more complex the problem, the more incentive there is to tell the truth. It's too complicated to lie.

The role of the NJA can easily be computerized if each of the negotiating parties agrees to submit its confidentially scored template and its reservation bid (perhaps different from its reservation value) to the computer. Indeed the computer can provide a software package that could assist each party to score the template. Going backward, the computer package could also help the parties in their joint efforts to construct a suitable template by posing a series of questions to each protagonist.

Postsettlement Settlement: Embellishment

Let's suppose that negotiating parties A and B, working by themselves, end their negotiations at point X in Figure 18.2. They know that X is jointly feasible, but they are not convinced that they can squeeze out further joint gains. Can a bigger pie be baked? Enter a new type of helper, H, whom we'll refer to as an embellisher, who proposes the following deal to the negotiators. "Look, I might be able to help you. But you will have to cooperate. I'm going to ask each of you for a lot of information about your tradeoffs, but I'll be careful to avoid questions about your BATNAs or RVs" (we hope the embellisher would use less jargon with most negotiators). H continues, "With the information I collect from the two of you, I will then be able to find out if there are joint improvements over your negotiated outcome X. If I think there are, I have two alternatives:

1. I could propose, for your consideration, a contract on the efficient frontier that dominates X and that I deem fair from what you have told me; but if either of you vetoes my proposal, then X will prevail.

2. If you don't want me to make a final proposal, I could suggest to you how you might continue your deliberations to squeeze out more joint gains. For example, my advice might suggest that the two of you spend more time working on, say, issues 5, 9, and 12, since there are joint gains to be realized involving those issues."

Now suppose A and B go along with the embellisher-helper, H; they help him with designing the template and they each work with him confidentially to score the template. The embellisher performs like a neutral joint analyst, except that now no embarrassing questions are asked about RVs. H then works alone, finds the efficient frontier, and ascertains that there are indeed contracts out there that jointly dominate X. Embellisher H then offers an efficient and fair point on the frontier that dominates X. Take it or leave it. What we mean by "fair" will be further explored in the next chapter. It could be that X is efficient and no jointly better outcome is possible. It could be that A and/or B might not have told the truth to H. Or perhaps A and B have changed their minds, and their preferences have shifted. In any case, A and B are in control at the end.

We ask the reader, "If you were a negotiator like A or B, would you use the services of the embellisher? Would you disclose your preferences truthfully to him?" We have done this in the classroom/laboratory setting, and most student-subjects agree to use the embellisher. Most of them sheepishly admit that they also told the truth about their preferences, because they didn't know how to lie. (How best to lie, if you are so inclined, is a formidable task and it's not easy to do. It's too complicated.)

There are some special cases where embellishment is especially appealing.

- *Financial embellishments.* In financial deals, parties may divide the complexity of the negotiations. They first explore deals with immediate monetary transfers. If they can get some sort of accommodation, they may then open a second round of negotiations to look for embellishments to the deal by introducing delayed incentives (such as bonuses), contingency transfers, and financial option deals.

- *Decompositions.* In very complex negotiations, the negotiators may divide the set of issues into those of primary importance and those of lesser importance. They may refuse external help on the primary issues, but if they can get an agreement on those issues, then in a more relaxed atmosphere they might agree to use helpers, including embellishers, on the secondary set of issues—especially if those issues are technical and relatively apolitical. But there is a downside here. In negotiating their way to a jointly acceptable outcome on the truly divisive issues, they may be tempted at some stage to break off negotiations and implement their BATNAs. But their RVs for the negotiations on the critical issues should, in principle,

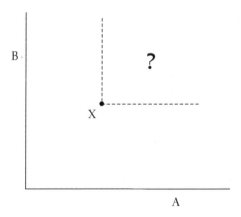

Figure 18.2. Embellishment. The arbitrator searches for possible agreements in the space northeast of X.

depend on what sweeteners can be achieved from the secondary class of issues. This is an argument for not attacking the primary issues first in isolation from the rest. It may even favor attacking the secondary issues first, or at least starting preliminary explorations of the second category of issues in tandem with the first group. We'll return to this topic later in Chapter 20.

Embellishment in the absence of agreement. Embellishment is a bit like nonbinding arbitration. The embellisher finds out the facts and then proposes a solution. But there is a whopping difference: in embellishment the parties do not have to disclose their BATNAs or RVs. The status quo point is the fallback contract point X and not the joint RV.

It remains to talk about the special, but critically important, case where A and B cannot by themselves agree on any feasible solution. How about embellishments in this case? The status quo point would now be the joint RV, and one of the advantages of embellishment disappears. If the default is the joint RV point, then the embellisher would have to probe into the alternatives for each disputant away from the bargaining table, and these possibilities may be too sensitive to share with any external helper. On the other hand, the protagonists, having failed to come up with any agreement on their own, may be more willing now to take another shot at negotiations—this time with an external helper.

Putting it into practice. There are some missing ingredients. For instance, how does the embellisher get paid? In the laboratory, where scores count, players A and B are each asked to give the embellisher 20 percent of their incremental additional gains. Who selects the embellisher? Who suggests the role of embellishments to the parties? Much of this has to be made accessible by specialized institutions that feature embellishment. One possibility is to add embellishment to the activities that mediators are prepared to perform.

Embellishment works better in the laboratory than in the real world. After a long bout at the negotiation table, the parties seem reluctant to go through all the work of scoring the template. Moreover, who will ever know if the final contract is efficient or equitable?

Case Study: Fair Division of an Art Collection

We can see how a nonevaluative intervention works by considering one of my experiences as a neutral joint analyst—although at the time I viewed my role as a proactive (analytic) mediator. I was approached by two brothers, G and R, who had inherited an art collection bequeathed to them by their recently deceased mother, Mrs. K. They were instructed to divide the collection "equally," without further guidance. The collection comprised about ninety-five individual pieces by some ten artists. The brothers had read my book *The Art and Science of Negotiation*, and since both knew me, they sought my counsel and help.

Brothers G and R were in their late forties at the time, and despite severe sibling rivalries, they and their families maintained, and wished to continue to maintain, cordial relations. They knew that they had to have outside help in order to divide the art collection without seriously impairing their tenuous, but highly prized, relationship. It also helped that the families lived a thousand miles apart. A few years before, G and R had worked together on settling their father's business affairs and nearly broke apart irrevocably. On that occasion, they also had outside help, but to no avail. They were all the more skittish about the present task because they had grown up with the paintings and had emotional ties to them.

I had dinner with them one evening, and they talked about the nature of their problem, the size and value of the collection, and were frank about their fears of negotiating together, alone, on a face-to-face basis—they knew it would tear apart the families. I also found out that there were two independent, incomplete appraisals of the paintings, but both brothers emphasized that the monetary appraisals were not critical, because each valued the paintings much differently. "After all," they told me, "this is a matter of aesthetics, tradition, and emotion and not a matter of money." Nevertheless, I concluded that the collection was then worth over $1 million; since then its value has probably quadrupled.

The negotiated process.   I was intrigued and decided that I had a lot to learn from this exercise; therefore I suggested a ridiculously low consulting rate which they gladly accepted. That was an easy negotiation. I was in it for the challenge, the experience, and the chance to help the brothers, both of whom I liked.

At the dinner table at that first meeting, I told them how I proposed to proceed. I would keep them apart and by some means or other I would collect confidential information about how they each felt about the individual paintings. I would subsequently suggest allocations to them that either one could veto, but I hoped this would not be necessary. I asked them how they felt about sharing the works of an individual painter, like Edouard MacAvoy. Would it make sense for one of them to get all the work of one artist and the other brother all of a different artist? It was clear that they each wanted a share of each of the main artists. I suspected that each anticipated a huge regret factor if the other brother owned all the works of an artist whose reputation suddenly "took off." It was important that they each had a share of each artist, but the division could be unbalanced artist by artist. That is, it was workable if G or R got a better deal with one artist than with another, but it would be best if it all balanced out in the aggregate. Balanced how? Not in appraised value, but each should feel that he got his fair share and that I had not favored one brother over another.

I told them that I wanted to prove to them and to myself that I could accomplish what they wanted me to do, but that it was up to them to act professionally and responsibly without too much posturing. I told them that I would try to represent their deceased mother. They were slightly amused at this, but I invoked her name on several occasions: "G, you're being unreasonable; how would your mother feel if she knew that you were behaving that way?" I also told them that one of my objectives was to see whether I could improve their personal relationship, and if, at the end, they felt I accomplished that, I would then like a token bonus.

I emphasized that any proposed partial allocation I might suggest along the way for their review was to be considered tentative—nothing would be settled until all was settled. I also told them that if I made a suggestion and one of them did not like it because their tastes had changed during the time of my analysis, then they should honestly tell me and I would try to make alterations, but never at the expense of the other party. As it turned out, I had to alter the allocations several times.

The outcome.   I asked each to tell me about his preferences in any way he wished. My inquiries were very unstructured at first. I did not demand any quantitative scoring or even full ordinal ranking. Initially I wanted to get a feel for the gestalt, what they really felt strongly about. Later on in the process I had to get more specific, asking for rankings of a few of the paintings at a time and asking questions such as: "Would you rather end up with this particular painting of artist A or these three paintings of artist B?" It was clear that some of their preferences were not well formed—and that made my job easier. In fact, the brothers' wide divergence of opinion made the task a piece of cake. Joint gains come from differences; and there were plenty of differences.

Thinking back to the Janet and Marty allocations, we recall that the extreme-efficient contracts were made by dealing from the top using the ratios of Janet and Marty's scores, the J/M column in Table 13.5. In this allocation I dealt from both the top and the bottom in the following sense: I sought paintings that reflected their wide differences of opinion, so the ratio of G and R scores would be very high or very low. I never probed deeply enough to get their quantitative scores, but I learned enough to convince myself that if I had, the ratios would have been extreme. For these extreme ratios I started making suggested allocations to each. I used a spreadsheet to keep track of my tentative allocations and monitored along the way to spot any imbalance that might be growing as far as appraised values were concerned.

I don't know whether I was right in suggesting partial proposals at various points in the process for the distribution of paintings. I did this because I knew they would be delighted with the initial results, and by this means I would get their cooperation for later requests I made concerning some of their more intricate preferences. To give you an idea, in dividing a subset of paintings that could be evaluated on a 100-point scale for each, my allocations netted each brother scores of 80 or 85. On subsets of paintings by some artists for whom the allocations were not obvious, I assessed a set of quantitative scores for each brother that I believed reflected their values, but I never did this formally with them.

Of the ninety-five items, most were paintings. There was some disagreement about whether five of these items, jewelry and baskets, should have been included in the collection to be divided. But I viewed this disagreement as minor, since the five disputed items were worth at most $3,000. Overall I had allocated ninety items to the two brothers, who were delighted with my proposals since each got much more than he expected. That part was a lark, but the disposition of those five minor items—minor to me—consumed more time and grief than all the rest. Each brother dug in his heels and would not budge. The two engaged in vigorous positional bargaining with unreasonable ultimatums. One brother threatened to sabotage the entire main allocation unless his voice prevailed. I ended up scolding them: "How would your mother feel if she could see the way you are acting? Her gift was meant to be a joy to you both, not a source of family friction." After a bit more symbolic posturing, the brothers settled and, as a bonus, took me to a fancy restaurant where the rule of the evening was that anything could be discussed except Mrs. K's art collection.

This is a case that some would say borders on proactive mediation. I prefer cataloguing it as neutral joint analysis and therefore belonging in the chapter entitled "Arbitration: Conventional and Nonconventional." The affected parties had a chance to reject my proposals, and I bent over backward not to judge who deserved more; I played a nonevaluative role.

Core Concepts

Traditional arbitration. In conventional arbitration, the external helper, H, plays an evaluative role: H first examines the facts of the case and then imposes a contract solution that becomes binding on the parties. Typically the protagonists are disputants engaged in a distributive bargaining problem. Since they cannot decide on a compromise value, they are required by law or some other prior arrangement to have an arbiter resolve the problem for them. Four key ingredients may be underlined:

- A single variable (such as monetary compensation) is in dispute.
- When there is an impasse, it is obligatory that they submit their dispute for resolution.
- The intervenor is expected to make evaluative judgments.
- The arbitration is binding.

Variations. Negotiators may consider using arbitration as a tool:

- voluntarily (when costs of litigation or possible damage to the relationship make it worthwhile to have someone else decide quickly)
- suggestively (without binding themselves, but simply to get an expert opinion that can be used as a guide)

Final-offer arbitration. In final-offer arbitration, each disputant submits a sealed final offer and, after fact finding, the arbiter must choose one of these final offers. No other choices are possible. To maximize the chance of one's offer being accepted, one should make a reasonable offer—or else the arbiter won't accept that offer—and the motivation is to obtain reasonable bids. But, as is shown in the appendix, it is more complicated if one considers not only the probability of winning but the size of the winnings.

Neutral joint analyst. A jointly invited, third-party helper, playing the role of a nonevaluative, neutral joint analyst (NJA), can guide and coach the parties on how to get the benefits of collaborative analytic negotiation. Although parties, in a confidential interview with an NJA, might be willing to disclose truthfully their value tradeoffs, they may balk at disclosing their BATNAs. A double-auction method can avoid this stumbling block.

Embellishment. Parties can seek the services of an embellisher, a special kind of helper, to investigate joint improvements of a contract. The embellisher elicits information about the structure of the negotiation (for example, the template), the parties' values (such as what the confidential scores would have been), and does efficiency analysis à la Chapter 14 to find efficient contracts (if any) that dominate the existing one. The embellisher acts very much like a neutral joint analyst, except the embellisher, knowing the tentative agreement, does not have to get involved with the stickiest part of negotiations, the determination of BATNAs and RVs.

Appendix A18: Decision and Game-Theoretic Analysis of Final-Offer Arbitration

All's Not Well with the Procedure

What are some of the strategic aspects of final-offer arbitration? Analysis of a sample negotiation will help here.[1] Suppose that management (M) and the union (U) are at an impasse. They have negotiated without success, knowing full well that they will have to submit the determination of the basic wage rate (a single number) for final-offer arbitration. Management submits a sealed final offer, m; the union submits a sealed offer, u. The arbitrator then selects one of these two offers, depending on which value seems more appropriate. How shall we formalize this?

Assume that the arbitrator, after determining the facts, has some ideal value, a, in mind. The arbitrator will elect whichever final offer, m or u, is closer to a. If we imagine m, u, and a to be plotted on some linear scale—say, dollar value (see Figure A18.1)—it will be easy to see which offer more closely approximates the ideal. It is possible, though, that the arbitrator might have different psychological measurement scales on either side of his ideal; m might be close to his ideal in terms of dollars, whereas u might be closer in terms of some other value. But this complicates our task prematurely. Let's just suppose that in terms of one linear scale, the arbitrator selects the offer that is closer to his ideal. Following is a discussion of three special cases of this situation: first, in which the value of the arbitrator's ideal is known; second, in which there is a commonly perceived probability distribution for the ideal; and third, in which there are differing probability distributions for the ideal.

Value of the Ideal Is Known

Suppose that both M and U know the value of a before they submit their final offers. What is the best m against any given u? If, on a linear scale, m is quite a bit closer to a than is u, then the discrepancy between m and the ideal could even be slightly greater and m would still be the superior choice. So it is not true

---

1. The following discussion is based on the work of Chatterjee (1979).

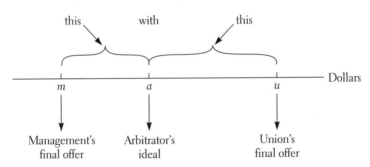

Arbitrator compares:

Figure A18.1.   The arbitrator's choice. In final-offer arbitration the arbitrator chooses the final offer *m* or *u* closest to his ideal, *a*.

that the best retort for M against u is $m = a$. The best retort is to choose a value for m below a that is only slightly closer to a than the distance that u is above a.

To what extent can the parties be sure of their payoffs? If M chooses m below a on the linear scale, then M can guarantee a payoff no worse than $a - (a - m)$. That value would be realized if u were selected at a distance above a that is equal to m's distance below a. Hence, to optimize M's security level (that is, to maximize M's minimum potential payoff), m should be set equal to a. Likewise, to optimize U's security level, u should be set equal to a.

Let's now look at possible equilibria. If U selects $u = a$, then M's best retort is to select $m = a$; conversely, if M selects $m = a$, then U's best retort is to select $u = a$. Hence, the pair $m = a$ and $u = a$ are in equilibrium and, from the analysis above, it is clear that they are the only equilibrium pair.

If a were known, would the players choose their final offers equal to that value? Probably not all; but as the players become more experienced, the known value, a, exercises a strong attraction.

Commonly Perceived Probability Distribution for the Ideal

To keep our case specific and simple, suppose that the annual wage rate for a starting fireman and the arbitrator's ideal value as perceived by M and U—which we will designate a—could be any value from 28 to 32 (in units of thousands of dollars). Both M and U perceive (and each knows that the other perceives) that all ideal values from 28 to 32 are equally likely for the arbitrator. The median and mean are both 30. Should M and U submit offers that are close to 30?

As a prelude to the analysis, we shall first determine the best retort for U against a known value of m—say, $m = 29$. Calculations for selected values of u

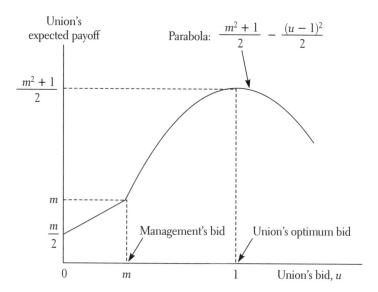

Figure A18.2. Union's expected payoff as a function of u for fixed m where all values of a between 0 and 1 are equally likely.

are shown in Figure A18.2.[2] For example, if u were set at 30, then the outcome could either be 29 (at M's offer) or 30 (at U's offer), depending respectively on whether a were less than or more than 29.5. (For this continuous range of a values, we don't have to worry about a being exactly 29.50000.) The probability that a is less than 29.5 is three-eighths, or .375. So under our assumptions if u is chosen at 30, then U will be exposed to a lottery with payoffs 29 and 30 and probabilities of .375 and .625, respectively. An appropriate summary index for a risk-neutral U is the expected value: $.375(29) + .625(30) = 29.625$.

From Table A18.1 we observe the rather surprising fact that the best retort against $m = 29$ is $u = 32$ for an expected value of 30.125.

It's not difficult to show that to maximize expected monetary value, the best retort for U against any assumed value of m is the response $u = 32$. This is a strong and remarkable result. Analogously, the best retort for M against any assumed value of u is the response $m = 28$. The pair $m = 28$ and $u = 32$ are in equilibrium, but this is not nearly as strong as saying that $u = 32$ is a best re-

2. *Analytical elaboration.* Let's say the $\bar{a}$ is rectangularly distributed between 0 and 1 (no loss of generality). It can readily be shown that U's expected return, as a function of u for fixed m, is:

$$\bar{U}(u|m) = \begin{cases} \dfrac{u+m}{2} & \text{if } u < m \\[2mm] \dfrac{(m^2+1)}{2} - \dfrac{(u-1)^2}{2} & \text{if } u \geq m, \end{cases}$$

which is depicted in Figure A18.2. Against m, the optimum response is $U^\circ_m = 1$ (for all $m \leq 1$).

Table A18.1   U's expected-value payoffs for selected values of u against $m = 29$ when all values for a from 28 to 32 are equally likely

Value of u	Possible outcomes	Probabilities	U's expected value
30	29	0.375	29.625
	30	0.625	
31	29	0.5	30
	31	0.5	
32	29	0.625	**30.125**
	32	0.375	
33	29	0.75	30
	33	0.25	
34	29	1	29
	34	0	

Note: Boldface indicates maximum in column.

sponse whether or not M plays its equilibrium value of 28; and $m = 28$ is a best response whether or not U plays its equilibrium value of 32.

If the commonly perceived distribution for the arbiter's ideal value is not rectangular (that is, if all values are not equally likely between some lower and some upper value) but is a more natural, symmetrical, bell-shaped, distribution centered at a value, a (see Figure A18.3), then against any assumed m the optimum retort U_m is higher than the mean of the distribution of the arbiter's uncertain value—surprisingly higher. And as m approaches the mean (central value), U_m^o drifts further to the right. So it is not true that if M makes an offer close to the center of the distribution of a, that U should reciprocate. Intuitively, the higher the value of m, the more U can afford to gamble. But also if U suspects that M is risk averse and will choose an m-value close to the mean, then U can afford to gamble with a higher u-value. All this, of course, depends on U's being risk-neutral. If U is risk averse, then there will be an attractive force toward the center of the distribution of a. Conversely, if U is risk prone, U^o will be higher still.

An analogous story holds if we look at M's optimum retort, M_u^o against an assumed value of u.

Differing Probability Distributions for the Ideal

Some theoretical models show that with complete exchange of information, M's and U's probabilistic perceptions of the distribution of a should be identical. Empirically this does not turn out to be true, however, and very often the distributions are displaced in directions favoring each protagonist (see Figure A18.4). All of this, of course, is speculative, and it is doubtful whether baseball players,

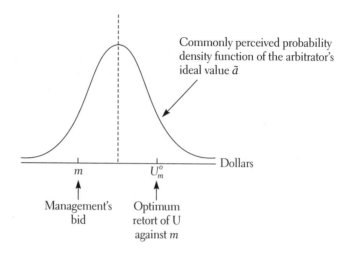

Figure A18.3.   U's probability distribution of a and U's optimum bid against m.

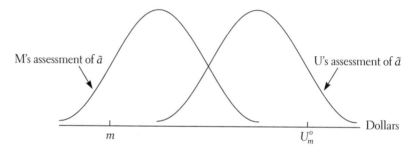

Figure A18.4.   Different probability distributions. The case where U and M have different assessments of a and U_m^o is the optimum retort against m. Vertical scale is such that the area under each curve is 1.0.

ball club owners, firemen's unions, or city managers formulate probability distributions. But the distributions shown in the figure would probably be reasonable approximations, especially if both parties voluntarily chose to submit to final-offer arbitration.

If U were to consciously calculate the best U_m against m, then U would use its own assessment of a, and as is shown in Figure A18.4. there would be a vast discrepancy between m and U_m (tempered somewhat by risk aversion). But now a further complication is introduced: U might suspect that M's assessment of a will be displaced to the left of U's own distribution, so that U might expect the possibility of very low m-values. Contrary to common wisdom, with the anticipation of low m-values and great uncertainty about a, U does not have much security and must be careful. With differing perceptions of a and differing perceptions of perceptions, with risk aversion and differing perceptions of the other party's risk aversion, with anticipated limited rationality and with expected miscalculations—in short, with full reality—this is a tremendously complicated problem.

Final-offer arbitration should have great appeal for the daring (the risk seekers) who play against the timid (the risk averse). It may be true that the proportion of cases going to final-offer arbitration is smaller than the proportion going to conventional arbitration. This is often cited as an advantage of final-offer arbitration. Of course, the logic is marred a bit, because conventional arbitration preceded by a round of Russian roulette would still do better.

Is it easier for an arbitrator to administer conventional or final-offer arbitration? In both cases he presumably would have to determine the facts. In conventional arbitration, he would have a continuum of choices; in final-offer arbitration, the adversaries—and in this case they really are adversaries—present the arbitrator with a dichotomous choice. It might be an easy choice: if the final offers are close together, it may not make much difference; if they are far apart and one seems ludicrous, again it's an easy matter. But if they are far apart and equally ludicrous, the task may be extremely difficult: the arbitrator might want to settle in the middle, but he can't. He does not have the luxury of being able to make fine distinctions in his judgments. On the other hand, if the arbitrator selects m or u, he does not have to publicly announce how he would have decided for every potential, embarrassingly difficult pair of offers.

19

What Is Fair?

We introduced the concepts of equity and fairness in Part III in the context of an idealized joint collaborative analysis with a mutual commitment to FOTE. Let's continue this theme and pursue the ideal case in which parties AAA and BBB have agreed on a template to be resolved, each has scored the template privately, and each has identified confidentially its BATNA and corresponding reservation value. They now have to agree on a final contract.

An external helper, if privy to all this information, could compute the results shown in Figure 19.1. The figure shows the joint RV value, the efficiency boundary, the maximum feasible values for AAA and BBB, as well as AAA's and BBB's potential (the difference between the maximum feasible value and the RV). This is the canonical representation of the problem, employing a FOTE analysis, carried to the extreme. Given their own privately held information, AAA and BBB know the general shape of Figure 19.1, but neither has sufficient information to draw the figure to scale.

The external helper makes the following proposition: "If each of you confides your private scoring of the template and RV to me I will not only be able to draw Figure 19.1 to scale, but I could specify an arbitrated, binding, final contract. Do you want me to do that?" Before committing themselves, the parties would naturally want to know just how this selection would be made. At this stage, they both risk subjecting themselves to an appallingly bad allocation.

Given the data in Figure 19.1, how should our external helper, acting as a neutral arbitrator, make the selection of a final outcome? What is fair?

This problem is almost isomorphic to the famous bargaining model proposed and analyzed by John Nash (see Box 19.1). There is a subtle difference in the formulations regarding the units of measurement on the axes. In the Nash model they are utilities, whereas in the model of negotiations the units are in terms of strengths of preference.

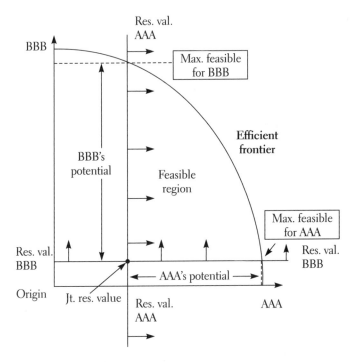

Figure 19.1. Feasibility and efficiency.

Since the problem of fairness in the context of negotiations (see Figure 19.1) and that of trades in the Nash model are analytically the same, much of what is discussed in this chapter is distilled from the literature about the Nash problem, with a slight modification of vocabulary. Instead of talking about trades T', T'', we'll talk about the joint evaluations of contracts C', C''. The Nash no-trade point becomes the joint reservation value (see Figure 19.1 and the figure in Box 19.1).

Units of Measurement

The units of measurement on the axes of the Nash bargaining model are utilities (see figure in box), whereas the units of measurement on the axes of Figure 19.1 are in strength of preference units, and we have, admittedly, been a bit sloppy about the properties of those scales. Let's take the point of view that the negotiating parties are risk neutral in terms of units of measurement, so that we can also interpret the units on Figure 19.1 as utilities. We'll return to this distinction at the end of this chapter when we consider an extreme example of the difference between the units of measurement.

Box 19.1 The Nash Bargaining Problem

John Nash proposed a variant of the fairness problem (Nash 1950). His formulation is rendered as follows: consider two individuals, 1 and 2, who "are in a position where they may barter goods but have no money to facilitate an exchange." Although they lack money, they can perform randomized experiments involving the distribution of the goods they hold.

Each individual comes to the market with an initial bundle of goods, and a trade takes place if and only if each consents to it. By a trade we mean an actual reapportionment of the joint bundle of goods held by them. Let T, T', and so on denote different possible trades. In the class of all possible trades, there is one that is distinguished, namely, when no trade actually occurs—the status quo. This we denote by T^*.

Nash goes on to assume that for each possible trade T, each player assigns a utility value that reflects the desirability of the trade. Thus the joint evaluation of trade T is the utility pair (u, v), giving the respective utilities of individuals 1 and 2 for T. Denote the utility pair for T^* by (u^*, v^*). In this way, each trade T is represented as a point in the plane. If T is represented by (u, v) and T' by (u', v'), a randomization between T and T' is represented by a specific point on the line segment (u, v) and (u', v'). The justification for this rests in the meaning of "utility."

We've now set up the fair-division challenge. The Nash bargaining problem boils down to the following: in the figure below, let R denote the set of all points representing trades and randomization of trades; let (u^*, v^*) represent the special status quo point. If no trade occurs, the payoff is (u^*, v^*). A trade takes place if and only if both players agree upon a unique point of R, which then constitutes the payoff. Player 1 wants easterly points and 2 wants northerly ones.

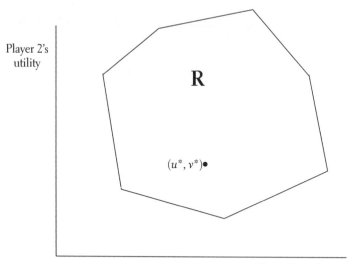

Player 2's utility

R

$(u^*, v^*)\bullet$

Player 1's utility

If, in the negotiation problem, we identify A with player 1, and B with player 2, the joint RV with (u^*, v^*), and the region of possible negotiated outcomes with the region R, then the two problems are isomorphic. There is an important and extensive literature in economics and game theory on the Nash bargaining problem that is relevant to our challenging question: what is fair?

Fairness Proposals

Equity Analysis: Setting up the Problem

Let's start with the following abstract but simple problem. Consider the problem captured in Table 19.1 and Figure 19.2. AAA and BBB, operating with a FOTE orientation, have identified six extreme-efficient contracts, labeled A to F with joint payoffs as shown in Table 19.1. These points are plotted together with the critical ratios that give the marginal rates of substitution involved in going from point to point. The parties have thought hard about their best alternatives to negotiation, and in the scoring system they have adopted for the template, they have assigned reservation values: AAA will settle for no less than 45 points; and BBB for no less than 15 points. As we see in the figure, there is a feasible region that meets these constraints. With this level of information in full view, AAA and BBB should somehow select a point on the efficient frontier. But which one?

Turn your attention to the right-hand side of Table 19.1. First let's consider an illustrative point in the feasible region; we arbitrarily choose the point with coordinates (70, 30). Since the *reservation values* are 45 and 15, respectively, the *excesses* for the two parties are 25 and 15, respectively. The next line, *maximum feasible*, contains the maximum value achievable for AAA, 90, given the constraint that this contract must meet BBB's reservation value. Correspondingly, the maximum feasible value for BBB is 72. Those numbers, 90 and 72, can be read off Figure 19.2. If the graph were not readily available, as it will not be in some more complicated situations, we might have to solve two separate linear programming problems with SOLVER to find those maximum feasible values.

The difference between the maximum feasible value and the reservation value will be called the *potential*. For AAA the potential is 90 − 45, or 45; for BBB the potential is 72 − 15, or 57. To get the values in the next line, *proportion*

Table 19.1 Equity analysis

	AAA	BBB	Critical ratio		AAA	BBB
A	0	100		Value	70	30
B	10	95	2.00	Reser. value	45	15
C	40	78	1.76	Excess	25	15
D	60	55	0.87	Max. feas.	90	72
E	80	30	0.80	Potential	45	57
F	100	0	0.67	Prop. of Poten. (POP)	0.56	0.26
				Min. POP	0.26	

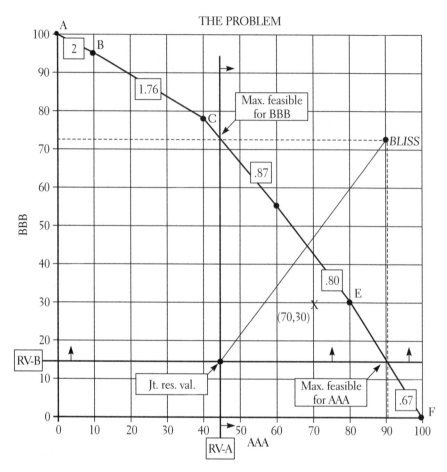

Figure 19.2.   Graph of equity analysis.

of potential, we divide the excess for each party (for the given point) by the potential for the party. Thus AAA's POP is 25/45 or .56; and BBB's POP is 15/57 or .26. The minimum of the two POPs is .26.

Alternative Proposals

We now introduce some contenders for what is most fair. Let the joint reservation value be at (0, 0) and the maximum feasible for each party at 100, so that the potential for each one is 100. This is not so special, because we can always change the value scales to achieve this normalization with no loss of generality. In fact, we use the illustration shown in Figure 19.3 because it highlights the salient differences between the alternative proposals for fair division.

The Maximin or Equipop Point

We've already introduced one contender: the efficient point that maximizes the minimum POP, which we dub the "maximin value." It gives each party the same POP value. In Figure 19.3, the maximin point has coordinates (67.5, 67.5). With just two negotiators, the maximin point yields equal POP payoffs to the two parties and hence we can (for the case $N = 2$) label this as the *equipop* point. With more than two negotiators, as we shall see in Part V, there is a difference between the maximin and equipop points and the difference may not be trivial.

The maximin point is on the ray joining the joint RV and the so-called *bliss point*, which gives each party its maximum feasible score.

The Mid-Mid Point

If we start out, in Figure 19.3, by giving each party half of its POP, we're led in this case to the point with coordinates (50, 50). Now using this point as a new reservation value and repeating the process by halving the new POP, we're led to the point on the frontier that is dubbed the "mid-mid point" and that in this case has coordinates (62.5, 75).

The Maimonides Solution

Another solution, we'll call it the "Maimonides" solution, has been proposed by Robert Aumann, a brilliant game theorist and biblical scholar, who traces its roots back to the Jewish philosopher Maimonides. The computation runs as follows. There is no dispute that AAA should get 50 points to begin with. The dispute is over the remaining 50 points. Let's split this, giving each player half of his or her remaining potential. That leads to the point with coordinates (75, 50).

The Nash Solution

The last contender we identify was originally suggested by Nash and helped him win a Nobel Prize in economics. Nash argued in a mathematical, axiomatic style made popular by Kenneth Arrow, whom we'll meet later. Nash first articulated what would be reasonable desiderata (or axioms) for a solution to fulfill and then proved that the only way to simultaneously satisfy these desiderata is to maximize the product of the scores of the two parties.

If we took a vote among all the economists and philosophers who have

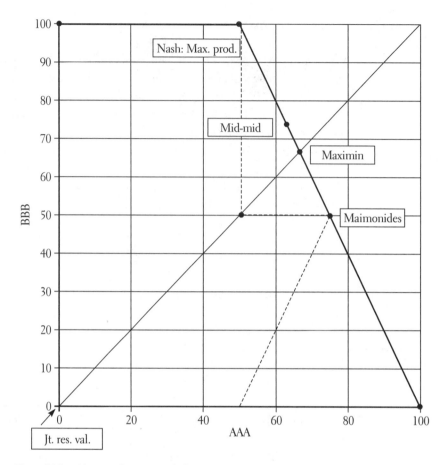

Figure 19.3. Alternate fairness proposals.

thought seriously about this problem, we suspect that the Nash solution would be the winner. But we're still not convinced. As external helpers we usually propose maximizing the common POP—that's the easiest to explain to our negotiators in most (but not all!) contexts—but our favorite is the balanced-increment solution, which we'll meet later. In the next section we discuss a context that some think favors the Nash solution.

An Example: Never on Thursday (or Oot vs. Tath)

Oot and Tath are two lucky beneficiaries of a windfall. They are given 100 vouchers for meals at a very desirable restaurant. The vouchers are dated and are not transferable. They can only be used by Oot or Tath and the dates are the next 50 Tuesday and Thursday nights. But there is a complication: Oot can go *only on Tuesdays* whereas Tath can go on *Tuesdays and Thursdays*. (Hence the

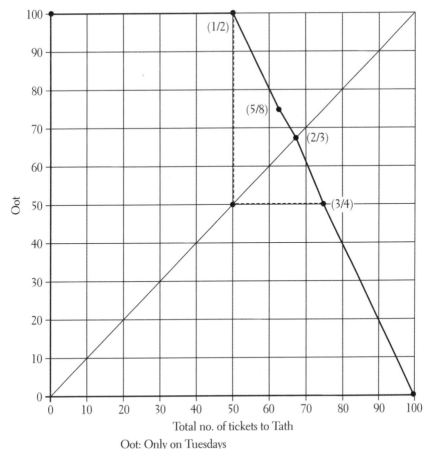

Figure 19.4. Equity analysis for the division of tickets.

acronym names.) Having struggled with their problem for a while, they come to you for a suggestion about how to divide the meal vouchers fairly between them. What do you suggest?

In Figure 19.4 we give Tath a single point for every voucher he receives; Oot gets two points for every Tuesday voucher he receives. We do this to make the range of payoffs for each from 0 to 100 points. Thus the region in Figure 19.4 is the same as in Figure 19.3. We are not concerned about this normalization of units, because we will not make any interpersonal comparison of them.

1. Maimonides says: "Give the Thursday vouchers to Tath; they are not in contention. For the 50 Tuesday vouchers in contention, split them evenly." This

means ¾ of the vouchers go to Tath, and the solution point is 50 for Oot and 75 for Tath. This yields a joint score of (75, 50).

2. The procedure that maximizes the common POP—or equivalently, maximizes the minimum POP—gives all Thursdays and ⅓ of Tuesdays to Tath, netting Tath ⅔ of all the vouchers. Oot gets ⅔ of the vouchers he is interested in. This yields a joint score of (66.7, 66.7).

3. The mid-mid procedure gives Tath all Thursdays and ¼ of the Tuesdays, so Tath nets ⅝ of all the vouchers and Oot gets the remaining ⅜. This yields a joint score of (62.5, 75).

4. The Nash procedure gives all Thursdays to Tath and all Tuesdays to Oot. They both get 50 tickets but Tath gets 50 percent of his potential and Oot gets 100 percent of his potential. This yields a joint score of (50, 100).

Who's the fairest of them all?

The good news is that for most nonartificial cases with two parties, it doesn't much matter which fairness solution is adopted. They all tend to be close together. But when we generalize these results to many-party cases—meaning three or more parties—there will be striking differences between the arbitrated solutions.

Rationale for the Nash Solution

Suppose we are seeking to decide what's fair, and we're currently considering a point (x, y) on the efficient frontier. A suggestion is made that AAA give up Δx for a gain by BBB of Δy. The Nash procedure says, in effect, that the tradeoff will be considered fair if and only if the proportional gain $(\Delta y/y)$ for BBB is more than the proportional loss $(\Delta x/x)$ for AAA. The implication of this rationale for fairness indicates that the optimum occurs when the product of the payoffs are maximized.

Proof. At the optimal (x, y) coordinate on the efficient boundary, we would require

$$-(\Delta x/x) = +\Delta y/y, \quad \text{or} \quad x(\Delta y) + y(\Delta x) = 0, \quad \text{or} \quad \Delta(xy) = 0,$$

which is the differential necessary condition that (x, y) be chosen so as to maximize the product xy.

Numerical Illustration of the Nash Rationale

If the joint payoff (40, 60) is under consideration, and a decrement of 2 for AAA yields an increment of 2.5 for BBB, then this movement is rejected by the Nash criterion because the proportional loss for AAA (that is, $\frac{2}{40}$ or .05) is greater than the proportional gain for BBB (that is, $\frac{2.5}{60}$ or .417). On the other hand, a movement from (38, 62.5) to (40,60) would be encouraged by Nash, since the proportional gain by AAA (that is, $\frac{2}{38}$ or .0526) is greater than the proportional loss for BBB (that is, $\frac{2.5}{62.5}$ or .04).

If we started, however, from a position where both parties received the same amount—for example, 60 each—then, according to the Nash procedure, we should move along the efficient frontier if Δy is larger than Δx. So let's imagine that we start from the point that maximizes the common POP—the maximin point. Then, according to Nash, we should certainly move from there if the marginal rate of transfer, given by the critical ratio at that point, is not unity. Now that's a result we can think about. We might opt to remain at the maximin point, even if the slope of the frontier at that point is not unity. Of course, if we're at the maximin point of (60, 60), BBB might self-servingly argue in favor of moving to (58, 62.5), but AAA might rightly object. In the present context, the arbitrator would decide.

The critical question, however, is if we're in a state of ignorance regarding who might benefit from a move from the maximin point, would the parties wish to move from it? That's what the debate is about. We still think there is something nice about the equality of the POPs.

Independence of Irrelevant Alternatives

In his seminal book *Social Choice and Individual Values* (1951), Kenneth Arrow, Nobel laureate in economics, introduced the principle of the independence of irrelevant alternatives (IIA). Here's what he says: Suppose that a decision maker must choose, according to his or her preferences, a best alternative from a prespecified set, S, of alternatives. Now suppose new alternatives are added to set S to form T. The set S is a subset of T. Now Arrow emphasizes the following: if the best in T is also in S, then it should also be the best of S. Or, stated somewhat differently: the best of T should either be the best of S or one of the new alternatives. Let's consider a concrete problem.

An Example Illustrating the Independence of Irrelevant Alternatives

At the weekend, a traveler to a new city goes into a restaurant for dinner, scans the menu, and after considering an expensive filet mignon, opts for a duck dish. The waiter volunteers the information "Today, the chef is also offering a few specials," and he includes alternatives such as foie gras and frog's legs. None of these specials appeals to our traveler, but he remarks, "Oh, in that case I would like to change my order from duck to steak."

Does this make sense? All the available dishes (including the specials not on the menu) constitute the set T. Those on the menu are the subset S, of T. The rule of the independence of irrelevant alternatives says: if steak is best for T, and steak is available in S, it should also be best in S. Alternatively, we might say, if duck is best in S and new alternatives are added, then the best for the enlarged set should be duck or one of the new alternatives. But steak is not a new alternative.

An argument against the IIA principle: our traveler prefers the duck to steak because, among other things, the steak is quite a bit more expensive than the duck. When the traveler learns that the chef is also offering frogs legs and foie gras, his assessment of the quality of the cooking goes up. His assessment of quality is partially determined by the items available on the entire menu. There is information in knowing that some exotic dishes are available. So perhaps it's worth spending the extra money to order steak over duck.

Arrow's Impossibility Theorem

Arrow used the principle of the independence of irrelevant alternatives in his Impossibility Theorem. There is no way, said Arrow, of combining individual preferences to get a group preference that meets certain desiderata (including the independence of irrelevant alternatives axiom). Arrow's groups comprise three or more members. We'll expand on this result in Chapter 24, on voting.

Nash uses the same independence of irrelevant alternatives desideratum to show that maximizing the product is the only legitimate arbitration scheme. Here's the argument for the special case in Figure 19.5. The decision maker is the arbiter. She considers two regions, R and R', with the joint RV at the origin as shown in Figure 19.5. Set R, with defining points {A, B, D, E}, is a subset of R', with defining points {A, C, E}. The best alternative in R' is the point (5, 50). That follows from requirements of efficiency and symmetry—all of our contending arbitrated solutions would single out point (5, 50) in R' as best. Now, since (5, 50) is also available in R, the IIA principle says that it should also be best in R. That's the Nash solution.

Convincing? For us, not quite, for a reason related to the evidence provided by the traveler who switches his choice with the new information about the extended menu. The IIA principle may not be compelling when the decision maker is an arbiter. Let me explain. When new points are added to the feasible region, we are not convinced that the best arbitrated point should remain fixed or shift to one of the new points. If (5, 50) is deemed best in R, then lopping off prospects that worsen the aspirations of BBB should shift the arbitrated value in favor of AAA despite the fact that (5, 50) remains available. The aspirations of the parties should depend on the shape of the entire feasible region, rather than

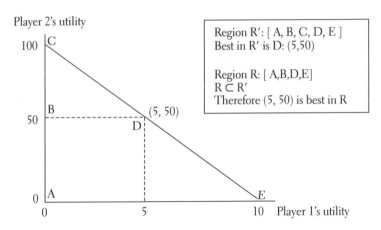

Figure 19.5. Nash's rationale.

be a local property of what happens differentially to a given point on the frontier. Still, there is something elegant about the Nash point.[1]

The Balanced-Increment Solution

Some notation will help. Consider the graph depicting the feasible payoffs and let B(x) be BBB's coordinate on the efficient frontier corresponding to an x value for AAA; and let A(y) be AAA's coordinate on the efficient frontier corresponding to any y value for BBB. See Figure 19.6. For any (x^*, y^*) feasible point denote its corresponding bliss point by $(A(y^*), B(x^*))$ and consider the ray from (x^*, y^*) to its bliss point. The maximin or equipop point starts out with the joint reservation point and follows its ray to where it pierces the frontier. The mid-mid point traverses this same ray, but goes only half the distance and then recalculates a new bliss point and a new ray, and again goes half the distance and continues the process until the frontier is attained. Instead of traversing the ray for half the distance and then recalculating, we can go a quarter of the distance and recalculate, or one-tenth the distance and recalculate, and we can push to the limit. The limit point approached on the frontier will be labeled the *balanced-increment point*.

1. In a paper I (HR) wrote in 1950 and informally published in a University of Michigan working series, I proposed a solution called the *balanced-increment solution*, which is a variant of the mid-mid solution, which in turn is a variant of the maximin point. In the literature my name is erroneously attached to the maximin point, when I've always viewed that solution as less desirable than the balanced-increment solution. The maximin point, however, has the merit of being simple and, once identified, hard to get the parties to change.

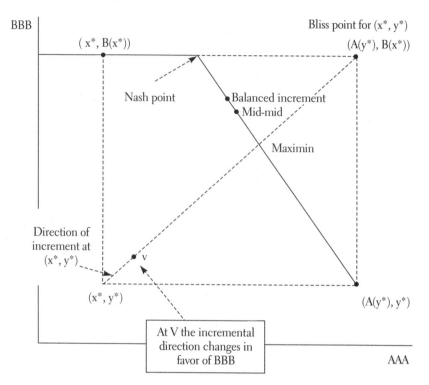

Figure 19.6.   The balanced-increment solution.

The balanced-increment solution is very sensitive to the shape of the entire efficiency boundary. Imagine starting with some region and grafting onto it some added possibilities or deleting some possibilities, resulting in a perturbed efficiency frontier. We feel that the arbitrated solution should be sensitive to any perturbation of the frontier. The Nash solution is insensitive in the extreme to perturbations on the frontier; take the case where a feasible region is given and points are deleted from this region that do not include the original Nash point. Then the Nash solution for the modified region keeps the same solution point, whereas the balanced-increment solution is very sensitive to any perturbation of the region. The independence of irrelevant alternatives assumption justifies the Nash argument, but in the context of negotiations, changing the feasible boundary is not irrelevant, because it changes the perception of what is fair.

Back to Units of Measurement and Utilities

As Paul Samuelson has quipped, when discussing the hole in a donut, you should select a donut for review with a large hole in it. Similarly, when discussing the significance of differences between units of measurement, one should select for review an extreme example that will highlight the difference. Should it make any difference to you as an arbiter that one of the parties is wealthier than the other? How, for example, would you divide up one million nontaxable dollars between a rich man and a poor man? Suppose that they can't decide between themselves and that you have been appointed as arbiter. Assume that they have invited you to propose a binding allocation of the million bucks.

In Figure 19.7 I (HR) exhibit how $1 million could be shared between the two recipients. The point (.15, .85) labeled A in Figure 19.7, referring to an allocation of .15 million to the poor man and the rest to the rich man.

The units of measurement in Figure 19.7 are in dollars. Now let's convert to utilities, assuming that the rich man is risk neutral and the poor man very risk averse. For example, let the poor man's certainty equivalent for a fifty-fifty crack at $1 million be $.1 million (that is, he would be indifferent between receiving $.1 million outright and taking his chances with the lottery); and let the rich man's certainty equivalent for a fifty-fifty $1 million prize be $.45 million.[2] Figure 19.7, in dollar units, is converted to utility units in Figure 19.8. I purposely chose the details of the poor man's utility function so that the efficient frontier (in utility units) is symmetric, which means that all of the procedures we have discussed (Nash, maximin POP, mid-mid, balanced increment) for calculating the arbitrated solution suggest point A, with coordinates (.8, .8). The division that achieves that common .8 utility might give $150,000 to the poor man and $850,000 to the rich man. At point A the rich man and the poor man will each get an amount that they think is just equivalent to a .8 chance at their best outcome of $1 million. In some sense that's equitable: they get the same equivalent chance at their best prizes. Is this fair? No!

I could have concocted more extreme numbers (that is, chosen a donut with an even bigger hole) and come out with a split of $50,000 for the poor, risk-averse, desperate fellow and $950,000 for the risk-neutral playboy. Why should their attitudes toward risk matter? Why should I care if the poor man is risk averse? Isn't that irrelevant?

Many game theorists argue that attitudes toward risk do enter the picture, because they are involved in the dynamics of bargaining. The rich man says, "If I don't get $850,000, there will not be any deal." The poor man can settle for $150,000 now or take a chance at possibly getting more—but possibly getting nothing. In that setting, each side is making threats or backing off from threats; it's a highly uncertain environment. So the argument goes that even if the problem seems deterministic, probabilities get into the act and utilities can't be far behind. The fact that the rich man is more risk neutral gives him much more bargaining power. The rich man could squeeze hard: he could say, "$150,000 is

2. *Certainty equivalent.* For any lottery, l, we define the certainty equivalent of l, denoted by CE(l), as the amount certain for which the decision maker is indifferent between l and CE(l).

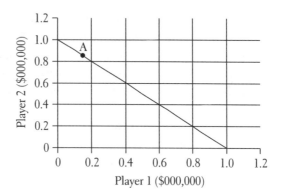

Figure 19.7.   Monetary payoffs to a rich man and a poor man.

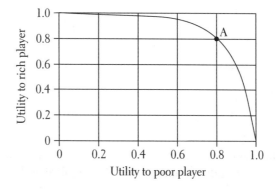

Figure 19.8.   Utilities to a rich man and a poor man from the same monetary payoffs.

more important to you than $850,000 is to me, so let's settle here or not at all."
The threat is credible. But my trouble is that I don't like it when people use
power that way, and as arbiter I would have strong prejudicial interests of my
own. As arbitrator, should I dictate a solution that reflects that kind of raw
power?

As arbitrator, should I try to predict how the negotiators would settle their
controversy without my services, and then try to do better for each? I think not.
Should I give preferential treatment to a party who acts more irresponsibly and
irrationally, because this gives that party more power? I think not. Should I
imagine how reasonable negotiators should behave and then impose that solution on them? That's coming closer, but it's not very operational.

"Stop quibbling," you may be thinking. "Just how would you select a 'best'
fair point? How would you settle the split of the million bucks?"

If the setting were one in which I could appropriately take power into consideration, I would fudge a little. I would give each the same monetary benefits.

If the monetary allocations are taxable, a case could be made for giving the same amount to both in after-tax dollars, which means giving more to the rich man in pretax dollars. This would be understandable, but it wouldn't reflect our values as arbiter. Certainly I would not base my analysis on utilities, because the more risk averse the poor man is, the less he'll get. A lot will depend on how I became the arbiter. Who selected whom? Just how did I get into the act? I can imagine circumstances where I would want to give the whole million to the poor man plus fine the rich misanthrope for being such a mean character.

"How about if the rich man wanted to give the poor man a half million at the outset, and the poor man wanted more because of the principle involved? What if that were the reason they came to you?" In that case I'm being asked to play an evaluative role, and a lot will depend on just how the pair was given the prize to share in the first place. After all, when I suggested an allocation of the art collection between the two brothers, it never occurred to me to favor the poorer of the two brothers because this could correct an imbalance in their wealth, or favor the brother who had a greater appreciation for art, or favor the brother with a better house to display the paintings in, or favor the brother with the larger family because more would share the beauty of the paintings, and so on.

The context of the dispute and my role as arbitrator in that dispute are sometimes of paramount importance. Sometimes the raw abstraction may abstract away too much context to be of direct use; but sometimes the abstraction can protect the arbiter from playing God, a role that should be reserved for God or one of Her agents.

Core Concepts

This chapter has focused on the choice of an equitable point on the efficiency frontier. In earlier discussions we had already considered several candidates for best solution, with the two leading contenders being the point that (1) maximizes the minimum POP; and the point that (2) maximizes the product of the excesses.

We have now added a few other respectable alternatives such as the Maimonides solution, the mid-mid solution, and the balanced-increment solution. Except for some vagueness about the units of measurement (utilities versus strengths of preference) the problem of "what is fair?" is isomorphic to the problem on the bargaining model formulated by John Nash. The Nash solution (to maximize the product of the two players' utilities when the status quo point is normalized to be zero) is a mathematical consequence of the famous principle of the independence of irrelevant alternatives first promulgated by Kenneth Arrow. We have offered some reservations about that principle in the context of fair arbitration.

The chapter then turned to the division of a one-million-dollar prize between a rich and poor man, a problem that looks entirely different if examined in dollar or in utility units. We concluded that in such a stark example it is critical to know how the arbiter got involved in the first place, her mandate, and something about the origins of the problem.

20

Parallel Negotiations

We turn now to conflicts and disputes in the world at large. What advice can we give to collaboratively minded participants in tough, intractable conflicts? For instance, what can be done to help resolve the Israeli–Palestinian dispute, the Greek–Turkish dispute on Cyprus, the Nationalist–Loyalist dispute in Northern Ireland, or the Kosovo Albanian–Serbian conflict? We offer some modest suggestions on how analysis may—and we repeat *may*—make a difference.

Orientation

Types of Parallel Negotiations
Our analysis focuses on barriers to successful negotiation and how to facilitate agreement through the use of *parallel negotiations*. We begin by categorizing negotiation channels as front channel, back channel, or Track II. We don't mean to suggest that real negotiations fall into a single category—there is almost always some overlap—but the distinctions will help clarify what types of parallel processes can be effectively implemented.

Front-channel negotiation. Front-channel negotiations involve government officials, who usually talk in private with one another, but are obliged to openly discuss concessions, demands, agreements, and victories with political constituents and the general public. This is the classic form of diplomatic negotiation. What happens if these front-channel negotiations reach an impasse? What if one or both sides view negotiation as a politically nonviable alternative? Assuming either one of these conditions holds, top decision makers can pursue two paths, singly or simultaneously. First, they can establish secret "back channel" negotiations with their high-level counterparts. Second, they can devise "Track II" exploratory talks involving a broader range of constituents. We'll consider these two approaches separately, although they may occur together in the same conflict.

Back-channel negotiation. Back-channel negotiations involve participants with significant influence over decisions that affect relations between the two parties. The existence of such negotiations assumes that a front channel already exists. Knowledge of back-channel negotiations is strictly limited to those few people who "need to know." This means that front-channel negotiators are almost always unaware that another line of communication exists between the parties. Figure 20.1 presents a two-way breakdown of the distinction between front and back channels and Track II negotiations. Textbook examples of back-channel diplomacy include: secret talks over SALT I during 1968–1973 between Secretary of State Henry Kissinger and the Soviet ambassador to the United States, Anatoly Dobrynin; and the Oslo talks between the Israeli government and the Palestine Liberation Organization (PLO) during 1992–93. In both cases back-channel negotiations took place in parallel with front-channel negotiations. The talks involved a very small number of highly influential decision makers, and were held in strict secrecy. Both successfully reached an agreement, whereas their front-channel counterparts did not.

Track II negotiation. Track II negotiations, like back-channel talks, are typically pursued in parallel to front-channel negotiations—although this need not be the case. These negotiations differ from front- and back-channel efforts particularly in the status of the participants: they involve nonofficial negotiators—for instance, academics, business representatives, or medical doctors—who cannot formally commit political constituencies to joint agreements. Rather than producing immediate consequences, the benefits of Track II talks accrue at a societal level, leading, it is hoped, to a long-term change in attitudes. These Track II negotiations can be secret, publicized, or open to the public. Good examples include joint scientific committees involving U.S. and Soviet researchers during the cold war, the Arab-Israeli dialogue held in Toledo, Spain, in the 1980s, and the work of former undersecretary of state Hal Saunders in Tajikistan in the 1990s.

Blurred boundaries in practice. We'll see that the boundaries of this simple categorization are blurred in practice. For instance, do secret meetings between two ex-foreign ministers who may return to office constitute Track II or back-channel negotiation? Are highly respected academics, who also wear the hat of senior policy advisers, considered decision influencers? Even though the distinctions break down in practice, they help us think more clearly about designing methods to overcome barriers to negotiation.

NEGOTIATOR INFLUENCE ON DECISION

		LOW	HIGH
TRANSPARENCY OF PROCESS	CLOSED	TRACK II	BACK CHANNEL
	OPEN	TRACK II	FRONT CHANNEL

Figure 20.1. Two-way breakdown

Barriers to Negotiation

We list below eleven barriers to negotiation. This list, while not exhaustive, illustrates some of the most common concerns that inhibit negotiation between antagonists.

1. High political cost of initiating negotiations that turn out to be unproductive (or of even simple recognition of your enemy).
2. Posturing, positional bargaining, and excessive claiming
3. Inability to explore options through brainstorming
4. Inability to use FOTE collaboratively
5. Front-channel negotiators are not natural compromisers
6. Power imbalances
7. Lack of trust and insecurity of contracts
8. Ripeness of the conflict
9. Divided selves: Blocking coalitions of extremists
10. Ignoring the potential value of future joint collaborative opportunities that may result from a settlement (referred to as "option value" by economists).
11. Logistics of getting to the table

We shall describe how back-channel diplomacy or Track II (exploratory) negotiations can be employed to address the barriers we have specified.

A Hypothetical, Abstract Problem of Intractability

Let's use the abstract example of Alba and Batia as the motivating case for the development of our analysis. Consider the following common characteristics of a difficult bilateral dispute:

- Alba and Batia are two countries (entities) engaged in a long-festering, seemingly intractable, dispute.
- Both sides are nonmonolithic, each having a militant extremist faction.
- Both sides have legitimate historical reasons for their palpable hatred of each other.
- It is politically costly for either side to initiate negotiations.
- Any hint of starting a negotiation is accompanied by a need to proclaim a set of initial claiming demands.
- There is a profound lack of trust.
- Any formal, official negotiations must be conducted with full press coverage.
- Negotiators are evaluated by their constituencies and leaders according to how tough they are and whether their side has gotten a "better deal" from the other side.

Why Parallel Negotiations Help

High Political Cost of Initiating Unproductive Negotiations

The leaders of Alba—and Batia as well—are afraid to call for negotiations. To do so would be a sign of partial capitulation to their hated enemy. They are engaged in a slow, escalatory game of bravado, and a call for negotiations would constitute blinking under pressure. Furthermore, Alba's leaders honestly don't know if anything can be accomplished by negotiating. Negotiating carries a high political cost—even higher if negotiations are not perceived to be successful. And although each leader is remaining intransigent, extremist factions on each side want them to be even tougher. Alba is ruled by a coalition, and any sign of weakness will put strains on the tenuous internal alliances. Yet Alba's leaders might still be deftly talked into negotiations if they thought there would be an overall improvement of the situation. Alba has a choice: to negotiate or not to negotiate, and the dominant uncertainty is whether negotiations, if begun, would be successful.

Let's draw an analogy. A business firm has the choice of remaining with an old product or switching to a new one. Will the new one be successful? The problem is pervasive: to keep with the status quo or to move into uncharted waters. One prescribed, partial remedy is obvious and widely employed: take a sample, try a pilot study, run an experiment, collect partial information that will help the decision maker gain better insight into the nature of the uncertainties. In the negotiation sphere is there an analogous information-gathering strategy?

"Experimental non-negotiations."   Let's suppose that through the intervention of a third-party helper, unofficial surrogates of the contending parties quietly meet together, not to negotiate but to "dialogue." This is a euphemism for negotiating with a small n rather than negotiating with a capital N. The leaders of Alba and Batia don't know about the dialogues, or they choose not to know about them; they maintain a stance of maximum deniability. But through such dialogues, as each side gains information about the other side's goals and expectations, the high cost of initiatiating negotiations may be somewhat reduced.

Troublesome Logistics

It's tricky for a dedicated intervenor to get the parties, or even surrogates of the parties, to a conference table. He or she—let's make it a she—has to find appropriate surrogates and get them to meet without tipping off the ever-vigilant press. Where to meet? Who pays the expenses? What to do once the surrogates meet? What will be the protocol if the press gets wind of the attempt at informal negotiations?

The facilitator will have to remind the surrogates about the nature of the exercise. They are not engaged in an attempt to find "the solution" to the impasse but to ascertain whether possible solutions exist that beat the reservation values of the contending parties, a tall order in itself. This calls for a different type of negotiation: exploring the efficient frontier without claiming.

The leader of the Alba side may truly not know whether negotiations, once started, will lead to an impasse, as depicted in Figure 20.2 with efficiency frontier A, or be potentially successful, as with frontier B. It would be helpful if he or she could get a better grasp of these realities. Secret, unofficial explorations with maximum deniability might be highly useful. Such non-negotiations (either official or nonofficial) are held not in parallel to front-channel negotiations but in lieu of them.

Posturing, Positional Bargaining, and Excessive Claiming

As we have repeatedly pointed out, real negotiations involve a delicate balancing of moves to *create* joint gains with moves to *claim* more-than-favorable shares. In our discussions of the negotiator's dilemma, we have indicated that the tactics employed for claiming a larger part of the pie often interfere with the tactics for creating a larger pie for both parties to share. In tough, highly publicized negotiations between official representatives of belligerent countries such as Alba and Batia, negotiators must talk to the other side, but also to their home constituencies. There is a need to posture when negotiating in a fishbowl. Negotiators are forced to do a good deal of positional bargaining, and they are not free to engage in speculative brainstorming. Track II or back-channel negotiations, because they are held in secret, make it easier to focus on creating rather than claiming.

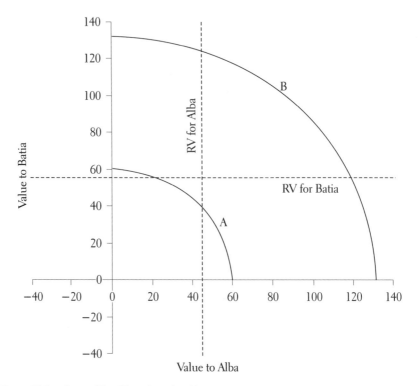

Figure 20.2. Cases of feasible and nonfeasible contracts.

Inability to Brainstorm

In the case of intractable conflicts with entrenched groups and extremists on both sides hollering for no concessions, it is extremely difficult for high-level officials to brainstorm constructively. Even if the press is not invited to observe the deliberations, leaks are commonplace. (For a case where leaks were minimized and brainstorming sessions maximized, see Box 20.1.) Leaders are loath to float ideas that may be unpopular at home, especially if any floater is distorted by the press or by political opponents. Brainstorming on issues like national security is considered a nonstarter, because "security" is not generally thought to be something people negotiate about. Such highly emotional, public negotiations promote positional bargaining with an emphasis on claiming.

In back-channel negotiations, claiming tactics are less prevalent because of secrecy, informality, and less need for posturing for the folks back home. With fewer people at the table, and obstructionist constituents temporarily excluded from the process, negotiators have more room for maneuver. Track II negotiators can similarly benefit from these factors. In addition, it is usually their task not to identify a single compromise contract, which requires a lot of claiming tactics, but to explore the efficient frontier. With no formal mandate to demand or to concede, they are free to investigate, and consequently there is much more of an

---

Box 20.1   The Philadelphia Convention, 1787

Long gone are those idyllic days that prevailed in 1787 at the Philadelphia Convention, when representatives of twelve states of the union (New Jersey was absent) met in secret session. The press knew that these negotiations were taking place, but the leaks were few (for fear of the wrath and opprobrium of their leader, George Washington). When Washington sat in his chair on a dais, the deliberations were formal and recorded. But he periodically vacated his chair, purposely turning the session into a Committee of the Whole when nothing was recorded and all was informal. Brainstorming prevailed. Frequent straw polls were taken during these informal sessions, but votes were not for attribution.

---

emphasis on creating. It's much easier in parallel Track II deliberations to approximate the conditions of joint collaborative analysis.

Inability to Use FOTE Collaboratively

In formal negotiations the leaders of Alba and Batia speak for the record with intended audiences at home as well as abroad. In formal negotiations the leaders are play acting. Speeches are made for public consumption, and for propaganda purposes the negotiators can't appear to be too chummy. In Track II and back-channel negotiations, the parties can relax and talk off the record. They can get to know one another as each other as individuals and establish a rapport. They quickly realize that they are intimate enemies who know most all that is relevant about the other side—each reads the press of the other, each has an extensive spy network that reaches far into the affairs of the other—and they learn that they might as well talk in a collaborative FOTE style: there is nothing to hide. Further, they quickly realize that they are joint problem solvers and, unlike in formal negotiations, they don't have to exaggerate to win some concession, as is so necessary in claiming. In some circumstances they might at first interact in a POTE style—not lying to each other but keeping quiet about some delicate information that A mistakenly thinks that B doesn't know about, but later realizes that B is indeed aware of it. And B now knows that A knows that B knows . . . Even intimate secrets are common knowledge and POTE slides into FOTE.

Front-Channel Negotiators Are Not Natural Compromisers

In the forum of formal, official negotiations, the main actors are politicians and diplomats trained to be tough, to be cautious, and not to explore irresponsibly.

Many are trained in military academies, where being tougher than the other guy is prized. In Track II diplomacy, the actors are usually idealists, starry-eyed academics, who love to brainstorm. They also may be more skilled in interest-based negotiation for mutual gains. It's true that heads of state have academically inclined advisers that accompany them in real negotiations who might brainstorm in back rooms, but the atmosphere is entirely different in the two types of negotiations.

In back-channel negotiations, exploring creative options for mutual gain is still inhibited by the ingrained conservatism of the participants. Progress is contingent on the establishment of a personal rapport between the handful of politicians or officials involved in the process. Without personal chemistry, back-channel talks are unlikely to produce fruitful joint agreements—the process is highly dependent on a meeting of minds. Protected from the glare of publicity and the scrutiny of extremist constituents, back-channel negotiators have a good chance of forging a joint problem-solving relationship—becoming a coalition in the middle with potential blocking coalitions on either side. The challenge then becomes selling their creative agreement to the excluded, more radical constituents and persuading political opponents.

Power Imbalances

Power imbalances are an often-cited barrier to negotiation. The dominant view is that negotiations do not make sense if there are power imbalances between the contending parties. Why should the more powerful party, Alba, negotiate with weaker Batia? Alba is already "winning" without negotiating, so why risk "losing"? A zero-sum mentality. Power is a multifaceted concept. It can mean the added power of information about the other party's reservation value and tradeoffs; or the power of better negotiation skills; or still other forms of power. But what most people have in mind is the power of an advantageous alternative to negotiations—a very good BATNA and an associated reservation value that can be backed by force. With such an attractive alternative to negotiation, why negotiate?

We address the problem of power imbalances in Figure 20.3, which depicts symbolically the standard two-party negotiation problem. Each axis is defined in terms of a composite satisfaction index for the negotiating parties Alba and Batia. The line B'B' depicts Batia's reservation value; Batia needs a value north of line B'B' *to* meet what it could accomplish with no negotiations. The line A'A' is Alba's reservation value; Alba needs a value east of line A'A' to meet what it could accomplish with no negotiations. Alba and Batia seek agreements depicted in the dotted region northeast of point J', but both Alba and Batia are only qualitatively aware of the realities of Figure 20.3 and lack knowledge of crit-

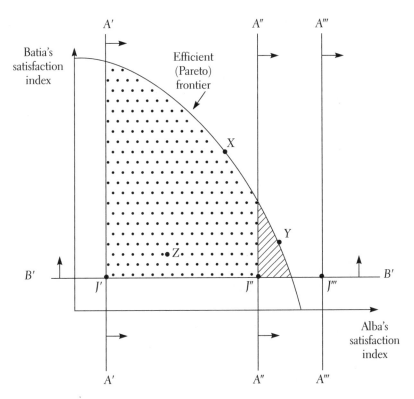

Figure 20.3.   Feasibility, efficiency, and the power of a good alternative.

ical details. In looking for agreements, the parties might "satisfice" on a Pareto inferior point Z without realizing that they are leaving potential joint gains on the table. Point Z is inefficient, but perhaps equitable; X is both efficient and equitable; Y is efficient but not equitable. Now for power.

Let's examine this question by assuming that Alba's fortunes have just changed—its "power" has increased—and that now, instead of requiring an agreement to the east of reservation value line A′A′, it must be east of line A″A″ (or even east of A‴A‴) (see Figure 20.3). Of course if the reservation value line is A‴A‴, then the zone of possible agreement is empty and no acceptable negotiated agreement can be found.

But let us keep in mind that Alba (and Batia, for that matter) does not know the details shown in Figure 20.3. Even though we now posit that Alba is a powerful player with a good reservation value, Alba may not know if the reality is of the type A″A″ or A‴A‴. If it is A″A″, it might still be hard for the parties to find a mutually acceptable agreement—an agreement possibility northeast of point J″. But it may be worth a try. Incidentally, if Alba has a strong reservation value (like A″A″), Alba might want to be quite open in order to convince Batia of its veracity, because otherwise Batia may be thinking that reality is something like A′A′.

On the other hand, if Alba postures as if the line is A'''A''' rather than the reality A''A'', it may forgo the possibility of getting value added from the negotiation.

If the reality is A''A'', there is still motivation for Batia to negotiate a modest gain for itself. Weaker parties often overestimate the stronger party's advantage in negotiation. The stronger may be able to annihilate the weaker on the battlefield, but the stronger cannot or may not want to exercise that power. So the advantage may not be as large as the weaker party perceives it to be.

Lack of Trust and Insecure Contracts

In Chapter 4 we talked about trust in connection with the repeated social dilemma game (or prisoner's dilemma). The two players would be maximally rewarded if they achieved a lasting pattern of cooperate-cooperate (that is {C,C} pairs) even though there would be a myopic gain for each side to defect (that is, using D rather than C). We saw how difficult it was to get into a C,C pattern, and once one player defected, how difficult it was to reestablish the cooperative pattern. We talked about the fragility of trust and how important trust is. But we learned more. Empathetic trust is neither a necessary nor a sufficient requirement for cooperation. Each party might hate the other—indeed, each may take some satisfaction in hurting the other—but choose not to defect from a C,C pattern for fear that it would lose in subsequent rounds.

There is a positive message here for seemingly intractable disputes: it is possible to get joint gains without empathetic trust. This is not to say that empathetic trust is unimportant; it is. But in repeated-play games it is possible to have mutual, operative, working trust even though the parties remain bitter enemies.

In parallel negotiations, where negotiators work intimately together over long periods, empathetic trust often develops between them even though neither can promise that any agreement he or she makes will be accepted back home. The conversation may take the following form. Alba: "I know what you are asking is reasonable and privately I agree with you, but I would have to deny it publicly. Look, I realize that you're going to have trouble selling what I'm suggesting to your extremists. What can my side do to make life easier for you?"

Securing insecure contracts.   The secret is to get into an agreement that yields good payoffs to each side and where each side refrains from taking momentary advantage of the other, for fear of subsequent retribution. (Think of the case where at each round in a repeated prisoner's dilemma game, each side does reasonably well with {C, C} and knows that a switch to D will bring only short-lived gains.) But mistakes can be made, and an insecure pattern of repeated {C, C} choices may be broken when a D pops up. There may be a change in the internal politics of Alba, and new leaders may not understand the delicacy of the unstable dynamics. So it may be prudent for the parties to think ahead to formulate ways of stabilizing their tenuous balancing act. They might start an escrow account, controllable by some outside group, that will compensate the injured party at the expense of the defector. Or perhaps some benefactor, Candia, may withhold its philanthropic support from the country that defects. Or perhaps if Alba defects, it will negatively influence its prospects for favorable trade with some international body. It pays to be imaginative in securing insecure contracts. Of course, the less empathetic trust there is, the more stabilizing sanctions must be built into the deal.

For the Palestinians and the Israelis, for example, back-channel talks in Oslo enabled them both to overcome periods where implementation was itself politically risky. Instead of using the threat of breaking off relations, the parties continued to work quietly and secretly through what were at that point the most difficult years of coexistence.

Ripeness of the Conflict

In the summer of 1992 I participated in a workshop held outside of Athens, Greece, for high-level diplomats of the Conference on Security and Cooperation in Europe (CSCE) countries. The workshop was organized by the not-for-profit, Cambridge-based consulting firm the Conflict Management Group, and I was a member of its team. We consulted the diplomats on various issues, including the appropriate time to start negotiations. But first, some information about the Valletta Accord.

CSCE dispute settlement mechanism: The Valletta Accord. In February 1991, at Valletta (Malta), the CSCE adopted a mechanism for the peaceful solution of disputes. If the parties are unable, within a reasonable period of time, to settle their dispute in direct consultation or negotiation, any party to the dispute may request the establishment of a CSCE Dispute Settlement Mechanism, which invokes a series of possible external interventions: mediation, conciliation, good offices, fact finding, arbitration, and adjudication.

Note that before the CSCE mechanism is invoked, the conflict has to mature or ripen, the parties must have been frustrated by nonproductive, unassisted negotiations, and one disputant must trigger the mechanism. For some steps in the mechanism, all disputing parties must agree.

The CMG group invited the participants of the Athens workshop to address another possible approach. Incipient disputes may be easier to manage and to solve at an early stage, before tempers flare in public, before escalatory actions are taken, before positions are frozen. "Would it be helpful," we asked, "if dispute counselors could offer their process assistance in a very low profile, non-public, informal, strictly confidential, exploratory, nonbinding, nonjudgmental way to the disputants at an early stage?" There are very good arguments that negotiations have a better chance of succeeding if the participants feel an urgency about getting an agreement. But at the same time, negotiations at an earlier stage may prevent a conflict from escalating and positions from hardening.

Divided Selves: Blocking Coalitions of Extremists

The Alba negotiators, whether the official negotiators or the back-channel officials, must keep in mind that any agreement has to be ratified back home. With any external agreement, there will be winners and losers, and the losers may be able to undermine or block the agreement. So it behooves the negotiators to keep this in mind. This general question is addressed in Chapter 26 on pluralistic negotiations, but a few words may be in order here. There may be righteous indignation when some faction feels left out of the process, after secretive meetings have taken place and agreements have been made without taking its views into consideration.

So let's imagine that the Alba negotiators fear that faction A′ might upset their agreement with Batia. They have threatened, preached, and cajoled, but A′ remains adamant in its opposition. What can be done? We suggest four possible remedial alternatives, some of which can be combined:

1. Tilt the negotiated contract to the benefit of A′.
2. Add new issues to the template being negotiated with Batia and resolve them in favor of A′.
3. Logroll this negotiation with other negotiations with Batia or with other parties in such a way that the combined outcomes favor A′.
4. Creatively compensate A′ internally by giving its members something they want. (See, for example, the Panama Canal case in Chapter 25, which describes how President Carter creatively compensated the military, then playing the role of A′.)
5. (Not remedial.) Engage in a political battle with the extremists. Sometimes—and this must be carefully weighed—A might enlist the aid of the other side, B, in the internal fight with A′.

Ignoring Future Joint Collaborative Opportunities

Alba and Batia are neighboring countries that have deep-rooted, antagonistic relations. They have fought borderline skirmishes in the past and each has memories of the barbaric acts of their enemies. Now they are at a standstill. There is little hope that they can resolve their territorial disputes, and this intractable stalemate blocks any constructive joint dialogue between them.

A facilitator invites a few nonofficial but highly prestigious individuals from both sides to a neutral site to talk about everything other than the differences that divide them. They are brought together not to negotiate their differences, but to investigate the possibilities of implementing joint projects that could be of mutual gain. These projects might be: joint economic ventures, trade, worker exchanges, environmental projects, tourism, educational endeavors, sports competitions, construction of joint medical facilities, cultural exchanges, and so on.

The BCP project. I (HR) call this effort to persuade opponents to explore the possibility of joint collaboration the art of "finessing controversy." In teaching how this might be done, I initiated a pedagogical, pilot project at the Kennedy School of Government at Harvard University. The setting was the controversy between Bolivia, Chile, and Peru, three countries that are locked in a perpetual dispute over Bolivia's lack of access to a port on the Pacific. Wars have been fought and diplomatic contacts seem to be frozen. I invited representatives from these countries who were taking graduate courses at Harvard to meet with me for a constructive dialogue. We dubbed this the BCP project.

As facilitator of this exercise, I had to remind the participants that they were not involved in a mock negotiation to resolve their differences; rather, their task was to be creative in seeking joint opportunities. I repeatedly ruled out of order interventions that placed blame on the others for past atrocities, and urged them to concentrate on future possible collaborations. After a couple of sessions, the participants entered into the spirit of the project and became ingenious in developing plausible schemes for cooperation. Rather than engaging in claiming activities, they more often said something like, "I realize my suggestion might produce negative reactions in your country. What might my country do, to help you to convince your fellow countrymen that we both will gain if we undertake proposal Q?" The dialogue progressed from establishing an inventory of possible collaborations to strategies for implementation.

I was encouraged by this pilot example to try to make a contribution to the region. But the BCP project floundered from lack of follow-up. We needed more prestigious participants who were influential but not official leaders in their country, and this would require fund raising that I was not prepared to do. Maybe someone will eventually try to implement such dialogues that "finesse controversy." Success at finessing controversy could also help improve the atmosphere for tackling the divisive issues that plague the region. Thus the project could act as a consensus-building strategy.

We can't ignore the fact that the opposite might prevail. The primary controversy might be so intense that no auxiliary compromises can realistically be implemented until "peace" is achieved. Another negative: progress on ancillary issues may take the heat off addressing the fundamental divisive issues. More research is needed.

Option value of future collaborative opportunities. In decisions about whether to start or continue negotiations, it is important to keep in mind that if an overall agreement is achieved, there may be a stream of subsequent follow-up agreements that could contribute to the well-being of all. These auxiliary, potential, positive add-ons provide an *option value* for any agreement. Even though these potential ancillary agreements may not be on the table during negotiation of the divisive issues that separate the parties, they should be kept in mind in deciding whether to start or quit these discussions.

If these potential ancillary agreements are not included in the primary negotiations or given their proper option value, the efficient frontier will be biased downward. This is a barrier both in the decision to start and in the decision to continue negotiations. Because these ancillary agreements can affect the overall desirability of a final agreement, it may be wise to explore their potential joint gains in parallel Track II negotiations. Look at Figure 20.2 again to see the value of factoring in future collaboration when trying to reach agreement. Including the option value in the analysis may push frontier A toward frontier B, resulting in new-found feasible alternatives.

How It Is Done

Premediation Briefing Reports and Track II Shuttle Analysis

One of the impediments in facilitating an unofficial, low-key dialogue between hostile factions is finding suitable surrogates on both sides to come to some distant place both at the same time and to devote the necessary time to devise and brainstorm. And then there are all the expenses. A much more modest, achievable program is for a single investigator—perhaps a doctoral student writing a thesis—to go to suitable surrogates seriatim, not unlike shuttle diplomacy, and to write a premediation briefing report (PMBR) that might be suitable for a mediator who wanted to be thoroughly briefed about the nature of the dispute. The report itself, if properly leaked to the authorities in the antagonistic countries, could help initiate negotiations. The report could also be used by the main negotiators to jump-start negotiations and to help the two sides develop a suitable template for their consideration.

Think tanks concerned with fostering more and better negotiations might want to subsidize the writing of these premediation briefing reports that result, in part, from facilitated, shuttle, Track II prenegotiations. In Box 20.2 we list possible chapters that could be included in such reports.

Box 20.2   Contents of a Premediation Briefing Report

1. Brief history of the conflict (past attempts at negotiation, partisan reviews of the history)

2. Identifying those who have influence (biographies of the key parties, how they negotiate, their ideologies)

3. Examining the internal political structure (who decides, who must influence whom, internal vulnerabilities)

4. Identifying interests and values (fundamental values and derived interests elicited from surrogates)

5. Envisioning the future with an agreement (long-run advantages of a settlement—for example, development)

6. Analyzing alternatives to negotiation (scenario if no agreement is reached, best alternative for each party)

7. Exploring options (joint devising with surrogates, creating options without claiming)

8. Constructing a negotiation template (listing possible issues to be negotiated and resolutions for these issues)

9. Identifying legitimate criteria (objective standards of fairness, precedents)

10. Evaluating the template (eliciting tradeoff and preference data from the participants)

11. Determining a range of possible appealing agreements (to lure the parties into negotiating by making joint gains salient)

12. Identifying external parties who may contribute resources (identify outsiders who gain; garner compensation and subsidies)

Case Study of a PMBR: The Energy Taxation Dispute in Finland

Since the early 1990s Finland has imposed environmentally related taxes on energy production to discourage CO_2 emissions. The energy taxation dispute between segments of Finnish industry and environmentalists was intense, and Joanna Pajunen, in her doctoral dissertation for the Helsinki School of Economics and Management Sciences, prepared a PMBR of the conflict using surrogates. For her surrogates she used two representatives each from industry (chosen from the Confederation of Finnish Industries and Employers) and from the environmentalists (chosen from the Green Parliament Group and the Council of the Green League). Her dissertation was patterned after a working paper distributed at the International Institute for Applied Systems Analysis on the PMBR process. An outline of her dissertation tells the story.

In Part One, the Introduction prepares the reader (or would-be mediator) with both the history of the dispute and the fundamentals of negotiation and intervention analysis, leading to a discussion of PMBRs in general.

Part Two, Applications, describes how Pajunen led her surrogates (1) to formulate an objectives hierarchy for each side of the conflict; (2) to develop a negotiation template for the problem; (3) to identify their preference tradeoffs; (4) to identify and evaluate each side's BATNA and RV—all of which enabled her (5) to find efficient compromise (equitable) solutions. Looking forward to an implementation stage, she also generated single negotiating texts, or SNTs, for actual debates in the Finnish Parliament. Throughout this stage her surrogates did not enter into face-to-face negotiations.

After the PMBR exercise was completed, the surrogates wanted to pursue a next stage: to negotiate among themselves in a mock Track II negotiation. This led to the next part of the thesis.

In Part Three, Implications, Pajunen recounted how she acted as a hands-on, directive, proactive mediator for negotiations between the surrogates. She used her analysis to structure a starting SNT that was revised four times. After a final agreement was reached, the participants urged the researcher to publicize the results, and the thesis documents how the word was spread and how the whole enterprise affected the Finnish budget negotiations in the fall of 1996.

In summary, Pajunen wrote: "This study was a clear indication of the applicability and usefulness of the premediation concept in a conflict situation where the disputants have not before tried to resolve their disagreements through negotiations. Consequently, we can sincerely recommend the application of the analysis in other conflicts, be they environment-related, intra-organizational or any other type."

A PMBR can also serve as a conflict assessment report, which structures the dispute, provides background information, does partial analysis, indicates the merits of negotiation, provides guidance for negotiators and possible intervenors, and develops a first draft of materials (such as designing a template) that could be useful in negotiations.

Case Study: Peru-Ecuador Conflict 1995

Consider these Track II exploratory negotiations facilitated by a team of negotiation scholars from Harvard University.

The conflict.   In 1941 Peru and Ecuador went to war over a disputed, tropical, underdeveloped area on their borders. The Rio Protocol of 1942 declared Peru the victor and ceded the land to Peru. In 1960 Ecuador, reacting to new geographical revelations, denounced the Rio Protocol, and skirmishes over this almost useless bit of territory ensued, blossomed, and erupted in early 1995 into a minor war. Periodic discussions between Peru and Ecuador were nonproductive and merely provided a platform for each side to make unacceptable demands.

Facilitated joint brainstorming (FJB).   In April of 1995, the Conflict Management Group and the Harvard Negotiation Project, led by Roger Fisher, invited eleven influential, nonofficial participants from Peru and Ecuador for a weeklong workshop to "explore how Peru and Ecuador might help defuse the tension between their countries." Fisher dubbed the exercise *facilitated joint brainstorming (FJB)*. I participated in the exercise, but as a minor player.

The facilitators invited individuals from the two countries who were known to have the ability to influence decision making in their own countries. Getting the right participants to the meeting table was critical and not easy. They were individuals who had held high offices in the recent past (for example, a former foreign minister, a former head of the joint chiefs of staff) and to lure them to Cambridge it helped that many of them had been graduate students at Harvard in the distant past. They were invited and they came. Their expenses were paid.

We intervenors called ourselves "facilitators," but in fact we acted more like proactive mediators—except that we were careful to emphasize that this activity was not to be considered a negotiation. The invitees were prepared to give opening vitriolic, patriotic pronouncements, but the facilitator implored them to try a new sort of game. They were invited to dwell not on the past but rather on what constructive steps they could take in the future to defuse the belligerent atmosphere. The participants were seated around a table, with each party having immediate neighbors from the "other side." In the opening presentation they were paired, Peruvian and Ecuadorian, and asked to take fifteen minutes to talk to each other in private so that each could introduce the other to the assembled group. Two opposing military men reluctantly left the room with obvious disbelief but returned after a short while with new-found attitudes. They both learned that they were proud parents of a severely retarded child whom they adored. They were still enemies, but now compassionate ones.

The facilitator played teacher and the message was:

- Peru and Ecuador are not involved in a zero-sum game.
- Joint gains can be achieved by brainstorming about creative options based on the truthful shared interests of the two sides.
- They were not there to negotiate, so that claiming should be kept to a minimum.

The facilitator explained what was meant by brainstorming or devising. The participants were encouraged to propose half-baked ideas to the group, and if group members were in a brainstorming mode, they were instructed to look for positive aspects of the proposal. It took some time to get the participants to enter into the spirit of brainstorming. Imagine trying to do this during real, official negotiations with the press looking for juicy stories.

The goal of the facilitation team was not only to generate creative options for the resolution of the dispute between the antagonists but to get them to think differently about the role of negotiations; instead of thinking about negotiations as hard-nosed, adversarial win-lose interchanges, they were gradually won over to thinking about interest-based negotiations for mutual gains. I led the group in some simulated exercises, far from the context of their territorial dispute, to illustrate the need for collaborative problem solving.

The group was surprised that the facilitators, at the end of the sessions, discouraged any attempt by the participants to draft a compromise document for the solution of their dispute. Each of them left with a different attitude and loads of ideas about how joint compromises might be made. Team members were pleased at the outcome but were not sure that there would be any lasting effect.

Aftereffects.   A year later my wife and I visited Lima, Peru, and met with several of the eleven participants. I asked whether they thought the week at Harvard had been profitable and whether the interchange was having any effect on Ecuadorian–Peruvian relationships. It was true at the time that the dispute had been settled and relations between them seemed to have generally improved, but could the Cambridge exchanges take any credit? To my surprise, they emphasized that not only had it been beneficial to them but it had had a positive effect on the relationship between the two countries. At a dinner my wife sat next to one of the military men, who was now back in the government as a chief military adviser on the Ecuadorian side, and he said to her, "Tomorrow I'm off to negotiate with the Peruvians on arms control, and I've made lots of preparations for 'interest-based negotiation for mutual gain.'" A couple of the Peruvians were writing a book on creative opportunities for interchanges in trade, health, and tourism, with extra emphasis on joint environmental opportunities—even opportunities in joint development of the formerly disputed territories.

It's now another couple of years later and Peru and Ecuador have signed an agreement settling their differences (*The Economist*, October 31, 1998). In a moving ceremony at Harvard in the spring of 1999, President Jamil Mahuad of Ecuador personally conferred his country's highest medal of honor on Roger Fisher for his contributions to peace in the Andes. The foreign minister of Peru sent a congratulatory concurrence.

Core Concepts

This last chapter on interventions in two-party disputes has dealt primarily with the role that informal, unofficial, parallel negotiations can play in resolving intractable disputes. We identified eleven barriers to the negotiation of lingering disputes and show how parallel (Track II) negotiations can be used to scale various impasses.

A very successful facilitated joint brainstorming (FJB) exercise between high-level surrogates from Peru and Ecuador was used to illustrate the power of this type of intervention. We believe that it has wide applicability as a constructive response to long-smoldering conflicts. The FJB, however, requires the presence of distinguished, influential surrogates working together around a table. A poor man's version of this, requiring fewer resources, can be achieved by having an analytically inclined, academic researcher shuttle back and forth between surrogates of the disputing sides. The goal is to develop a deeper understanding of the dispute and perhaps to discover creative options for later, more official negotiations. The researcher can also produce a premediation briefing report (PMBR) about the dispute that can be used later by a facilitator or mediator.

Many Parties

We now extend our inquiry into many-party territory. By "many parties," as in some primitive tribes, we mean more than two.

Everyone knows that reaching a decision becomes harder the more people are involved. Negotiations among multiple parties—or even decisions, which must be jointly taken by a group whose members are ostensibly on the same side—are often long and unhappy affairs.

Up to now we have been talking about relatively simple bilateral negotiations (with the addition of a third-party helper). Once we move out of that setting, many different geometries can be considered:

- A group of separate, individual negotiators
- Bilateral negotiations with multiple participants on each side
- A group of advisers preparing one side for negotiations
- A permanent decision-making or advisory group
- An ad hoc decision-making or advisory group

Each of these groups poses particular challenges. But there is also a core of similar difficulties that will have to be managed in all of those contexts. Chapter 21 starts by listing in what ways some groups do worse than individuals. It then attempts to understand the underlying reasons for this behavior. After examining losses caused by cognitive overload, poor coordination, poor communication, and poor motivation, the chapter considers general prescriptive advice designed to partially ameliorate some of these causes. The advice considers such matters as membership in the group; the use of a facilitator or chair; the need for an ongoing visual documentation of the deliberations; the role of brainstorming and devising; a focus on purpose and choice of a problem-solving framework; the decomposition of tasks and formation of subcommittees; and the allocation of time.

Chapter 22 then considers groups that strive for consensual agreement. Each of the members wishes to act in unison, but strains in the community exist. They do not resort, however, to voting or coalition formation. To a large extent this discussion generalizes the two-party material in Chapters 11–13, which has a strong FOTE flavor. As was the case for two parties, we once again examine the fair-division problem—but now with more than two parties—before launching into our full analysis of feasible, efficient, and equitable contracts for the many-party problem.

Chapter 23 has a game-theoretic flavor. It starts out by discussing the dynamics of coalition building in various contexts. We next consider the highly structured problem where every coalition is valued at some specific, fully known, total monetary worth (with common knowledge of this input data) and players must then join coalitions and decide how to divide the resulting joint revenue. It's a wild game that defies any prediction of outcome. Still, some modest advice can be given. We examine principles that can be employed to divide up the joint proceeds of the coalition of the whole in ways that reflect the power of various subcommittees. We end our consideration of coalitions by giving partisan advice to one party engaged in two separate intertwining negotiations; each sets up a BATNA for the other.

Chapter 24 examines the use of voting procedures for group action. Common schemes such as majority rules are fundamentally flawed, because they can lead to intransitivities of group preferences. We review Arrow's famous Impossibility Theorem that proves that there is no way for three or more (that is, many) individuals to combine their individual, ordinal preferences to obtain a group ordinal preference without violating some appealing desiderata. Many-party negotiations sometimes result in the identification of a few viable contracts for adoption, and guess what? They vote. One way out of the dilemma is to demand a richer set of inputs from the voters. They are asked not only about their ordinal preferences but about the intensities of their preferences. They are asked for cardinal orderings. However, easier said than done—especially with the realities of insincere voting. We introduce a case study that examines how a group of scientists selected a trajectory for the Voyager mission to outer space.

Chapter 25 examines cases where at least one side in a negotiation is nonmonolithic. We imagine a two-party (external) negotiation across a table that is complicated by the existence of an intense internal negotiation on each side of the table. Contracts that are negotiated across the table may result in winners and losers on one side of the table, and the losers may try to block the negotiations unless they are duly compensated by internal transfers. Some call these "bribes." The challenge is to synchronize internal and external negotiations.

In Chapter 26, we extend the material in Part IV on interventions in two-party deals and disputes to many parties. We examine a hypothetical case of a community that has to negotiate with a family that wants to use part of their extensive landholdings to erect a shopping mall and some high-rise office space. The quid pro quo is that the town will get money from a higher tax base and some needed land that can be used in many different ways. A facilitator/mediator is more active with the many-sided interests within the town than with the division between the town and the developer. We conclude with a case study of energy negotiations in the European Union. A series of facilitator/mediators vainly sought a feasible solution (for all the participating countries), before Luxembourg stepped up to the plate and found the way toward mutual acceptance.

We end our book on a philosophical note by going back to the two-person prisoner's dilemma game (or social trap) and introducing more players. We consider cases involving the commons problem (from sheep grazing to over-population of our planet); the overutilization of resources (like fishing and forests); the pollution of the atmosphere and the despoiling of the seas; free riders (both individual, noncharity givers and countries that want everybody else to curtail their green-house emissions); the Not-in-My-Backyard (NIMBY) syndrome; environmental injustice and why disadvantaged communities might exact enormous compensations for the acceptance of an undesirable facility. And finally, all this is brought down to the personal level: how much are you willing to sacrifice for the good of the whole?

21

Group Decisions

Our book as a whole focuses on how to use quantitative methods and game-theoretic analysis to illuminate negotiation problems and obtain better outcomes. One could call it science in the service of the art of human interaction. In this chapter, we turn that relationship around. We look at traditional negotiation advice, organizational behavior, and sociology to illuminate the problem of how to use quantitative methods—or any other difficult analysis—in the hurly-burly of a group of negotiators. Art in the service of science, perhaps.

We proceed by first examining the problems that people face when they try to make decisions in groups (or, as some would have it, the reasons to shun group decision making when possible). We next look at the benefits that are sometimes available from group decision making. And finally we make suggestions regarding the norms, processes, and techniques that provide the necessary infrastructure for a group wanting to jointly use quantitative methods. The discussion will be relevant to several kinds of groups:

- *Established groups making joint decisions:* Imagine a university faculty deciding whom to hire or what to do with a generous bequest, a board of directors taking action, or a business partnership involving several owners. The individual decision makers may have different interests and perceptions, but they have many joint interests as well.
- *Unitary decision makers with advisers:* The advisory group for a unitary decision maker may be ad hoc (that is, advisers invited to address a particular problem) or may be a standing committee designed to perform advisory functions for a class of decisions as they arise (think of the U.S. president's Cabinet). Then there are combinations of ad hoc and standing committees. For a dramatic example, think of President John F. Kennedy as the decision maker at the time of the Cuban missile crisis. The president was advised not only by his regulars but by other special, knowledge-

able invitees. In some decision configurations the principal decision maker may need to have his or her decision ratified by a council (a board of directors, Congress, voters).

- *Negotiations among many unitary actors.* Negotiations among many actors are the preoccupation of this part of the book and will be discussed in ensuing chapters—how to map the space created by multiple intersecting BATNAs, how to maximize satisfaction among multiple actors with complex value functions, and so on. In this chapter we consider the problem of keeping a fractious group organized and productive—in order to have any chance of achieving those other goals.

In each of these cases several individuals will have to come together to exchange information, find or invent alternatives, argue the merits of their positions, and concoct a workable solution to their joint problem. We wish to examine how this is done and how it could be done better. Subsequent chapters look at quantitative methods adapted to the particular challenge of group decision making. But first let us examine some of the intricacies encountered in group decision making.

Behavioral Realities

In what ways do committees perform badly? Why is this so? After examining these questions, we turn to prescribing ways to improve committee behavior.

Problems with Group Behavior

The problems common to group interaction are familiar to anyone who has spent time working in an organization:

- People all talk at the same time. No one can make himself heard or, worse, hear anybody else.
- People don't listen carefully. They think about their next input.
- They forget what was said; and no record is kept.
- Discussions are disorganized, go around in circles, and easily get sidetracked.
- Too much time is spent on trivia and not enough on substance.
- Often the discussion breaks down into several parallel meetings, reaching inconsistent understandings.
- Some parties disengage from interactions. Some withdraw. Some free ride.

If these sorts of complaints are prevalent, then the group is in trouble and needs help. But figuring out what to do requires some analysis.

Why Groups Do So Poorly?

What makes multiparty interactions so unproductive compared with those of individual decision makers? One reason is simply that the problems groups face are often more complex. Finding a substantive solution is more difficult. Imagine a family that wants to pick a spot for vacation. Mom wants to go somewhere warm. Dad wants to be out in the country in a quiet wilderness spot. It's not so hard to think of likely locales. But now add some more parties, who will impose additional restrictions on the decision. Brother wants to be able to go shopping for camera gear. Little Sister wants horseback riding. Big Sister wants chic nightlife. Grandmother wants someplace within driving range—she hates flying. Now it doesn't seem so easy to make everyone happy.

Still, with more people, the group has more brainpower and more personhours at its disposal. Shouldn't the extra intellectual firepower more than compensate? Lamentably, the presence of more people usually hurts more than it helps. Most groups organize themselves (if it can be called organization) in ways that diminish their power to solve problems. Once she knows everyone's interests, the mom in our problem can probably find an answer that meets the criteria faster if she doesn't have to talk about it with the others (assuming they trust her enough to let her find the solution). We will say a few words about some additional reasons.

Coordination loss. Social psychologists have long noted that the effort put forth by a group is often less than the sum of what the members could do as individuals, resulting in what is known as "coordination loss."[1] For example, a group pulling on a rope, as in a tug-of-war, does not pull as hard as one would expect from the participants' individual ability. One reason is that it is hard for a group to pull in exactly the same direction. The result is that those who pull in slightly different directions cancel out much of each other's work. Similarly, those at a meeting may fail to agree on the purpose of their gathering. They don't agree on what they should be talking about at any particular moment—what they need to decide, what information to share, and so on.

Communication overload. A group has more people, but the amount of airtime available for people to talk does not increase. Yet each still wants to press

---

1. See Thompson (1998), p. 145, for a discussion of what she terms "coordination loss," "motivation loss," and "conceptual loss."

her point, and people overfill the available space. The unhappy result of this communication overload is that some "dumb down" their points, reducing their input to the simplest possible positions. A few people typically dominate the discussion—those who are most comfortable shouting over the others. Each additional person brings new issues to discuss. The conversation is easily sidetracked, and time is wasted on irrelevancies. The same arguments are rehearsed again and again, without persuading anyone.

Cognitive overload.   It is true that a group has more total reasoning power available. But each individual's capacity to process information doesn't change. With more participants there are more inputs and a greater variety of inputs. Because no one person can keep track of all the information and process it effectively, cognitive overload results.

Interpersonal conflict.   It inevitably happens that some people rub each other the wrong way. Their interpersonal styles may conflict, or they may prefer to think about problems in different ways. They may compete for attention and status.

Disengagement.   Returning to the example of the tug-of-war, we find that another reason groups fare poorly is that some members are tempted to slack off and let the others do the work. Economists refer to "free riders," social psychologists talk about "social loafing."[2] Each player knows that if he works hard at organizing the discussion and reaching an integrative agreement, the benefits will mostly go to others. And the more difficulty the group is having, the less sense it makes for anyone to invest time and energy in improving it. No one assumes responsibility. Some withdraw. They may actually stop attending, or they may check out mentally. The result of this disengagement is often no agreement, or one that favors the loudest and the most stubborn.

Benefits of Group Decision Making

With so many difficulties flowing from group interaction, one may well ask, "Why bother?" Perhaps the best course is simply to have all decisions made by individuals. And yet there are some compensating benefits that would be lost if that were the case.

Resources.   A group has more manpower available and can use it with greater flexibility than an individual. Groups especially have a wider range of exper-

2. For a useful discussion, see Latané, Williams, and Harkins (1979).

tise—regarding the subject matter or even in skill at managing a group—and more specific knowledge relevant to the problem at hand.

Arousal. Being in a group often makes people aware of being observed. Whether because of a desire to impress their colleagues or simply as an outlet for their feelings of discomfort, some people work harder when others are around. They don't do as well at learning new skills under such circumstances, but they do perform the things they already know with greater zeal (Zajonc 1965). Interaction with well-liked teammates may also raise morale and, with it, productivity.

Ownership. There is the hope that participating in decision making will lead people to accept and support the decision.

Prescriptive Advice

What should negotiators or decision makers do to not only improve the quality of their outcomes but reduce the effort and stress required to reach them? The broadest answer is to make a greater investment in the infrastructure of the group—adopting norms, establishing heuristics to guide collective thinking, and dividing roles and responsibilities in a way that will help the group as a whole. This is hard work, and requires both time and energy. Our belief is that for groups that need to work together, and even for negotiators facing an important dispute that will require a significant effort to resolve, that investment makes sense.

We organize our prescriptive remarks in two major categories: membership and organizing the substance of the discussion. Under the latter, we include choosing purposes, using a framework, using labor efficiently, managing the conversation, and watching time.

Membership

The most important rule for group membership is this: invite the people you need—and no more.

Each additional person involved imposes costs: communication gets more difficult, and the task of managing the process gets harder (see, for example, Hackman and Walton 1986). But adding people can bring benefits: more ideas, increased expertise, the ability to contribute greater resources to an agreement. How should one decide whether or not to include a candidate? A general rule is to invite a person only if you can articulate a clear reason why his or her presence would help.

Individuals may make a useful addition for two kinds of reasons: (1) they have something to contribute to making a better decision, or (2) any solution will need their approval or support. In the first class we can distinguish among those who have:

- Expertise in dealing with this sort of problem
- Information about this particular problem
- Skill in managing negotiations and decisions

Sometimes multiple people are available to contribute needed information or ideas. In that happy circumstance you can also try to select people who will help the meeting run more smoothly. Your choice might be based on:

- Negotiation style (willingness to use FOTE, to be collaborative on distributive issues)
- Ability to lead subgroups
- Time and energy available to commit to the project.

Involving those whose approval is needed presents a dilemma. If too many are asked, they can make the meeting unwieldy. We know of a professional partnership that tried to rewrite its compensation policy in a meeting of more than one hundred. Even had they had the best formal process and highly skilled facilitation (which they didn't), the meeting still would have been a disaster. It is difficult for more than a handful of people to draft any complex document together. It is better to have a subgroup take care of drafting.

The experience led some of the leading partners to write a policy among themselves. They presented what they'd come up with when the other partners arrived at the next meeting, and asked for an immediate vote for approval. The plan, a good one, was resoundingly rejected. If those who must approve an agreement have no opportunity to have a say in its contents, they will likely view it with suspicion. And if it is sprung on them without adequate time to consider it, they won't want to risk agreeing to something they don't understand.

The solution is to disaggregate three elements: (1) contributing one's ideas and interests, (2) drafting, and (3) approving. Don't try to have everyone actually at the meeting where it is created—but give all participants a chance to offer their input indirectly before they are asked to approve. A formalizing structure allows one to break down the problem for input—the partners could have been asked to each offer a memo with their views on the firm's objectives and alternatives in creating a compensation plan. Or a group with similar interests might send a representative to the table to transmit members' suggestions.

Don't stop thinking about membership once the work starts. Keep in mind that as the group learns more about the problem under consideration, its needs may change. New participants with different skills and information might help; others might find that their input is no longer required. We will return to the question of organizing manpower in a subsequent section on disaggregating tasks and assigning them to subgroups. But first we will need to discuss frameworks that groups can use to break up a task and make it easier to tackle.

Organizing the Substance of the Discussion

After deciding who will come, the next question is how to use group members' time and effort in a coordinated fashion. An individual will often benefit from taking some time to organize a process to guide his thinking. For a pair of negotiators, agreeing on a process is even more important. For a group, it is indispensable.

Choosing purposes.   Too often, the participants in large negotiations or problem-solving sessions have different purposes—and still worse, they don't even know it. Imagine a business student couple who have difficulty choosing a city to live in after graduation. They go to marriage counseling. One spouse thinks that the desired result is a choice of destination. The other wants them to decide whether to stay married or get divorced. They will be more successful in solving either problem—or both—if they can agree which they are working on at any given time.

The first task for a group is to establish common purposes. What would the members like to get done together? In the language we introduced in Chapter 2, what is the problem they will be working on? To some, "problem" implies a reactive response to some difficulty. We might easily substitute "opportunity" to provide a more proactive, forward-looking flavor. In either case, we are talking about the gap between the current state of the world and some imagined future state that we would like better. A common purpose makes coordination easier. It can also increase motivation—each will work harder if she is convinced that there is a worthwhile end to this toil. We offer a few general guidelines for determining a group's purposes.

Articulate ideas about the purpose before the meeting. A group will make more progress if everyone comes prepared to make a useful contribution. Circulate a memo to all the invitees with some preliminary thoughts about what needs to be decided and what information could help that decision. Each can do some individual preparation before coming, and possibly consult any constituents who will be concerned with the outcome.

State the purpose in terms of a tangible product. Some purposes are all too easy to express without saying anything useful. "We're going to meet to talk about improving productivity." Talking about something might be useful, or it might not. Creating a definite product, like a list of five alternatives that can be investigated further, is more likely to be helpful.

Set nested aspirations. Having a target to shoot for can motivate a group and give it a common object to steer for. But it can also limit productivity—when we achieve the stated target we tend to slack off. We satisfice. The effort devoted to any one project can easily expand to fill the entire time. And if we only set one goal and fail to achieve it, the participants are not going to return to the same group with much enthusiasm. A way to avoid both problems is to set a series of aspirations.

It is hard to know whether any particular one is too ambitious to accomplish in the time available. If we only get the first done, that will at least be something. And if we get that done with time to spare, we can go on to the next.

Solicit input on purposes. Group members are more likely to work hard toward a purpose if they have helped to create it. They are more likely to see expectations for their performance as reasonable if they have had a hand in setting them. When possible, give everyone an opportunity to offer his or her ideas about the purpose for a meeting. Doing so may risk an open disagreement over the purpose—but an open disagreement is easier to deal with effectively than a buried one.

In these preliminary meetings about purpose and potential product one might ask such generic questions as:

> Why us?
> Why now?
> Why this problem?
> Who should be deciding?
> Who will be affected by any decisions we make?
> Where does this problem occur elsewhere?
> How do we find out about the experiences of others working on similar
>   problems?

Remember: Keep the option of recycling backward if need be! The group should not get trapped by adopting an inflexible agenda. As the group understands its problem better, it may be useful to rethink its basic mission and recycle.

Use a framework to coordinate group thinking.   In Chapter 2 we discussed useful precepts for an individual to follow in making choices. We recognize that most people do not use this method, and yet decision making is still often fairly effective. Good decision makers may subconsciously do a lot of that rational calculus without formalizing each step of the analysis (much as outfielders know where a fly ball is going without knowing any of the ballistic equations that a physicist would use to predict it).

But what about a group mind? The decision-making group has no such mysterious synthesizer; therefore, we would argue that there is a greater need for the group to think systematically about the processes it will follow. The individual might get away without adopting a systematic approach to decision making, but the group cannot be so cavalier.

It is even worse when dealing with very complex problems, like the ones groups typically face. One person may be able to keep a complex problem organized in her head—the way some people can play chess blindfolded. Most will benefit from some formal structure. When one is working with a group, formal structure becomes a necessity.

Using a framework helps reduce coordination loss, by keeping everyone focused on the same subject at the same time. It reduces conceptual loss as well. Using a chart to lay out a structure for the problem at hand frees up room in the participants' minds. And even the people who can play chess blindfolded play better when they can see the board—it cuts down on the effort spent keeping track of the pieces. A formal organization is like the chessboard—looking at it helps you keep all the pieces straight in your head. It doesn't channel the resolution to any particular result—parties can and will disagree at every step. But they will be disagreeing about the same thing at the same time, so that their reasons for disagreement can be identified and handled.

An industrial process breaks a manufacturing job into discrete tasks, which allows more people to usefully work on the same job. A group of decision makers can break their job down into discrete thinking tasks. They can then take on the subtasks one at a time or "subcontract" them out to subgroups.

Choosing a problem-solving framework. Some professions have well-defined structures that guide them in working together. Management consultants have models to guide their analysis of a company's strengths and opportunities. Pilots have preflight checklists to guide them in prepping their aircraft. Finding the framework appropriate for your meeting will depend on the task and on the level of sophistication of the participants. Several approaches are adaptable for general use.

Good frameworks will have a few characteristics in common. They will be simple and easy to use. The easier they are to understand, the more mental energy is left for the substance of the task. They will not limit the group's thinking—in the sense that they will not control the outcome. Rather than discouraging disagreement, a good framework will encourage people to disagree more efficiently. It will help them identify exactly where their thinking diverges, and will suggest ways in which they can gather more information or ideas that will help resolve their disagreement.

Flexibility and learning. One hallmark of a good framework for analysis is its flexibility. We don't intend our prescriptions to be followed in a linear fashion. Just as in the PrOACT way of thought, one may start out with an original understanding of what the problem is about but learn, in the process of undertaking other tasks, that the original conception of the problem was too narrow; one must cycle back and change the definition of the problem. The cycle of analysis is not

$$Pr \rightarrow O \rightarrow A \rightarrow C \rightarrow T$$

But there should be feedback loops at every stage back to earlier stages as one gains a deeper understanding of the circumstances and events. The way the problem is first conceived may itself be the problem. The issue is not always to get from "here" to "there" but to learn where "there" is (and perhaps where "here" is as well). In pursuing any analysis, one should expect to be surprised along the way and prepared to cycle back in the reformulation of tasks. This is even more true of group analysis. Group members should anticipate that they will reframe their projects to reflect their deeper understandings. Remember that unexpected insights in the course of doing analysis may be the impetus for breakthroughs, accidents may be more important than planned events, and disappointments may lead to successes. Keep these feedback loops in mind as we suggest what appears to be a linear list of steps for a group to follow.

The PrOACT Way. A group with similar interests, at least nominally on the same team, facing a joint decision, can use the PrOACT system laid out in Chapter 2 (identifying the *Problem*; clarifying *Objectives*; generating *Alternatives*; examining *Consequences*; making *Tradeoffs*) for analyzing individual (or unitary) decisions under certainty. In cases of uncertainty, each alternative has an associated risk profile. In the literature on risk analysis, one speaks of *risk assessment* (of the uncertainties), *risk evaluation* (of consequences), and *risk management* (a combination of structure and synthesis). The Separation Theorem of decision theory asserts that the tasks of assessment (of probabilities) and evaluation (of consequences)—can be done separately and then synthesized. In performing a risk analysis, the group might want to decompose the problem into its constituent parts and then recompose the analyses of the separate parts. One advantage for the group in breaking apart assessment from evaluation is that it establishes a place for the input of scientific experts, who are specialists in only one aspect of the problem.

Other problem-solving frameworks. Another simple framework to help concentrate the "group mind" is the Circle Chart.[3] This framework posits two divisions among kinds of thinking: descriptive versus prescriptive, and concrete

3. For a fuller explanation, see Fisher, Sharp, and Richardson (1997).

versus theoretical. The two cuts make four "quadrants": data, diagnosis, prescription, and action steps. Another framework, specifically tailored for negotiation, is organized around "seven elements": interests, options, criteria, alternatives, commitment, communication, and relationship.[4] Or one can get even more rudimentary and adopt Ben Franklin's suggestion of comparing two alternatives by writing down the pros and cons of each (Hammond, Keeney, and Raiffa 1999, pp. 88–90). No matter which framework is used, it will be a better help in structuring fruitful interactions than no framework at all.

Using labor efficiently: Organization of tasks and subcommittees. The secret of good analysis is to divide and conquer. Working with groups, the trick is to:

- decompose the problem into constituent tasks (subcommittees)
- assign the right people to each of these tasks (subcommittees)
- recompose or synthesize the work of the separate committees

Decomposition of tasks. Deciding which subcommittees to form depends both on the nature of the problem and on the composition of the group: participants' abilities, their distribution of expertise, their individual interests, and how the anticipated product of these subcommittees might conceivably aid us in solving the problem.

Each subcommittee should have a clear initial statement of its purpose; the product that it is hoped it will produce; and the time frame. In the process of its investigations, the subcommittee should of course refine its purpose, but it should inform the parent committee of any substantial restatements of its task.

Assignments to subcommittees. Divide roles to meet members' interests. Keep in mind that the division of roles is in itself a negotiation. People have preferences in their position within a group. A team composed of members who are happy with their roles will likely outperform one made up of malcontents.

For each task or for each subcommittee you plan to form (whether as a group leader or as part of a group making collective decisions), you should identify the talents and expertise needed. Who has special knowledge of the problem? What are the skills that the group will need? Who has interests at stake and needs to approve the outcome? The decision of which subcommittees should form depends on the composition of the group: members' abilities, their distribution of expertise, and their individual interests.

Recomposition and synthesis. Let's assume that we have now identified and

---

4. See Fisher, Kopelman, and Schneider (1994); compare with another negotiation framework that focuses more on psychological dynamics that can affect the distributive aspect of negotiation (Shell 1999).

structured the problem and assigned tasks to subcommittees. They report back
to the main committee. The time has come for the committee as a whole to put
the parts back together again and to decide or make its recommendations. This
is often done poorly. First of all, it's hard to do. Second, usually not enough time
is allocated for this group task. Third, often the wrong people are asked to do the
synthesis. Narrowly focused experts may be too concerned about only one aspect
of the problem. Fourth, the reports of the subcommittees may not be integrated
into an overall evaluation, because it's not clear how this should be done. The
rhetoric and advocacy appeals of special interests then begin to dominate the discussion. It's almost as if no fact-finding has been done, no subcommittee reports
have been submitted, and the closing debate could have taken place right at the
beginning.

A good facilitator is indispensable at this phase. Good analysis can help. At
one extreme is the use of a completely formal scheme in which synthesis is
achieved by an optimization algorithm. But it is admittedly difficult to formalize
all aspects of a complex problem. One suggestion is to keep track of those realistic concerns that are omitted in the formal analysis and have the group discuss
how these excluded considerations might affect the formal conclusion.

It may not be possible to achieve a unanimous report. But the group members might record where they agree and where they disagree, and try to structure
their disagreements.

Managing the conversation. There remains the task of moderating the
group's discussion from minute to minute to keep it efficient and productive.
Tomes can be written about this subject, but we confine our remarks to those
few ideas that seem most crucial to the task of quickly organizing a new decision-making group:

1. The need for a good facilitator (or discussion organizer and leader)
2. The need for a good scribe (recording what is being said and what has
   been said)
3. The need to generate ideas

Facilitation. The facilitator can be thought of as a traffic cop whose job is to
keep the discussion moving smoothly.[5] She—or he—can direct discussions by
choosing topics and making sure everyone has a chance to contribute. She can
suggest subgroup divisions.

A facilitator takes on much of the work of managing the group in order to

---

5. Another problem of usage here: consultants who help groups conduct meetings and make
decisions are called facilitators, even though they do tasks associated with both "facilitators" and
"mediators" as those terms are used in discussing third-party intervenors in bilateral disputes (see
Chapter 17).

free the other participants to work on the substance. And the success of the meeting will to some degree depend on how well she plays that role. What sort of person should we be looking for?

We want someone who knows something about how a group can be productively organized. She (or he) should have some ideas about how to handle the problems dealt with in this chapter. It is less important that she agree with the prescriptions than that she has some plan of how to proceed in mind.

We want someone with a good head for organizing a complex problem. She should be able to keep the issues straight, and know what questions are important to ask. She need not be able to solve the problem, but she should be able to see its outlines and know how to direct people toward subproblems that they can solve.

She should have good interpersonal skills. She will need to deal with tough personalities and defuse tense situations. She should gently guide people back to the task at hand when their comments are off point.

Negotiators are often suspicious about the motives of a facilitator. If any of the parties see the facilitator as biased against them, they will resist her guidance on the process—even if her suggestions actually do no damage to their interests. It isn't a good idea to have a facilitator who is known to have personal interests at stake in the negotiation that are at odds with those of any of the participants. Close personal relationships, or business entanglements, with any of the participants are also likely to raise the hackles of the rest.

If it has been decided that the facilitator will be selected from among the parties in the negotiation, it is a good idea to choose the one with the least at stake. Another approach is for the facilitator to delegate advocacy of her substantive interests to another participant with similar interests, and devote herself to worrying about the process. This requires a good bit of trust on the part of the facilitator of the person handling her affairs.

Documentation. People often think of the person who takes notes at a meeting as the one who contributes least. Yet much can be accomplished by a skillful scribe. He—or she—needs to be good at reducing complex statements to a pithy essence that can be jotted down for all to see. He needs to be good at checking his understanding against the contributor's meaning.

An ongoing visible record, especially when the group is using a formal decision framework, keeps a group moving in unison. The medium can be a chalkboard or a mass of flip charts, or a network of computers on which each participant sees the progress displayed. Each player knows that her contribution is being incorporated into a record that will be useful later on—so no one feels unnoticed.

A visual representation relieves the participants of the mental effort of keeping the discussion organized in their heads. The depiction will show the relationship between the different thinking tasks—letting everyone know where the discussion came from and where it is going. This saves time that might be spent explaining the problem or repeating what has already been said. A good record shows anyone coming back from an absence (whether she was called to the phone or simply lapsed into a daydream) what the topic of discussion is and what ideas have already been discussed.

A visual record also helps solve a dilemma that troubles some participants using a framework. On the one hand, if the organization of the discussion is too rigid, it may inhibit creativity. Sometimes a bright idea is sparked by a discussion of another topic. On the other hand, if the discussion is too free-flowing, it is hard to get anything done. Some get confused by conversations that jump around. The brilliant idea on one topic causes us to lose solid workmanlike thinking on another. Recording with a framework allows a group to get the benefits of both approaches. There is a clear rational progression for those who prefer it, and there is also a way to catch ideas that come out of order. Visual recording tends to improve the quality of oral communication. The scribe can gently prod a verbose participant to rework his rambling comment into an informative sentence. Someone who interjects a non sequitur can be pushed to explain the relation between her comment and the previous discussion. Over time this process tends to produce more concision and more consideration of how one uses the group's time and attention.

Generating fresh ideas. As we discussed in Chapter 2, the more alternative courses of action considered by decision makers, the better the payoff to that decision is likely to be.[6] Some alternatives may be available "off the shelf"—well-known solutions to similar problems. Others might be invented by the decision makers themselves. The most common method for doing so is "brainstorming" (see Osborn 1957). The basic elements of the practice are this: a group sets aside a period in which it will only invent ideas; criticism of the ideas is not allowed (since otherwise fear of their colleagues' poor opinion may prevent some group members from offering suggestions); in fact, unworkable ideas are encouraged in the hope that they will suggest other, better ones; and the ideas are not ascribed to any individual, but only to the group as a whole. The group should seek an environment where it is safe to openly express half-baked ideas and where it is taken for granted that only some of these ideas will mature into serious contenders for later deliberation.

There is some evidence that "nominal groups" (composed of individuals working in isolation on the same task in parallel) do better than actual brainstorming meetings in terms of both the quality and the quantity of ideas (Diehl and Stroebe 1987). Researchers find that the players lose more from the brainstorming interaction than it generates new value. But that is not the choice that actual decision makers and negotiators face. A nominal group takes a great deal of organization and resources to set up. The practical choice is probably be-

---

6. In addition to the work of Hammond, Keeney, and Raiffa (1999), see also Shaw (1981). Note that "alternatives" in decision theory parlance are "options" according to negotiation writers.

tween five negotiators at a brainstorming meeting and five negotiators each sitting at home watching television.

- *First, generate ideas in isolation.* Set aside a period in which group members work alone—whether in their own offices or simply in silence around a conference table. Each is assigned to come up with as many ideas as possible for a short period—before interaction with others distracts them or other people's ideas displace original ones in their own mind (Valacich, Dennis, and Connelly 1994). At the end of this stage the participants can share their ideas, ideally recording them in a visible location.
- *Second, generate ideas together.* The group can now generate synergy from its members' thoughts. Members should try to find new ideas that have not been presented, and to improve and tinker with the ones that have. They will only nominate a certain number of courses of action, without making any decisions among them. One of the keys to generating better ideas is simple: don't stop when you find a good one. It often happens that the first idea that anyone has is the only one that gets considered. Working in a group can be stressful—and some latch on to the first idea mentioned as a way of getting the meeting done.
- *Third, evaluate, criticize, and improve.* Just as groups fail in creating new ideas, they often commit the opposite error: failing to attack weak ones. In 1971 Irving Janis coined the term "groupthink" to describe the tendency to demonstrate loyalty to the group by supporting the majority view—whatever the merits of the plan. We often use others' opinions as a proxy when we are short of information. And we fear ostracism if we go against the majority (or the boss). The solution is to set aside a period in which to look for holes in the possible decisions the group is considering. Ask members to assume that there is something wrong with the idea under consideration—the challenge is to find what it is. Make clear that there is no penalty for attacking an idea. One problematic way to do this is to institutionalize the role—as the Catholic Church did by creating the office of devil's advocate to argue against the canonization of potential saints. There is a tendency to ignore the views of someone in such a permanent role—after all, one wouldn't want the devil to prevail. A better solution is to have participants periodically switch roles, sometimes generating ideas and other times evaluating them.

Watching the time.   Most committees have a terrible time managing time. Here are a few insights that can improve their performance.

- Make initial time allocations for different tasks and review these periodically.
- Appoint a timekeeper to keep track of the clock for the group.

- Leave adequate time for synthesis at end.
- Subcommittees may need the discipline imposed by a deadline.
- Be aware that there is a behavioral bias toward underestimating the time it takes to do certain tasks.
- The group may have to negotiate for more time.
- Group members may be able to use time delays for their own competitive advantage in negotiations.

The Individual Decision Maker with a Group of Advisers

Let's consider the case of an important decision maker (such as the president of an enterprise or even of the United States) who has a staff of advisers and can request additional help from experts. What process should President G follow in addressing a hypothetical problem—or opportunity?

The Advocacy Model

The leader, President G, listens attentively to the rhetoric of various advocates of different positions or actions. She may seek out advocates of various extreme positions and listen to a debate between them as a way of making up her mind. The rhetoric concentrates on the merits of the speaker's preferred alternative and marshals those arguments that favor that alternative. Just as a judge listens carefully to both sides of a court case, including cross-examinations, the leader who has to make a decision does likewise.

One advocate argues in favor of alternative Q, stressing how well Q fares on some subset of evaluative objectives, ignoring those other objectives for which Q does not do so well. Another advocate belittles Q by examining objectives that favor his preferred alternative W. Rarely do advocates come together and report on those matters they agree on and try to probe where and why they disagree. They make it hard for the ultimate decision maker to make tradeoffs across evaluative objectives or to make a decision that synthesizes alternatives. The decision maker often adopts the advice of the advocate who makes the slickest presentation.

Structured Analysis: An Alternative to Advocacy

Suppose G has a tough decision to make as president of her organization. She doesn't have the deep understanding of the problem she needs in order to choose wisely at this point. She has, however, the services of a staff whom we'll designate as either (narrow) experts or (broad) advisers. The experts are special-ists who have a deep understanding of some aspect of the problem but don't have a grasp of the big picture; they are not the synthesizers. In addition G has a small number of close advisers who will be encouraged to review the problem in its entirety, work with the experts, and make balanced, integrative judgments or recommendations at the level of action. Some individuals may play a dual role, having deep expertise about some aspect of the problem as well as being able to synthesize the disparate parts into an integrated whole.

So G works with a few of her broad advisers to disaggregate the problem; they break the problem down into manageable parts so the experts can do their work and provide the synthesizers with valuable information. President G may choose not to articulate all of her evaluative objectives, some of which might have to do with her secret personal ambitions. This is one reason she prefers to make the final decision. But it is still a mind-boggling choice, and she needs help in making it. Unfortunately, as is true of most important executives, even if she could do her own analysis, she doesn't have the time.

She participates enough to identify some key uncertainties and to comment on the list of objectives for the organization—perhaps masking some of her per-sonal goals. She leaves it to her advisers to structure some viable alternatives, to do an assessment analysis of the key uncertainties, to generate a set of plausible evaluative objectives and to examine how each alternative fares on each of the evaluative objectives. She encourages her advisory staff—not the narrow ex-perts—to make a few simplifying tradeoffs to make the problem more transpar-ent and tractable—like discounting future monetary streams, and pricing out concerns about the environment, about image, and about labor goodwill.

So let's imagine that a given problem has been disaggregated, the experts have added their inputs, and each trusted adviser has presented his or her view-point about which action should be taken and why. But these trusted, high-level advisers disagree. Now what? Instead of engaging in an advocacy process, we suggest that G investigate, perhaps with the aid of a methodologically oriented facilitator, whether her advisers can agree on why they disagree. Can they get agreement on any lesser, second-level issues? Is it a question of the objectives? Is it a question of deep differences with regard to the uncertainties involved, or per-haps differences on basic values and tradeoffs? If the advisers can agree on why they disagree, they may wish to probe further: can that level of disagreement be decomposed further into subparts for which there may be subagreements?

The president's small cadre of trusted synthesizers may all prefer alternative Q to R, but she may opt for R. Why? Maybe she factors into the analysis previ-ously missing objectives; or maybe she finds out that while they all agree on Q over R, they do so for different reasons, and when she does her own synthesis of the assessments and evaluations, she comes up with R over Q.

Now President G has decided to go with alternative strategy R. There still remains the task of communicating this choice to her board of directors and to her constituency. She is now choosing not among alternative actions but among alternative stories to announce to her public, and this may call for a new array of evaluative objectives and a new set of advisers. But the first analysis leading to R should help inform the ensuing analysis. And so it goes.

Now for the downside: structured analysis takes time; it is hard to do; in the decomposition/recomposition process, issues may fall between the cracks. Structured analyses may be poorly done. Perhaps President G should experiment with structured analysis on some simple problems at first and build up the needed competence in her organization. Maybe some parts of structured analysis can be grafted onto the advocacy model to make a hybrid methodology.

Core Concepts

Groups have more resources available to use in decision making. But they suffer from "productivity loss"—the inefficiencies that result from the interactions of the people involved. We think that good norms, heuristics, and procedures can reduce the drag of productivity loss.

One way to reduce productivity loss is to minimize the number of people who attend decision-making meetings or negotiations. Take on only those additional members whose knowledge of the facts in this case, expertise in handling a class of problems, or skill at managing groups outweighs the cost of additional productivity loss.

Groups can structure their discussion around a cognitive framework that orders a problem, allows it to be decomposed and then recomposed, and provides a map that can be used to identify areas of disagreement. The PrOACT system is one good candidate, and there are others to consider.

Groups can work together on tasks that require input from many, and then break down into smaller groups, either for tasks that require more creativity or intricate problem solving, or simply to work on different problems in parallel.

The style in which meetings are run affects the amount of productivity loss. We recommend appointing a facilitator, keeping a visible record of the meeting (organized around the cognitive framework), and adopting both nominal and brainstorming techniques.

22

Consensus

Negotiations involving multiple parties take place under the shadow of different decision rules. Parliamentary negotiations are often governed by voting rules, with simple majorities, two-thirds majorities, and more complicated double-criteria as well. Talks between shareholders in a company can be railroaded by one party possessing a 51 percent stake, but they may still be constrained by a minority stakeholder—or coalition of stakeholders acting together—that holds at least 25 percent of the shares. In this chapter, however, we focus on one polar extreme for reaching agreements in multiparty negotiations—the rule of consensus.

Consensual Agreement

Imagine a set of N individuals (N > 2) who must choose a joint contract. Although they disagree on values and perceptions of uncertainties, they wish to act collaboratively. Any one of them can veto the whole deal. Each hopes an agreement will be reached, but each has a reservation value below which he or she won't settle. We assume that each has adequately prepared alone (determining interests, visions, alternatives, options, uncertainties, and so on) as described in Chapter 11. The group has prepared together in a FOTE manner and has jointly created a template to structure the give-and-take dynamics of later negotiations. By acting in a FOTE manner they agree to adopt the code of conduct enunciated by Milton R. Wessel in his 1976 book *The Rule of Reason* (see Box 22.1).

Each of the group members continues preparing alone by scoring the template—we'll relax this requirement later on—and each has assigned a reservation value to his or her Best Alternative to a Negotiated Agreement (BATNA).

Box 22.1   Rules of Reason: Milton R. Wessel

1. Data will not be withheld because they may be "negative" or "unhelpful."
2. Concealment will not be practiced for concealment's sake.
3. Delay will not be employed as a tactic to avoid an undesired result.
4. Unfair "tricks" designed to mislead will not be employed to win a struggle.
5. Borderline ethical disingenuity will not be practiced.
6. The motivation of adversaries will not unnecessarily or lightly be impugned.
7. An opponent's personal habits and characteristics will not be questioned unless relevant.
8. Wherever possible, opportunity will be left for an opponent's orderly retreat and "exit with honor."
9. Extremism may be countered forcefully and with emotionalism where justified, but will not be fought or matched with extremism.
10. Dogmatism will be avoided.
11. Complex concepts will be simplified as much as possible so as to achieve maximum communication and lay understanding.
12. Effort will be made to identify and isolate subjective considerations involved in reaching technical solutions.
13. Relevant data will be disclosed when ready for analysis and peer review—even to an extremist opposition and without legal obligation.
14. Socially desirable professional disclosure will not be postponed for tactical advantage.
15. Hypothesis, uncertainty, and inadequate knowledge will be stated affirmatively—not conceded only reluctantly or under pressure.
16. Unjustified assumption and off-the-cuff comment will be avoided.
17. Interest in an outcome, relationship to a proponent, and bias, prejudice, and proclivity of any kind will be disclosed voluntarily and as a matter of course.
18. Research and investigation will be conducted appropriate to the problem involved. Although the precise extent of that effort will vary with the nature of the issues, it will be concomitant with stated overall responsibility [for] the solution of the problem.
19. Integrity will always be given first priority.

Source: Milton R. Wessel, *The Rules of Reason: A New Approach to Corporate Litigation* (Reading, Mass.: Addison-Wesley, 1976).

We further assume that each party reveals its quantitative scoring simultaneously to the other group members or, perhaps more plausibly, to an analytically oriented external helper. Although we are painfully aware that these assumptions are rarely, if ever, achieved, we're pushing the limits of abstraction because there are lessons to be learned. We start with the simple fair-division problem to help prepare for later intricacies.

Fair Division with Many Parties

In Chapter 13 we discussed the fair-division problem for two parties. Besides being interesting in its own right, that problem helped motivate our analysis of more complex problems. Recall that the fair-division problem gives rise to an especially simple template: there is an issue for each item to be allocated, with each issue having just two possible resolutions. Furthermore, there are no complications about BATNAs and reservation values. Our analysis looked at the case which prohibited the use of monetary transfers to establish equitable allocations. In this chapter we generalize our discussion of fair-division problems to more than two parties, and we start out by considering the case with monetary transfers. We then prohibit those transfers and essentially generalize to more than two parties the discussion in Chapter 13 of collaborative FOTE analysis. In the next chapter we shall revisit the fair-division problem when a subset of parties, in an after-market, are allowed to make further trades, possibly including monetary transfers. In that discussion we shall be interested in how the existence of the after-market influences the initial allocation of items to be shared.

Fair Division with Monetary Transfers

Massachusetts, like other states, grants individuals the right to specify in a will how they wish to dispose of their property at death. If an individual does not write a will, the state will write one. The laws of descent and distribution on intestacy (determining who gets the property if there is no will) specify how the estate should be split among spouse and children. "For example, if A dies, wife B will take one-half, C will take one-fourth, and the two grandchildren will divide D's (their parent's) one-fourth equally" (Bove 1979).

It would be easy to divide the estate in equal or even in well-specified unequal shares, if it consisted solely of monetary resources. But how should one decide the disposition of items that cannot be easily sold and that have sentimental value for the inheritors? The problem is not as special as it might appear: similar problems are faced by husbands and wives in divorce settlements; by business partners in dissolving businesses; by victors in dividing spoils.

Let's examine a hypothetical situation and some of the ways in which it might be resolved.[1] A father leaves his estate of four indivisible commodities to be shared "equally" among his three children. Assume that the four commodities—A, B, C, and D—have the monetary values shown in Table 22.1, and that

1. This example is taken from Luce and Raiffa (1957), p. 366. The discussion given here is both more elementary and more extensive.

Table 22.1   Valuations of four commodities by three legatees (in dollars)

| | Monetary worth to each individual | | |
Commodity	1	2	3
A	10,000	4,000	7,000
B	2,000	1,000	4,000
C	500	1,500	2,000
D	800	2,000	1,000

the monetary worth to each of the children of any subset of the items is merely the sum of his or her monetary valuations of the individual items. Leave aside for a moment whether these monetary assignments have been strategically assessed by the individuals; assume simply that they are honest revelations and that the task is to suggest an allocation of the commodities to the children, with possible transfers of monetary amounts among them. There are three commonly proposed procedures for arriving at a solution.

Naive procedure.   Allocate each commodity to the person who values it most and collect its value for the pool of money to be shared. Thus commodity A goes to child 1 for $10,000; B goes to child 3 for $4,000; C goes to child 3 for $2,000; and D goes to child 2 for $2,000. The money collected is the sum of these amounts ($18,000), and each child gets one-third of this, or $6,000. The first child gets commodity A, less $10,000, plus $6,000—which nets out as A less a monetary payment of $4,000; the second child gets D plus $4,000; the third child simply gets B and C. Each gets a package that has been personally valued at $6,000.

Auction procedure.   Conduct the equivalent of an open ascending auction for each item; collect the payments; and share the proceeds equally. In this case, the first child gets commodity A not at $10,000 but at $7,000—the high bidder gets the commodity at the second highest price, since the auction would stop when the maximum price of the second-to-last bidder was reached and only the highest value was left. Commodity B goes to child 3 at $2,000 (not $4,000); C goes to child 3 at $1,500; and D to child 2 at $1,000. The pool would be the sum of these values, or $11,500, and each would get back $3,833.33. In this case, the first child gets A less $7,000 plus $3,833.33, or A less $3,166.67; the second child gets D plus $2,833.33; the third child gets B and C plus $333.34. Obviously, if the parties had to choose between the two proposals on purely selfish grounds, the first child would prefer the auction proposal, the second child the naive proposal, and the third child the auction proposal.

The Steinhaus-Knaster allocation.   A more complicated procedure for allocation was suggested by the Polish mathematician Hugo Steinhaus, and is known
as the Steinhaus fair-division procedure. Using the allocation shown in Table
22.2, we can see that, for example, individual 1's total evaluation of all four commodities is $10,000 + $2,000 + $500 + $800, or $13,300, and that his fair share
is one-third of this amount, or $4,433 (the initial "fair share" in the table).
Steinhaus would give the individuals total packages (goods plus transfer payments) that exceed their initial fair shares by the same amount. Here's the way it
works. The items are distributed efficiently: the first individual gets A, the second D, and the third B and C. Individual 1's excess over his initial fair share is
then $10,000 − $4,433, or $5,567. The second and third individuals' excesses
are, respectively, −$833 and $1,333, which makes a total excess of $6,067. As
long as the individuals differ in their initial evaluations, this total excess will be
positive—an important point. The total excess is divided equally: $2,022 to each
individual. The first individual should thus end up with an adjusted fair share
that is $2,022 above his initial fair share of $4,433, for a total of $6,455. This
is accomplished by giving him A and asking him for a cash contribution of
$10,000 − $6,455, or $3,545. Individuals 2 and 3 receive the same excess of
$2,022 over their respective fair shares. Note that the monetary side payments to-
tal zero: the market clears.

   Table 22.3 compares three of the above procedures. Individual 1 prefers the
auction proposal; individual 2 prefers the naive proposal; and individual 3 prefers the Steinhaus proposal.

Table 22.2   The Steinhaus-Knaster allocation

	Individuals		
	1	2	3
Item			
A	$10,000	$4,000	$7,000
B	2,000	1,000	4,000
C	500	1,500	2,000
D	800	2,000	1,000
Total valuation	13,300	8,500	14,000
Fair share	4,433	2,833	4,667
Commodities received	A	D	B and C
Monetary worth of commodities received	10,000	2,000	6,000
Excess	+5,567	−833	+1,333
Final division	A − 3,550	D + 2,858	B, C + 691

Table 22.3   Comparison of naive, auction, and Steinhaus-Knaster procedures

| | Items | Side payment (in dollars) | | Steinhaus- |
Individual	received	Naive	Auction	Knaster
1	A	−4,000	−3,167	−3,550
2	D	4,000	2,833	2,858
3	B, C	0	333	692

What is fair? One way to get one's thinking straight about such alternative proposals is to see how they would perform in simpler, more transparent situations. In the simplest case, there is a single indivisible commodity to be shared between two individuals.

A little elementary algebra will show the following. Let X and Y designate two players whose valuations of a given indivisible commodity are x and y, respectively, where $x < y$. The naive, auction, and Steinhaus proposals give the commodity to the Y player, and the payments by Y to X are, respectively:

$$x/2 \qquad \text{for the auction proposal}$$

$$y/2 \qquad \text{for the naive proposal}$$

$$x/2 + (y - x)/4 = [x/2 + y/2]/2 \qquad \text{for the Steinhaus proposal}$$

If the three inheritors X, Y, and Z have evaluations x, y, and z for a single commodity to be shared, and if Z has the highest evaluation, then the Steinhaus procedure yields to each player his initial fair share (one-third of his total valuation) and an incremental bonus of

$$\frac{2}{9}\left(z - \frac{x+y}{2}\right),$$

which seems quite reasonable.

Strategic misrepresentations with the Steinhaus procedure.   Let's examine the strategic problem a bit more deeply. Party 1 doesn't know how much the other parties might misrepresent their values. The more they misrepresent, the more dangerous it is for party 1 also to misrepresent. So although one can't really say that the Steinhaus scheme encourages honest evaluations, in many situations it may be the pragmatic thing to do. Honesty in this case is the super-cautious strategy (that is, the strategy that maximizes the valuer's minimum possible return—the so-called maximin strategy). It is also a good strategy against an extreme or naive exaggerator. Finally, it is the easiest and most socially desirable thing to do.

Fair Division without Monetary Transfers

Three parties: The problem and its analysis. The fair-division problem, without the use of monetary transfers, was used for two parties in Chapter 13 as an introduction to the more general case of integrative collaborative negotiations. The fair-division problem was easier to analyze than the general integrative problem, because its template is simple (two resolutions per issue) and it does not require an elaborate treatment of the alternatives to negotiation. The same prevails here where we deal with many parties.

Consider this problem: AAA, BBB and CCC are given ten items to share "equally." They choose *not* to use monetary evaluations or monetary transfers. Imagine that each scores the items by sprinkling 100 points over them, just as we did earlier for Janet and Marty. The data of the problem are exhibited in the box in Table 22.4, which exhibits a representative allocation (or contract). To be legitimate, the sum of the entries in columns E, F, and G have to sum to one in each row. In this allocation AAA is given items 2, 3, 4, and 8; BBB gets 6, 7, 9 and 10; and CCC gets 1 and 5. By using the sumproduct command, we can find the payoff values for the three parties on the spreadsheet: AAA gets 57.50; BBB gets 67; and CCC gets 38. For these values we list the sum, the minimum, and the product in preparation for SOLVER to make the calculations.

Table 22.4 Fair division: The problem and its analysis

	A	B	C	D	E	F	G	H
1								
2								
3						Allocations		
4	Item	AAA	BBB	CCC	AAA	BBB	CCC	Sum
5	1	18	2	20	0	0	1	1
6	2	16	4	13	1	0	0	1
7	3	22	2	15	1	0	0	1
8	4	16.5	1	15.5	1	0	0	1
9	5	17	4	18	0	0	1	1
10	6	2	30	10	0	1	0	1
11	7	4	10	4	0	1	0	1
12	8	3	20	3	1	0	0	1
13	9	1	14	1	0	1	0	1
14	10	0.5	13	0.5	0	1	0	1
15	Total	100	100	100				
16			Values					
17		AAA	BBB	CCC	Sum	Min	Prod	
18		57.5	67	38	162.5	38	146,395	

In using SOLVER, the variables to be controlled are in the region E5:G14. To be legitimate, these entries must be non-negative and sum to 1.0 for each row. Thus, for example, +E5+F5+G5 must be 1.0, and this should be true down to E14+F14+G14. Pure, as distinct from mixed, allocations add the requirement that entries E5:G14 be integers. We can now ask SOLVER to get to work and find the legitimate allocation that optimizes a specified objective.

Comparison of solutions. We exhibit in Table 22.5 five responses that SOLVER produced when asked to maximize: the sum, using mixed (m) values; the minimum, using pure (p) and (m) values; and the product using (p) and (m) values. CCC's return can vary from 38 to 52.3. We're inclined to use the min(m) calculation, but ex ante, not knowing the outcome of the comparison table, we can make a good case for the product(m). Key observation: the choice of what to maximize makes a difference!

Now we are primed to consider a collaborative FOTE analysis of the many-party integrative negotiation.

FOTE Analysis with Many Parties

A Concrete Abstract Case

An abstract case will highlight the essentials of negotiating according to the principle of consensus. Imagine that six players, AAA, BBB, . . . , FFF, have agreed on the scored template shown in Table 22.6. Full consensus is required, and if it is not obtained, any player can force the no-agreement state—which means that each player receives his or her RV. Each player has disclosed his or her scores to the analytical intervenor. It's not obvious whether a feasible solution exists. It does.

FOTE Analysis with SOLVER

In Table 22.7, the scored template is given in A5:H30. The variables at SOLVER's control are the contracts in I8:I30. These have to be legitimate, that is, they have to be non-negative and sum to 1 in blocks. For example, the sum of entries (shown in column J) must all be 1. For example, the sum of the entries in I8:I10, shown in J10, must sum to 1; and so on to the sum of the entries I26:I30 shown in J30, must also sum to 1. For any legitimate contract we instruct the spreadsheet how to find the payoff values for AAA to FFF. For example, the value for AAA in C33 of 42 is the sumproduct of the contract I8:I30 and AAA's scores in C8:C30. Each player has been given a confidential reservation value

Table 22.5   Comparison of solutions

Max	Values			Sum	Min	Prod
	AAA	BBB	CCC			
Sum(m)	54.5	87	38	179.5	38	180,177
Min(p)	54.5	57	48	159.5	48	149,112
Min(m)	52.3	52.3	52.3	156.9	52.3	143,056
Prod(p)	54.5	87	38	179.5	38	180,177
Prod(m)	47.5	87	44.6	179.1	44.6	184,209

Table 22.6   Scored template for all

(A)	(B)	(C)	(D)	(E)	(F)	(G)	(H)
		Parties					
Issue	Resolution	AAA	BBB	CCC	DDD	EEE	FFF
1	1	14	0	15	0	0	14
	2	8	22	20	4	11	8
	3	0	45	0	10	5	0
2	1	11	0	0	0	0	12
	2	7	25	0	0	20	8
	3	0	55	0	0	25	0
3	1	0	0	42	12	0	24
	2	5	0	35	8	2	18
	3	10	0	25	6	4	12
	4	17	0	0	0	9	0
4	1	35	0	30	0	10	40
	2	29	0	20	8	26	30
	3	20	0	10	13	40	23
	4	0	0	0	18	0	0
5	1	0	0	2	60	4	0
	2	5	0	4	45	8	2
	3	10	0	6	29	15	4
	4	15	0	8	15	12	7
	5	23	0	0	0	0	10
RV		35	75	50	60	45	45

that represents the scoring of its BATNA. The RVs are shown in C35:H35. The values of the illustrative contract less the RVs are exhibited in C37:H37 and labeled the *excesses*. The illustrative contract in I8:I30 is not only legitimate but feasible, since all the excesses are non-negative. However, for our purposes it isn't necessary to start out with a feasible contract; any legitimate contract will do.

The *maximum feasible* row (row 39) exhibits the maximum that each player can achieve with some feasible contract. Let us explain, for example, the entry 96.1 in E39, the maximum feasible for CCC. We called upon SOLVER to adjust the free variables (that is, the contract) in I8:I30, subject to the legitimacy constraints (that is, they are non-negative and sum to unity as required in cells J10, J14, J19, J24, and J30) and the feasibility constraints (that is, all excesses C37:H37 are non-negative) that maximize the value to CCC (cell E33). SOLVER took a second to solve this linear programming (LP) problem and yielded the answer of 96.1. Row 39 results from solving six separate LP problems.

The *proportion of the potential* (POP) is the ratio of the excess to the difference between the maximum feasible and the RV. For example, CCC's POP for this illustrative contract is 36 divided by (96.1 − 50) or .78, quite a high value. The spreadsheet is further prepared by exhibiting the minimum of the POPs, which is the number .08, shown in cell J41. The product of the excesses is shown in K35.

We are now ready to discuss the maximin and the Nash solutions—ones that are efficient and equitable.

Comparison of solutions.   Table 22.8 exhibits six different contending contracts of what may be termed "best." For each problem, we ask SOLVER to find a legitimate contract. If we demand a *pure* (that is, not mixed or randomized) contract, then we must require the free variables in I8:I30 of Table 22.7 to be integers and to be feasible (that is, all excesses are non-negative) to maximize either the *sum* of the values, or the *minimum of the POPs*, or the *product of the excesses*. The results are shown.

The problem with the sum criterion is the huge potential for inequity—so much so that many don't take it seriously as a candidate. But it's interesting to show. When we maximize the minimum POP, it will make a big difference whether we use pure or mixed contracts. With mixed contracts we can get all participants to achieve at least 28 percent of their potentials; with pure contracts, the 28 percent is reduced to 15 percent. If we give up trying to raise the minimum POP and use the Nash solution, then the min POP plummets down, but CCC and FFF are considerably enhanced. Clearly, it makes quite a difference whether we use the Nash or the maximin solutions.

Table 22.7   FOTE analysis with SOLVER

1	A	B	C	D	E	F	G	H	I	J
⋮										
5						Parties				
6	Issue	Resolution	AAA	BBB	CCC	DDD	EEE	FFF	Contracts	Sum of
7										
8	1	1	14	0	15	0	0	14	0	
9		2	8	22	20	4	11	8	1	
10		3	0	45	0	10	5	0	0	$x(1, r)'s = 1$
11										
12	2	1	11	0	0	0	0	12	0	
13		2	7	25	0	0	20	8	0	
14		3	0	55	0	0	25	0	1	$x(2, r)'s = 1$
15										
16	3	1	0	0	42	12	0	24	1	
17		2	5	0	35	8	2	18	0	
18		3	10	0	25	6	4	12	0	
19		4	17	0	0	0	9	0	0	$x(3, r)'s = 1$
20										
21	4	1	35	0	30	0	10	40	0	
22		2	29	0	20	8	26	30	1	
23		3	20	0	10	13	40	23	0	
24		4	0	0	0	18	0	0	0	$x(4, r)'s = 1$
25										
26	5	1	0	0	2	60	4	0	0	
27		2	5	0	4	45	8	2	1	
28		3	12	0	6	29	15	4	0	
29		4	15	0	8	15	12	7	0	
30		5	23	0	0	0	0	10	0	$x(5, r)'s = 1$
31										
32										
33	Value		42	77	86	69	70	64	Sum =	408
34										
35	Res. value		35	75	50	60	45	45	Prod =	2,154,600
36										
37	Excess		7	2	36	9	25	19		
38										
39	Max. feasible		62	100	96.1	89.9	92.8	75.2		
40										
41	Prop. of Potential		0.26	0.08	0.78	0.3	0.52	0.63	Min =	0.08
42										

Note: The entry in J10 is: +I8+I9+I10 or +sum(I8:I10).
  J14 is: +I12+I13+I14 or +sum(I12:I14).
  . . .

Table 22.8   Comparison of solutions

	AAA	BBB	CCC	DDD	EEE	FFF	Sum	Min. of POP
Res. value	35	75	50	60	45	45		
Maximize:								
Sum								
Pure	43	77	94	76	50	72	412	0.08
Mixed	43	77	94	76	50	72	412	0.08
Min. POP								
Pure	39	100	59	71	66	50	385	0.148
Mixed	43	90	88	68	69	54	391	0.279
Prod.								
Pure	42	77	86	69	70	64	408	0.08
Mixed	41.7	85.2	80.2	72.4	59	62.9	402	0.1024

POTE Analysis

Now let's assume, a bit more realistically, that the parties prepare the template together, but each scores it individually, including the evaluation of confidentially held RVs. Table 22.6 is *not* common knowledge. We're going to suggest how some very astute parties might actually negotiate in a POTE style—the truth but not the whole truth.

Negotiation template used as a decision aid.   One of the parties, AAA, exhibits the partially prepared Table 22.9 *without* the X's in column C. He makes the following speech: "Look, rather than telling you my interests, I'm willing to record what I ideally would want. I'm doing this not as a positional bargaining ploy, but to tell you something about my preferences in a manner you can record. I'll do that if each of you follows suit by recording your wish list. It's not going to be easy for us to find a contract on which we all can agree."

After some debate the others agree, and AAA puts an X next to each resolution he prefers. For example, he most prefers resolution 1 on each of the first four issues and resolution 5 on the fifth. Each of the other parties follows suit and each puts a single X next to the resolution he or she most prefers.

Party DDD outdoes AAA in providing information. She indicates her preferences but volunteers that, for her, the fifth issue is by far the most important. The parties agree mutually to disclose this additional information and CCC comes up with the idea (shown in Table 22.10) of using double and triple X's to indicate strength of preference.

Most groups engaged in this exercise share this type of information but do it

Table 22.9 The template as a decision aid to exhibit AAA's qualitative preferences

A	B	C	D	E	F	G	H
				Parties			
Issue	Resolution	AAA	BBB	CCC	DDD	EEE	FFF
1	1	X					
	2	—					
	3	—					
2	1	X					
	2	—					
	3	—					
3	1	X					
	2	—					
	3	—					
	4	—					
4	1	X					
	2	—					
	3	—					
	4	—					
5	1	—					
	2	—					
	3	—					
	4	—					
	5	X					

far less efficiently. Most important, they don't systematically record the information disclosed in a manner that all can simultaneously see and remember.

Friends and Enemies

Let's examine Table 22.10 more closely, keeping in mind who is *like* whom (a positive affinity) and who is *against* whom (an opposition). For example, party BBB and party EEE are in harmony. Party BBB feels most strongly about issue 2 and EEE doesn't feel strongly about this issue, but EEE agrees with BBB on what is best for it. Party BBB does not care about issues 3, 4, and 5 and therefore doesn't disagree with EEE on these issues. There is only a slight discord between BBB and EEE on issue 1. Yes, BBB and EEE could coordinate their action together. They have a strong affinity to each other.

In contrast, parties AAA and BBB have deep discord. On issues where BBB

Table 22.10   The table qualitatively rated

A	B	C	D	E	F	G	H
				Parties			
Issue	Resolution	AAA	BBB	CCC	DDD	EEE	FFF
1	1	X	—	—	—	—	X
	2	—	—	X	—	X	—
	3	—	XX	—	X	—	—
2	1	X	—	—	—	—	X
	2	—	—	—	—	—	—
	3	—	XXX	—	—	X	—
3	1	X	—	XXX	X	—	X
	2	—	—	—	—	—	—
	3	—	—	—	—	—	—
	4	—	—	—	—	X	—
4	1	X	—	XX	—	—	XX
	2	—	—	—	—	—	—
	3	—	—	—	—	XXX	—
	4	—	—	—	X	—	—
5	1	—	—	—	XXX	—	—
	2	—	—	—	—	—	—
	3	—	—	—	—	X	—
	4	—	—	X	—	—	—
	5	X	—	—	—	—	X

feels strongly, they disagree completely. Continuing in this vein, considering the affinity-opposition between parties, we can prepare a sort of sociometric figure showing affinities and oppositions and the strength of these. In Figure 22.1, we see that a three-way coalition could form among parties AAA, CCC, and DDD.

Who would make a neutral choice to conduct the meetings? Not AAA, because AAA is in strong opposition to BBB and in somewhat weaker opposition to EEE. The party that doesn't seem to be opposed to anyone else is FFF. He or she would make a suitable chair for further deliberations. Diagrams showing affinity-opposition connections can also be used in the discussion of coalitions, the subject of the next chapter.

Voting Dynamics

First vote.   The parties study the data in columns A to H of Table 22.10 and discuss what would make a good compromise. Looking at the X's for issue 1, the

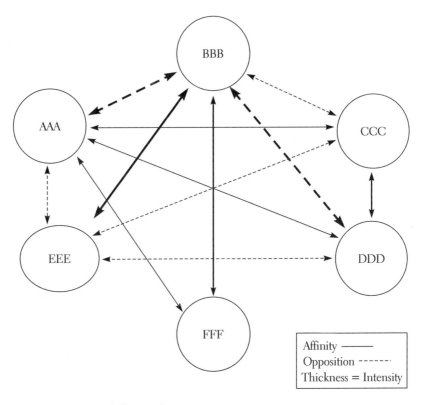

Figure 22.1.   Diagram of affinity and opposition.

group suggests starting tentatively with resolution 2. BBB would have preferred level 3, but you can't have everything (at least not right away). In the spirit of co-operation, they concoct trial contract 1: resolution 2 on each of the first four is-sues and resolution 3 on the fifth. These are shown in Table 22.11.

They vote on the acceptability of contract 1. Each values the contract using his or her private numerical scores and compares it with his or her reservation value. BBB and DDD vote No, in a vote that all simultaneously disclose. Sur-prisingly, most individuals vote truthfully and not strategically in these circum-stances. In the laboratory, most subjects still behave as if they all agreed on FOTE behavior. There are exceptions, of course, and these exceptions, interest-ingly enough, are most often economists or game theorists.

Contract 1 could be viewed as a single negotiating text—though it's also more than that—and the recorded votes on contract 1 tell, from Table 22.11, why it is not acceptable to BBB and DDD. How can it be sweetened for them in incremental steps that do not overturn the acceptability of the contract to the others?

To sweeten contract 1 for BBB we can push the resolution of issue 2 from level 2 to level 3 (see Table 22.12). Note how strongly BBB feels about this. This

Table 22.11   First vote

A	B	C	D	E	F	G	H	I
					Parties			
Issue	Resolution	AAA	BBB	CCC	DDD	EEE	FFF	Contract 1
1	1	X	—	—	—	—	X	0
	2	—	—	X	—	X	—	1
	3	—	XX	—	X	—	—	0
2	1	X	—	—	—	—	X	0
	2	—	—	—	—	—	—	1
	3	—	XXX	—	—	X	—	0
3	1	X	—	XXX	X	—	X	0
	2	—	—	—	—	—	—	1
	3	—	—	—	—	—	—	0
	4	—	—	—	—	X	—	0
4	1	X	—	XX	—	—	XX	0
	2	—	—	—	—	—	—	1
	3	—	—	—	—	XXX	—	0
	4	—	—	—	X	—	—	0
5	1	—	—	—	XXX	—	—	0
	2	—	—	—	—	—	—	0
	3	—	—	—	—	X	—	1
	4	—	—	X	—	—	—	0
	5	X	—	—	—	—	X	0
Vote no. 1		Yes	No	Yes	No	Yes	Yes	

change will make DDD somewhat happier. Let's see what happens with the change.

Second vote.   With the proposed change, BBB changes his vote but DDD does not change hers. Luckily, as shown in Table 22.12, we get all Yeses except for DDD. To make DDD a bit better off we could push issue 5 from level 3 to level 2. Let's see what happens (see Table 22.13).

Eureka! It worked. Table 22.13 shows unanimous agreement. Our single negotiating text no. 3 passed muster.

Scores on votes.   The votes and the scores for the parties, exhibited in Table 22.14, show that CCC and EEE have done extremely well in the sense that they both soared over their hurdle rates (that is, their reservation values). BBB, how-

Table 22.12 Second vote

A	B	C	D	E	F	G	H	I	J
					Parties			Contract	
Issue	Resolution	AAA	BBB	CCC	DDD	EEE	FFF	1	2
1	1	X	—	—	—	—	X	0	0
	2	—	—	X	—	X	—	1	1
	3	—	XX	—	X	—	—	0	0
2	1	X	—	—	—	—	X	0	0
	2	—	—	—	—	—	—	1	0
	3	—	XXX	—	—	X	—	0	1
3	1	X	—	XXX	X	—	X	0	0
	2	—	—	—	—	—	—	1	1
	3	—	—	—	—	—	—	0	0
	4	—	—	—	—	X	—	0	0
4	1	X	—	XX	—	—	XX	0	0
	2	—	—	—	—	—	—	1	1
	3	—	—	—	—	XXX	—	0	0
	4	—	—	—	X	—	—	0	0
5	1	—	—	—	XXX	—	—	0	0
	2	—	—	—	—	—	—	0	0
	3	—	—	—	—	X	—	1	1
	4	—	—	X	—	—	—	0	0
	5	X	—	—	—	—	X	0	0
Vote no. 1		Yes	No	Yes	No	Yes	Yes		
no. 2		Yes	Yes	Yes	No	Yes	Yes		

ever, ended up with 77, barely 2 points over his reservation value. Perhaps BBB could have played strategically and voted No on contract 2 in order to see if he could get more. But this would not be POTE behavior.

In the laboratory setting most participants tell the truth. We have tried to present the dynamics of these negotiations in a manner that captures how most groups play this exercise, although they rarely are as efficient as we have been in laying out the steps along the way. We can't emphasize enough the value of using a common decision aid so that all can see what's going on, rather than having each player keeping idiosyncratic and sketchy notes.

How good is this way of negotiating? It can be very effective. But as we shall see, it can also be disastrous. The process can easily be embarrassed by changing a few numbers.

Table 22.13   Third vote

A	B	C	D	E	F	G	H	I	J	K
					Parties				Contract	
Issue	Resolution	AAA	BBB	CCC	DDD	EEE	FFF	1	2	3
1	1	X	—	—	—	—	X	0	0	0
	2	—	—	X	—	X	—	1	1	1
	3	—	XX	—	X	—	—	0	0	0
2	1	X	—	—	—	—	X	0	0	0
	2	—	—	—	—	—	—	1	0	0
	3	—	XXX	—	—	X	—	0	1	1
3	1	X	—	XXX	X	—	X	0	0	0
	2	—	—	—	—	—	—	1	1	1
	3	—	—	—	—	—	—	0	0	0
	4	—	—	—	—	X	—	0	0	0
4	1	X	—	XX	—	—	XX	0	0	0
	2	—	—	—	—	—	—	1	1	1
	3	—	—	—	—	XXX	—	0	0	0
	4	—	—	—	X	—	—	0	0	0
5	1	—	—	—	XXX	—	—	0	0	0
	2	—	—	—	—	—	—	0	0	1
	3	—	—	—	—	X	—	1	1	0
	4	—	—	X	—	—	—	0	0	0
	5	X	—	—	—	—	X	0	0	0
Vote no. 1		Yes	No	Yes	No	Yes	Yes			
no. 2		Yes	Yes	Yes	No	Yes	Yes			
no. 3		Yes	Yes	Yes	Yes	Yes	Yes			

Table 22.14   Scores on votes and contracts

	AAA	BBB	CCC	DDD	EEE	FFF
			Scores on votes			
Vote						
1	Yes	No	Yes	No	Yes	Yes
2	Yes	Yes	Yes	No	Yes	Yes
3	Yes	Yes	Yes	Yes	Yes	Yes
			Scores on contracts			
Res. value	35	75	50	60	45	45
Contract						
1	59	47	81	49	74	68
2	52	77	81	49	79	60
3	47	77	79	65	71	58

The Two-Party Counterpart of the Voting Procedure

Let's examine the dynamics we followed as it applies to two parties. We begin with a proposed contract and let its joint evaluation be at point *a* in Figure 22.2. In this case BBB votes Yes and AAA votes No. So we change the single text and move to point *b*; still not acceptable to AAA. On the third trial we move to point *c*, which is in the feasible range. Success? Maybe. Point *c* may not be efficient or very equitable. In practice negotiators satisfice: when they find a feasible solution, they claim victory. No one is the wiser.

   If BBB's reservation value were higher, it is clear what could happen: when moving from *b* to *c*, the contract becomes acceptable to AAA but may no longer be acceptable to BBB. The process may not necessarily converge at a feasible value even if feasible values exist. Also, the end point depends in part on the starting point. Not very satisfying. A better way would be to adopt FOTE with analysis.

Comparison of solutions.   Turn back to Table 22.14. The last line shows the solution found by the sequential voting procedure. It's important to point out that we concocted the numbers so as not to embarrass the search process. By changing a few RVs, the sequential voting procedure could flounder or yield terribly inequitable results. This completes our analysis of this negotiation problem.

   If space permitted, we could push further to the case in which full consensus is not necessary. For example, we could require that only four rather than six

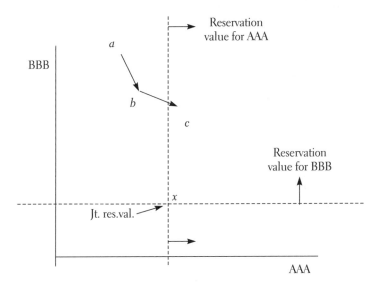

Figure 22.2.   The two-party counterpart of voting procedures.

excesses be non-negative. But in these cases we would have to be much more careful about our equity criteria. We could also talk about coalition formation and making side deals. We could also push in the direction of allowing trades among some players as a form of external compensation. This is the theme of the next chapter, where it will be easier to work with a much simpler example.

Strategic Claiming Behavior

Let's look a bit closer at the dynamics of the changing of proposed contracts. In going from proposed contract 1 to 2 (see Table 22.14), parties AAA and FFF both lost points, but they still (with POTE analysis) registered a Yes vote for proposed contract 2. Throw in a bit of realistic strategic claiming behavior, and we can easily see how AAA and FFF could howl against the proposed change from resolution level 2 to 3 on issue 2. In the proposed change from contract 2 to 3, issue 5 was changed from level 3 to level 2, much to the delight of party DDD. But parties AAA, CCC, EEE, and FFF were hurt—not enough to change their vote (using POTE), but enough so that a bit of claiming tactics could have ruled out this transformation.

In a laboratory setting, most sets of six players eventually come to an agreement like the one shown in contract 3, yielding player BBB a score of 77, just a bit over his RV of 75. Again throw in a bit of claiming behavior, and BBB might very well object, "I'm not going to settle for contract 3. I'm going to veto the whole proceedings unless we change issue number 1 from level 2 to 3!" Who knows what might happen. Throwing in a little claiming behavior can upset things in all sorts of unpredictable ways.

The dynamic that proceeds from a proposed contract to a modified proposed contract looking for more Yes votes is often the only act in town. Whether it results in an acceptable accord is often a matter of luck. Consider the dynamic process starting out with contract 1 in Table 22.11. Now let's play malevolently with the dynamics by changing the RVs of some of the players in such a manner that when we move from contract 1 to 2 to improve the roles of BBB and DDD, this tips the votes of, say, AAA and CCC from Yes to No. The changes may not be obvious even if the scored template of Table 22.6 (other than RVs) remains fixed. The RVs in Table 22.6 were chosen experimentally so that the dynamic adjustment process would work—not to embarrass it, as we could have. There is simply no way, in general, to ascertain whether there are feasible consensual outcomes for a scored template other than solving a mathematical programming problem.

Honest revelation?   Now suppose you, the reader, are a player in a negotiation that proceeds in a FOTE-like manner to the point where the negotiation can be summarized in terms of a scored template, as in Table 22.6, but with different numbers. Suppose, however, that the parties have not yet disclosed their confidential numbers. You know your own column of numbers, but not those of the other players. An analytical intervenor proposes to all the parties: "If each of you agrees to privately share with me your confidential scores and RVs, I'll find out if there is a feasible (fully consensual) agreement, and if so, I'll suggest to you the adoption of the Nash solution that maximizes the product of the excess payoffs. Anyone, at any time, can veto my suggestion, and if so, you will be on your own to try to find a compromise solution." The analytical intervenor is suggesting playing the role of a voluntary, nonbinding arbitrator. Would you accept this proposal? We would.

If you accepted the proposal, would you tell the analytical intervenor the truth about your preference scores and the truth about your RV? Many subjects agree to accept the deal and to tell the truth about preferred tradeoffs, but hesitate to tell the truth about their RVs. But how to lie? That's a tough one, and many subjects agree to tell the truth because they don't know how to lie.

Core Concepts

This chapter has focused on one polar extreme for reaching agreements in multiparty negotiations—the rule of consensus.

We assumed at first that the members all adopted a FOTE stance and pledged to adopt Milton Wessel's Rules of Reason (see Box 22.1). To ease into the complexity of the problem we lingered a bit on multiparty (involving at least three players) fair-division problems, where the template is especially simple and where BATNAs and RVs do not enter into the analysis.

We first considered the case where the parties assign monetary values to the items to be distributed and assume monetary transfers can be made between the parties. A Polish mathematician, Hugo Steinhaus, proposed a solution to this problem. Next, we no longer assumed the possibility of monetary transfers, and we used SOLVER to help generalize the discussion in Chapter 13 on the fair-division problem from two players to more than two players. Once again we compared maximizing (a) the sum of payoffs, (b) the minimum of payoffs, and (c) the product of payoffs; and this time with more players we noted how different the results can be. We deferred a fuller discussion of "what is fair?" until Chapter 24.

We then generalized the problem of fair division with more parties to the case where there are several issues, each with its own set of resolutions and where each party has its own confidential reservation value and additive scoring system. If the players, acting in a FOTE manner, are willing to disclose their scores and RVs, then a neutral analyst, using SOLVER, can compute efficient and equitable contracts. The stance moved from FOTE to POTE—that is, the parties are willing to exchange some information without lying, but not all—and we examined how many sets of players, in a laboratory setting, seek a feasible contract. There's no guarantee that the dynamic will find a feasible contract when one exists or that a feasible contract so identified will be efficient or equitable.

The chapter concludes with an appendix that suggests ways to determine efficient and equitable contracts that favor some members more than others.

Appendix A22: Consensus with Asymmetric Claims

Suppose the situation is as described in Table 22.6, with one wrinkle: the parties represent constituencies of different sizes. Party AAA may represent a constituency with twice the number of individuals as party BBB. For consensual agreement, the group seeks a contract that satisfies all the RV constraints; but in meeting that criterion, we might want the distribution of the excesses to be asymmetrically divided, giving more weight, so to speak, to AAA over BBB.

Some notation will help now. Let the RV of AAA be RV_A; of BBB be RV_B; and so on. For any contract K let the resulting score of AAA be $V_A(K)$; of BBB be $V_B(K)$; and so on. For any contract K, the excess for AAA is the difference $[V_A(K) - RV_A]$; for BBB it is $[V_B(K) - RV_B]$; and so on. The problem is to choose a contract K that is consensually agreeable—that is, all excesses are nonnegative—and that will optimize some criterion, such as the balanced Nash criterion, that is, to maximize

$$[V_A(K) - RV_A] \times [V_B(K) - RV_B] \times \dots, \qquad (A22.1)$$

or equivalently to maximize

$$\log[V_A(K) - RV_A] + \log[V_B(K) - RV_B] + \dots \qquad (A22.2)$$

To see one intuitive merit of maximizing the product form, consider the following illustration. Let

$$V_A(K) - RV_A = 20$$

and

$$V_B(K) - RV_B = 40.$$

If 20 points were added to AAA's score, it would be increasing AAA's score by the multiplicative factor of 2. If 20 points were added to BBB's score, it would be increasing BBB's score by the multiplication factor of 1.5. Thus using objective (1), or equivalently (2), we would prefer adding 20 points to AAA rather than to BBB.

If we want to reflect an asymmetry in the power of the parties, we could introduce non-negative weights W_A, W_B, . . . that sum to 1.0 and choose K to maximize

$$W_A \log[V_A(K) - RV_A] + W_B \log[V_B(K) - RV_B] + \ldots, \qquad \text{(A22.3)}$$

or equivalently to choose K to maximize

$$[V_A(K) - RV_A]^{W_A} \times [V_B(K) - RV_B]^{W_B} \times \ldots \qquad \text{(A22.4)}$$

23

Coalitions

There's a world of difference between two-party and many-party negotiations. We have been considering the rich class of negotiations in which each of the many parties is a bona fide participant in the negotiation process. The parties may be several members of a disputing family, or the many members of a firm's board of directors, or the many firms in an industry, or the many nations in a trade dispute. The parties may be of different types: a consumer interest group, a union, an environmental group, a firm, a state, a government agency—all in contention.

In negotiations with three or more parties, if you decide not to come to an agreement with all of your adversaries, you might still forge an agreement with a subset of the other parties. In other words, you can still cooperate with a coalition of some of the others. If there is only one other party, this complexity can't be formulated. But even if there are two other negotiating parties—say, B and C—you might consider what you could do with B alone, or with C alone, or with both. You must also contemplate what B and C could do without you. If you plan eventually to enter into negotiations with B and C, should you first approach B and compromise some of your differences with B before jointly approaching C? What should be your reaction if B and C collude before you can get into the act? Should you try to upset this coalition by trying to woo away B? How much do you have to give in to B so that B will not be vulnerable to enticements from C? How much can you inveigle from B by threatening to go to C and squeezing out B altogether?

The complexities can become surprisingly rich with just three players, even if we concentrate on the polar extreme where each party of players faces a world of certainty and where there is only one issue involved.

Cooperative Game Theory

Significant conceptual complexities arise when even a single new party is added to a two-party negotiation: coalitions of two parties can now form. Game theorists, starting with the seminal contribution of von Neumann and Morgenstern (1944), have investigated these complexities under the heading of "n-person games in characteristic function form," also known as *cooperative* game theory. This chapter sets the stage for our discussion of coalition formation by introducing the problem faced by three cement companies who can form a cartel. How should they split the synergies they would create? We imagine the case where all rewards and losses, tangible or intangible, are valued in monetary terms and money can be transferred from player to player.

The Characteristic Function Form

The basic problem:

- Let there be a universe of players, U, of n players (n greater than 2).
- For any coalition (subset), S, of players, let $V(S)$ be the monetary value of the coalition S.
- We assume that $V(.)$ is defined over all subsets of U.
- For any two disjoint subsets, S and T,

$$V(S \cup T) \geq V(S) + V(T).$$

The quantity

$$V(S \cup T) - [V(S) + V(T)]$$

is the synergy created when coalitions S and T form into the super-coalition $(S \cup T)$.

- The value of any coalition of a single player is 0.
- Preplay communication with binding agreements is permissible.
- All the above, including a complete specification of the V-function, is common knowledge.

Problem: Given the above with full knowledge of V, the players must decide which nonoverlapping coalitions to form, and how to divide the value of each coalition among its members.

Let's see how this structure plays out in some examples.

Case Study: The Scandinavian Cement Company

The Scandinavian Cement Company (SC) is the leading producer of cement in a nameless country. It has traditionally shared the market in a cartel arrangement—perfectly legal in that country—with two other producers, the Cement Corporation (CC) and the Thor Cement Company (TC). The cartel arrangement is about to expire, and the three companies are contemplating a formal merger.[1] The companies call in an independent consultant, Loran Chat, to prepare a preliminary analysis of the problem.

Loran Chat's analysis is summarized in Table 23.1. With the present arrangement—all firms separate, but with a cartel understanding—their earnings are 32 million, 23 million, and 6 million (net present value) monetary units for SC, CC, and Thor respectively. (For convenience, we'll call the monetary units dollars.) If they join in a total merger, they can do better than the sum of their earnings ($61 million): they benefit from synergies that add $16 million, for a total of $77 million. But Loran Chat also points out that there will be synergies involved if any two merge; for example, SC and CC together can command $59 million rather than $55 million (32 + 23) whereas Thor in this case would be reduced from $6 million to $5 million.

The SC representative argues that the $16 million synergy should be allocated according to size:

$$\frac{32}{32+23+6} \times 16 = 8.39 \text{ to SC}$$

$$\frac{23}{32+23+6} \times 16 = 6.03 \text{ to CC}$$

$$\frac{6}{32+23+6} \times 16 = 1.57 \text{ to Thor}$$

This proposal would result in the following payoffs:

$$32 + 8.39 = 40.39 \text{ to SC}$$

$$23 + 6.03 = 29.03 \text{ to CC}$$

$$6 + 1.57 = 7.57 \text{ to Thor}$$

The payoffs would total $76.99 million.

---

1. This armchair case is an adaptation of an adaptation of an adaptation. The original article was Lorange (1973). Lorange wrote a version of this case in a seminar that I conducted. My former research assistant, Kalyan Chatterjee, adapted Lorange's article for MBA classroom use. We now simplify further.

Table 23.1 Net present value of earnings for each merger

Type of merger	Earnings (in millions of dollars)
All firms remain separate	
SC	32
CC	23
Thor	6
Two merge, the third remaining separate	
SC, CC	59
Thor separate	5
SC, Thor	45
CC separate	22
CC, Thor	39
SC separate	30
Total merger	
SC, CC, Thor	77

"That's just not reasonable," argues the Thor representative. "I should end up with a lot more than $7.57 million."

"I don't see why," responds the SC representative. "We're all getting about a 26 percent increase in our worth because of the merger."

"I'll tell you why. According to Loran's figures, if my company, Thor, joins with CC the two of us can get $39 million—we would get more than you want to give us in the three-way merger. And in the case that Thor joins CC, SC would end up with $30 million and not the $40.39 million you want." Thor then turns to CC and says: "If you join me, we can command $39 million; you could take $30 million and I would take $9 million."

SC protests loudly. "You fellows are bringing in an irrelevancy. Are we in this together or not?"

"I'd rather go it alone than with the two of you," says Thor, "and only get $7.57 million. It's my company that's generating the synergy."

The CC representative enters the fray: "I think $7.57 million is a fair payoff for you, Thor, but $29 million is a bit low for me. Remember: if you don't join us, you'll end up with only $5 million."

"Yes, but you two will get only $59 million together, and I doubt that you, CC, will be able to get $29 million out of SC. Furthermore, if you two join as one entity and get $59 million while I get $5 million, then together we would total $64 million. So if we then all joined together, we could produce a synergy of $13 million (77 − 64) and it would then be fair to share that synergy evenly: half to your combined firm and half to me."

"Are you saying, Thor, that you want $11.5 million? If you are, you're being completely unrealistic."

The Core

And so the argument goes. Finally, they agree to ask Loran Chat what he thinks. Loran, being mathematically inclined, starts off by saying that he's being asked to find three amounts, X_{SC}, X_{CC}, and X_{TH}, that divide up the total of $77 million:

$$X_{SC} + X_{CC} + X_{TH} = 77. \tag{23.1}$$

These three amounts should, as a minimum, also satisfy additional inequalities

$$X_{SC} \geq 30, \tag{23.2}$$

$$X_{CC} \geq 22, \tag{23.3}$$

$$X_{TH} \geq 5, \tag{23.4}$$

$$X_{SC} + X_{CC} \geq 59, \tag{23.5}$$

$$X_{SC} + X_{TH} \geq 45, \tag{23.6}$$

$$X_{CC} + X_{TH} \geq 39. \tag{23.7}$$

Inequalities (23.2), (23.3), and (23.4) state what each firm can get alone against a coalition of the other two; inequalities (23.5), (23.6), and (23.7) state what pairs of firms can get if they form coalitions.

"The first thing," says Loran, "is to see whether we can find three numbers that will satisfy requirements (1)–(7). If we can, we will then try to describe all feasible sets of three numbers that do the trick. And after that we can talk about ways to decide, among these feasible triplets of numbers, if we have an embarrassment of riches."

Loran plots these inequalities in a rather strange way (see Figure 23.1). He uses a horizontal axis for X_{SC}, a vertical axis for X_{CC}, and equation (23.1) to account for X_{TH}. Requirements (23.2) and (23.3) are plotted directly. Inequality (23.4), when combined with (23.1), implies

$$X_{SC} + X_{CC} \leq 72. \tag{23.4'}$$

Inequality (23.5) is plotted directly. Inequality (23.6), coupled with (23.1), implies

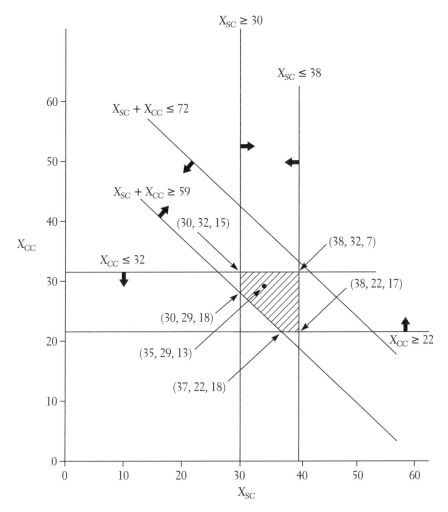

Figure 23.1. The feasible set of triplets that satisfies equations 23.1 through 23.7.

$$X_{CC} \leq 32; \qquad (23.6')$$

and inequality (23.7), coupled with (23.1), implies

$$X_{SC} \leq 38. \qquad (23.7')$$

Inequalities (23.6′) and (23.7′) are also plotted. The points that satisfy all inequalities lie in the shaded area, and each of the vertices of that region is labeled with three numbers: a value of X_{SC}, of X_{CC}, and of X_{TH}. For example, the most northeasterly vertex has coordinates 38 for X_{SC}, 32 for X_{CC}, and, because of requirement (23.1), 7 for X_{TH}. We see that lots of triplets of numbers are feasible, in the sense that they satisfy requirements (23.1)–(23.7). The set of feasible triplets comprise the *core of the game*.

The parties ask Loran to suggest a solution. "One possibility," he responds, "is to take some point near the center of the core. Estimating roughly, I would suggest 35 for X_{SC}, 29 for X_{CC}, and 13 for X_{TH}."

"I don't like your suggestion at all," says SC. "I represent the biggest firm and I get an increment of $3 million, while Thor is ending up with a $7 million increment."

"Let's compromise," says the CC representative. "We have SC's original suggestion and Loran's suggestion. I get 29 in each case. Let's split the difference. I suggest that SC get midway between 40.39 and 35, or 37.69; I'll take 29.02; Thor will get midway between 7.57 and 13, or 10.29. How's that?"

The SC representative scowls. "I don't like it, but for harmony's sake I'll go along."

The Thor representative smiles. "I don't like it either, but I don't know how to convince you that I deserve more. So I'll go along, too."

We'll come back to this story later. But first let's discuss a related problem that serves to highlight some complexities in the dynamics of coalition formation.

A Pure Coalition Game

Let's abstract away the context of the cement industry and consider a simply explained game (this is not the same as saying that it is a simple game) in which Loran Chat can find no solutions to the counterparts of equations (23.1)–(23.7)—that is, where the core is an empty set.

Game Description

Instructions. The game has three players: A, B, and C. You will be assigned one of these roles. Your aim is to join some coalition that commands a positive payoff, to negotiate how the joint payoffs should be split, and to try to maximize your own payoff (see Table 23.2). You will be scored according to how well you do: your payoff will be compared with the payoffs of others playing a similar role.

For example, if a coalition of A and C were to form, they would command a joint return of 84 units (see Figure 23.2). They might jointly agree to give 50 to A and 34 to C. Of course, C might want more from the coalition AC and might threaten A by courting the favors of B. After all, if B does not join any coalition at all (or remains as a one-party coalition), then B gets nothing. So B will be trying

Table 23.2   Payoffs in a pure coalition game

Coalition	Payoff
A alone	0
B alone	0
C alone	0
A, B	118
A, C	84
B, C	50
A, B, C	121

desperately to join A and C in a grand coalition ABC (commanding 121), or else to break up AC and join one of them.

The idea of the game is for you to maneuver about and eventually join a coalition that will offer you the best return. Of course, what you might demand from one coalition depends on what you can add to that coalition and what you potentially could obtain elsewhere. You should have no prior communication with the other two players (except for arranging for a meeting place) before the negotiations start. You are allowed thirty minutes for negotiations, but are free to complete negotiations sooner. All three of you should arrange yourselves in symmetrical positions at the beginning. If any two players want to arrange for a private meeting, the third must not interrupt for at least a two-minute period. *[End of instructions.]*

To start off, players examine the table of possible payoffs and devise the beginnings of a strategy. After being assigned roles, but before discussing the game with the other two players, subjects are asked to describe their strategy in writing. As they play the game, they record the outcome of the negotiations and the sequence of tentative agreements that were made along the way. After the negotiations have been completed, the three players discuss exactly what happened during the game.

[Reader: We suggest that you find two others to play this game with before reading further.]

Empirical Behavior

There are various ways in which players can jockey for inclusion in a coalition. Suppose that A rushes out and makes a private offer to B. "Let's join together without C and split the 118. Since I am obviously stronger than you, a reasonable split would be 78 to me and 40 to you."

"I don't think that's reasonable," B responds. "I don't care who my partner

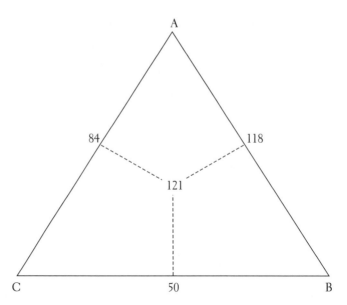

Figure 23.2. Coalition payoffs.

is, but my aspirations are far higher than 40. I can go to C, who is now out in the cold, and offer her 4, and take 46 for myself." "If you offer 4 to C," warns A, "I'll woo C away with an offer of 8."

"But if you do that," argues B, "then you'll end up with only 76, which is worse than the 78 you unreasonably demanded from me."

Offers that cannot be readily refused. With the above conversation as background, let's investigate how players can make offers that "cannot readily be refused" (see Table 23.3). What do we mean by "cannot readily be refused"? If, for example, C offers 42 to B, keeping 8 for herself, then B cannot go to A and try to get more than 42 without A's being vulnerable to an effective counteroffer from C that would both beat B's offer to A and yield C more than 8. Restated more slowly, if C offers 42 to B and if B threatens to go to A and request, say, 44 (leaving A with 74), then C in turn could go to A and offer him 75, which would permit C to keep 9 for herself—an improvement over her original 8. Thus C can say to B: "My offer to you of 42 is not readily vulnerable. If you are wooed by A who offers you more, I can 'out-bargain' you with A and you'll end up with nothing, while I will get my 8."

The weak band together. Here's a tactic that B can use. B muses at the very start: "I can make offers either to A or to C that cannot readily be refused, and in each case I would get 42. But A can make similar offers that would yield him 76 and C can make offers that would yield her 8. Yet all three of us cannot com-

Table 23.3   Offers that cannot be readily refused

| | Payoff | | | |
Offer	A	B	C	Total
Offer of A to B	76	42	—	118
Offer of A to C	76	—	8	84
Offer of B to A	76	42	—	118
Offer of B to C	—	42	8	50
Offer of C to A	76	—	8	84
Offer of C to B	—	42	8	50

mand 76 + 42 + 8, or 126. As a grand coalition we can only get 121. So it's critically important that I not be left out in the cold: it's imperative that I prevent a coalition between A and C. Should I approach A or C first? I think that I'm better off with C; and to make C really tied to me, I'll start off with a magnanimous offer: I'll offer her 10 units, 2 units more than she should expect from a two-way coalition that includes her. If C understands what I am doing and if she remains faithful to me, then we as a firm bargaining unit can then approach A. In that bargaining problem with A, there would be 71 points to share, (121 − 50 = 71), and our firm BC coalition should get 35.5 units of that. I'll suggest to C that we split this 35.5 units evenly between ourselves. So C will end up with 10 + 17.75, or 27.75 units, which should far exceed her reasonable aspirations. I'll end up with 57.75. Not bad, eh? Let's see how she responds."

C is favorably impressed and she agrees to the plan. The BC team then approaches A, who is shocked by their cold calculation. A refuses to negotiate for the 71 points that could be divided between himself and the BC coalition. "Once I start down that path," A ponders, "I'm a goner. My best bet is to try to woo C away from her partnership with B."

So A approaches C confidentially. "It just does not make any sense for you and B to share 50 between the two of you," he says to her. "I'm not going to join with you under those circumstances. If we brought in an impartial arbitrator, don't you think my fair share would be much more than B is suggesting that I get? How much is he offering you of that 50? I bet it's a lot less than half. If you agree to come with me, I'll give you 30 points. The principle and morality are all on our side. It was B that started the intriguing."

C now sees a possibility of getting 30 from A, rather than a secure 10 from B with a decreasing hope of an additional 17.75. But still, C has made an agreement with B. She wavers and says that she'll have to think about it. Quickly A goes to B and informs him that C is about to sign an agreement with him, but that there is still time for B to join with him. A offers to give him 45 of the 118 points they can command together. And so the jockeying continues.

Those who try to foresee outcomes in situations like this should not be too dogmatic about their predictions: anything can and does happen in such uncharted terrain.

Face-to-face versus terminal-to-terminal.   This coalition game was played by subjects under two very different interactive conditions. In an early version, subjects negotiated face to face. In a later series of experiments, conducted by Elon Kohlberg, subjects communicated via computer terminals; they did not know the real identities of their adversaries, and their messages tended to be much more circumscribed than those of the earlier set of subjects.

In the face-to-face-to-face negotiations, two of the three parties in each group occasionally talked to each other in the presence of the third party; other pairs arranged for private meetings. Over 90 percent of the triplets ended up in a three-way coalition, splitting the entire 121 units available. In about 80 percent of the contests that ended up with three-way coalitions, however, the players got involved in some two-way coalitions at some time during the negotiations. In the other 20 percent of cases that ended up in a coalition of the whole, the players never formed any two-way coalitions during the negotiations—they merely suggested successive changes in how the 121 total points should be divided. For face-to-face negotiations the average payoffs were roughly 69, 40, and 10 for A, B, and C, respectively—including the groups that formed two-way coalitions. Observe that $69 + 40 + 10 = 119$, which is slightly less than 121. This occurs because some groups never ended up with a three-way coalition.

A strikingly different set of statistics resulted when the interactions were computerized. Outcomes for sixty-seven triplets were recorded. Three triplets did not settle at all, and only three of the sixty-seven achieved a three-way coalition. Of the remaining sixty-one cases that involved two-way coalitions, twenty were between A and B, twenty-two were between A and C, and nineteen were between B and C. The average payoffs in the sixty-seven contests yielded 49 to A, 27.8 to B, and 5.7 to C—not a very efficient set of performances. On the average, all three parties fared far better in face-to-face-to-face negotiations.

How can we account for these differences? They are so striking that no statistical tests of extreme hypotheses need be conducted: they are not a statistical fluke. People probably find it easier to act tough if they are not looking at the other negotiators—if the "others" are anonymous. It's hard to squeeze out someone else from a coalition when that person is looking at you. Each of the parties seem to do far better (on the average) in the softer, more personal atmosphere of face-to-face-to-face negotiations; but the results were not conclusive. Perhaps the interactions via computer simply required more time. More experimentation certainly needs to be done. It would be interesting to include an intermediate case where negotiations are done by telephone via a three-way conference hookup. It might also be interesting to give subjects a choice of interacting face-to-face or less personally. On the evidence thus far, it would likely be to their advantage to choose personal contact.

Rationality, Fairness, and Arbitration

What would you do if you were asked to arbitrate this pure coalition game? What's fair? Subjects were all asked that question. One would-be arbitrator argued that each player alone gets nothing, whereas all three together get 121; so each should get one-third of 121, or 40.33. Others objected that this solution was unreasonable—that it ignored the power relations that accrued to the players because of two-party coalitions. The equal-shares advocate maintained that an arbitrator should not be concerned with that sort of power and intrigue. Most subjects, however, strongly believed that the payoffs for two-party coalitions should influence the division of the 121 total units—that the potential power of the negotiators should be considered by the arbitrator. We'll proceed with this assumption.

The core of the game. Paralleling the treatment of the Scandinavian Cement problem, several subjects tried to find sharing values X_A, X_B, and X_C for A, B, and C, respectively, that satisfied the requirements:

$$X_A \geq 0, X_B \geq 0, X_C \geq 0, \tag{23.8}$$

$$X_A + X_B \geq 118, \tag{23.9}$$

$$X_A + X_C \geq 84, \tag{23.10}$$

$$X_B + X_C \geq 50, \tag{23.11}$$

$$X_A + X_B + X_C = 121. \tag{23.12}$$

No one succeeded in finding a triplet (X_A, X_B, X_C) that satisfied requirements (23.8)–(23.12), because no such triplet exists. To prove this, we can argue as follows. Suppose that (X_A, X_B, X_C) satisfies requirements (23.9), (23.10), and (23.11). Adding these three equations together, we would have

$$2(X_A + X_B + X_C) \geq 118 + 84 + 50,$$

or

$$X_A + X_B + X_C \geq 126,$$

which contradicts equation (23.12). Hence we see that any allocation of 121 units among A, B, and C will have to violate requirement (23.9) or (23.10) or

(23.11). In this example, there is no allocation of the grand total that will simultaneously meet the demands of all two-party coalitions. The core is empty!

Some astute subjects argued that if the grand coalition commanded 126 units instead of 121 units, then there would be a triplet that would satisfy requirements (23.8)–(23.11), with 23.12 modified by the replacement of 121 by 126. The solution would be

$$X^o_A = 76, \qquad X^o_B = 42, \qquad X^o_C = 8,$$

where the superscript ^{o} is used to connote "optimal." The suggestion was made that a "reasonable and fair" solution would back off from these values to satisfy the 121 requirement. This is achieved by reducing each value by five-thirds, or 1.67. The resulting suggested triplet is then:

$$X^*_A = 74.33, \qquad X^*_B = 40.33, \qquad X^*_C = 6.33. \tag{23.13}$$

Subjects in earlier games learned an important tactical trick in negotiations: most people want to be fair, and they can be persuaded somewhat by fairness arguments. So it makes sense for you, as a negotiator, to step back from the fray and ask what an arbitrator might impose. In the course of negotiations, if you seem to be getting less than what you deem to be fair, then you can use this argument to support your position. (The obverse of this stratagem is more controversial: you should temper your aspirations toward fairness and should not try to get much more than your fair share.) One complication is that normally there is more than one seemingly fair solution. Astute negotiators, of course, will select those principles of fairness that favor their side. If several parties engage in these tactics, then a strange thing happens: instead of focusing on substance, the arguments shift to debates about fundamental principles—which is often a good thing. But the setting is somewhat corrupting, since the parties are persuaded by the implications for their own payoffs as well as by fairness in the abstract.

A few subjects, without any prompting, computed the fairness solution given in requirement (23.13) and used this to temper and guide their initial aspirations. Some used it quite openly and passionately when the negotiations were developing adversely from their vantage point.

The Shapley Value.   Another so-called fair solution for this negotiation exercise is known extensively in the literature as the Shapley Value, after the game theorist Lloyd Shapley. Consider a hypothetical model of the dynamics of coalition formation in which one player starts out singly, then is joined by a second player, and then by a third. With three players, there are six possible dynamic formations of the grand coalition of all three players. In the first line of Table 23.4, we see how a grand coalition forms in the sequence A then B then C. In

Table 23.4   The Shapley Value of the pure coalition game

| Order of players forming the grand coalition | Incremental value added by each player | | | |
	A	B	C	Total
ABC	0	118	3	121
ACB	0	37	84	121
BAC	118	0	3	121
BCA	71	0	50	121
CAB	84	37	0	121
CBA	71	50	0	121
Average	57.33	40.33	23.33	121

Note: The Shapley Value (for A, B, and C) is the vector quantity (57.33, 40.33, 23.33).

this sequence, A alone commands zero; when B joins A, B contributes 118; when C joins A and B, she adds 3 to bring the grand total to 121. In the last line of the table, C starts and brings zero; B joins C and adds 50; A then joins with C and B and adds 71. The Shapley arbitrated solution averages the contributions added by each player. Thus, according the Shapley's scheme, A would get a fair share (or arbitrated value) of 57.33, which is the average of the six numbers in the column under A. Notice how the Shapley arbitrated values (57.33, 40.33, 23.33) differ sharply from the values in equation (23.13), namely 74.33, 40.33, and 6.33.

A modification of the Shapley Values.   What would I do if I were the arbitrator? Even though the Shapley Value has some deficiencies, I (HR) am persuaded by many of its merits (see Luce and Raiffa 1957, pp. 245–252). But in this case I would suggest my own peculiar brew, which exploits a hodgepodge of the ideas we have touched on. Start with the analysis in Table 23.3 exploiting the idea of offers that cannot readily be refused. Add the possibility that any two-party coalition can bargain with the remaining party, and divide that synergy in half; take the half received by the existing coalition of two parties and divide that in half. Then average the results over the three different starting two-party coalitions.[2] All this is systematically done in Table 23.5. Suppose, for example, that we start off with the coalition AC, which commands 84 units. If A receives 76 and C receives 8 units, then this decomposition is not readily vulnerable to B's offers to A or C. This idea goes back to "offers that cannot readily be refused." Coalition AC alone commands 84, and B alone gets nothing. If, however, they join together, they create a synergy of 37 units. For this arbitration scheme we

---

2. We have not investigated how this would generalize to situations with more than three players.

Table 23.5   Another arbitrated solution of the pure coalition game

Starting two-party coalition	"Reasonable" payoff			Total
	A	B	C	
Coalition AB	76.0	42.0	—	118
Synergy	0.75	0.75	1.5	3
Total	76.75	42.75	1.5	121
Coalition AC	76.0	—	8.0	84
Synergy	9.25	18.5	9.25	37
Total	85.25	18.5	17.25	121
Coalition BC	—	42.0	8.0	50
Synergy	35.5	17.75	17.75	71
Total	35.5	59.75	25.75	121
Average	**65.83**	**40.33**	**14.84**	121

imagine that B is given 18.5 of this synergy and that coalition AC shares its 18.5 equally. So if coalition AC forms first, the 121 units are divided as follows: 85.25 to A, 18.5 to B, and 17.25 to C. The solution shown in Table 23.5 averages the partitions of the 121 units. Notice that in this case C gets 14.84.

We can now return to the Scandinavian Cement Company case and investigate other arbitrated solutions for that problem. The Shapley Values are (35.5, 28.5, 13), as shown in Table 23.6. My preferred arbitrated values, shown in Table 23.7, are (34.916, 28.416, 13.66). Both these solutions fall close to the center of the shaded region of Table 23.1.

The Game with One Strong Player

Let's look at an extremely simple example and compare the solutions obtained by using various methods. Assume that one strong player, A, and two weak players, B and C, are taking part. Their coalition payoffs are as follows: each player alone commands 0; coalition AB and coalition AC each command 10; coalition BC commands 0; all three together command 10. We see that A, the strong player, can play B against C; he needs only one of them. If we set up the following requirements:

$$X_A \geq 0, \qquad X_B \geq 0, \qquad X_C \geq 0, \qquad (23.14)$$

$$X_A + X_B \geq 10, \qquad (23.15)$$

$$X_A + X_C \geq 10, \qquad (23.16)$$

Table 23.6 Shapley Values for the Scandinavian Cement Company case

| Order of players forming the grand coalition | Incremental value added by each company | | | Total |
	SC	CC	TH	
SC, CC, TH	30	29	18	77
SC, TH, CC	30	32	15	77
CC, SC, TH	37	22	18	77
CC, TH, SC	38	22	17	77
TH, SC, CC	40	32	5	77
TH, CC, SC	38	34	5	77
Average	35.5	28.5	13	77

Table 23.7 Another arbitrated solution for the Scandinavian Cement Company case

| Starting two-party coalition | "Reasonable" payoff (in millions of dollars) | | | Total |
	SC	CC	TH	
SC, CC	32.5	26.5	5.0[a]	64
Synergy	3.25	3.25	6.5	13
Total	35.75	29.75	11.5	77
SC, TH	32.5	22.0[a]	12.5	67
Synergy	2.5	5.0	2.5	10
Total	35.0	27.0	15.0	77
CC, TH	30.0[a]	26.5	12.5	69
Synergy	4.0	2.0	2.0	8
Total	34.0	28.5	14.5	77
Average	34.916	28.416	13.66	77

a. These are the values that can be obtained by the company alone, remaining outside the two-company coalition.

$$X_B + X_C \geq 0, \tag{23.17}$$

$$X_A + X_B + X_C = 10, \tag{23.18}$$

then there is only one triplet of values that satisfies all of these, namely:

$$X_A^o = 10, \qquad X_B^o = 0, \qquad X_C^o = 0.$$

Is this a fair solution? The power resides in A; all A has to do is to get B or C to join him, and he can play one against the other. Think of A as the employer and think of B and C as workers. The obvious tactic is for the workers to unite and present themselves as a unified front to A, since without B or C player A is impotent. Players B and C should not squabble among themselves, because they're symmetrically constituted. It's easy for them to decide allocations: divide equally.

The core—that is, the set of triplets that satisfies individual and coalitional demands as given by requirements (23.14)–(23.18)—in this case contains only a single triplet (giving all to A). Yet this resolution does not have compelling predictive value: players B and C do join together in the laboratory setting.

The Shapley Values for this game are 6.67 for A, and 1.67 each for B and C. The counterpart to the arbitrated solutions in Table 23.5 would yield in this case 8.33 for A, and 0.833 each for B and C.

Now let's consider this same game structure with one dominant player (A, the "employer") and instead of two weak others introduce twenty-five weak others (B, C, D, . . . , Z). Assume that A and any single "other" can get 10 units. The core, which gives all to A and nothing to anyone else, seems to be a reasonable prediction, because it would be very hard for those twenty-five others to remain unified. Should a fair arbitrated solution reflect this reality? Should A get more and more as the number of others increases? The Shapley Values do this, but the core solutions do not.

It is not easy to suggest a compelling set of "fairness principles" that deserve to be universally acclaimed as the arbitrated solution. The more you think about this, the more elusive the dream becomes.

Moving toward Reality

As a reminder of how very restrictive our discussion about coalition games has been, consider the way in which the discussion specializes to two-party negotiations. Instead of players A, B, and C, we would have only players A and B. There is no loss of generality if we assume that each player alone commands zero and that as a coalition they command one unit of reward. The problem thus boils down to: how should A and B share one unit of reward? Obviously the focal point is .5 for each, which would be the Shapley Value. But the core in this case is embarrassingly rich: any division whatsoever of the unit reward—as long as each party does not get a negative amount—is a solution in the core. The two-player version of the pure coalition game is simply a distributive bargaining problem with openly disclosed reservation values—not a very interesting case. How very rich in conceptual complexity this trivial game becomes when we go from two to three or more players!

The fascinating part of two-party distributive bargaining arises from the fact that the negotiators do not know each other's reservation prices; indeed, they may have to work hard to determine their own. All these considerations are abstracted out in the simple coalition games. When Scandinavian Cement and the

Thor Cement Company form a two-way coalition and decide how they should divide up their spoils (the net present value of future profits), they are engaging in two-party distributive bargaining. It is the presence of that third company that brings a richness of detail to the situation. We can think of the three-party pure coalition game in part as a set of *interlocking two-party distributive bargaining games*, where each of the players in any such game has a reservation price that is determined to some extent by the other negotiations that can take place. To top things off, there is also the complexity of a three-way coalition. Matters get even more intricate when we include a fourth and fifth player.

Now let's add further reality to the potpourri. Increase the number of issues and let some of these be noneconomic, with nonobjective tradeoff rates between the levels of the different issues. The parties are not necessarily monolithic, and each party may not have a clear picture of its own value structure. There are uncertainties and asymmetries of information. In a case such as this, teams of analysts would have to work awfully hard with their clients, separately and collaboratively, in order to reduce the complexity of a real, multiparty, multi-issue negotiation problem to the format of a simple coalition game, in which each coalition has a numerical payoff made up of a decomposable commodity (like money) that can be traded. And after all this simplification takes place, after the players have really come to understand the strategic structure of interlocking coalitions, the bargaining dynamics can become especially bruising. To some extent, the complexity of the real situation softens the intensity of the bargaining dynamics. The parties are not clear about what is in their own interests, and their knowledge about the interests of others is likewise vague. Compromise is often easier to arrange in a situation of ambiguity. In this perverse sense, the complexity of reality yields simplicity: many real-world negotiations are happily not as divisive as starkly simple laboratory games, because in the real world it is difficult to see clearly what is in one's own best interest.

Intertwining Negotiations

Let's identify with player A who can remain by herself or form a coalition with either B or C but not both. If A were to join with B, they would have to resolve a complex negotiation involving several issues, each with several possible resolutions. Such problems were elaborated in Part III, which deals with two parties and many issues. The same structure would hold if A were to try to merge with C. Think of A, B, and C as hospitals that are being pressured to affiliate but where there are no obvious synergies between B and C. The issues to be resolved in an A&B alliance may be different than with an A&C alliance. To make matters a little more realistic and complicated, let us hypothesize the existence of player D, with whom C might choose to join. This problem differs from the standard coalition problems described earlier in this chapter in two important ways: (1) not all coalitions can form, and (2) each potential coalition requires an integrative negotiation to consummate.

Let's help player A to prepare. First A should reread Chapter 11, dealing with how A should prepare alone: articulating A's interests or objectives; learning about the interests and alternatives of the other parties. In this context A must consider whom to approach first, B or C. Does it make sense to have a three-way discussion? Probably not. We have to keep in mind that our reservation value when negotiating with B depends on what we might expect from a deal with C, and this in turn depends on what C can possibility attain from aligning with D. Player A has to think through a set of intertwining negotiations, the sequencing of which is only partially under the control of A. Furthermore, information learned in one negotiation will not only affect reservation values and aspiration levels of other negotiations, but will influence the way A will proceed in each of the intertwining negotiations.

Let's idealize. First assume that both B and C are willing to participate in FOTE-like negotiations—except, perhaps, for the revelation of their reservation values. Here's a way that A might suggest they proceed. Let A and B negotiate separately from A and C. For each of these interactions, assume the parties jointly design a template for negotiations. The templates for these two negotiations will in general be different. Assume furthermore that the parties score their templates, and exchange information to allow them to do full efficiency analysis. So far no discussion of reservation values need be involved. Now let the claiming aspect be resolved by an open, ascending auction mechanism, to be described.

Party A prepares two tables, one for B and one for C. Each table gives a listing of the extreme-efficiency contracts ordered from bottom to top in increasing preference for party A. A typical row in the table for B would list one possible set of resolutions for each of the issues in that template. Now the auction begins. B or C starts by selecting an opening bid, which consists of identifying a row in its table. This determination says that B is willing to settle with A at that contract. Now A turns to C and indicates how high C has to go in its table of extreme-efficient contracts to better B's bid. This requires A to make paired comparisons of contracts on the A/B template with those on the A/C template. If C tops B's bid by offering a better bid, then A determines how high B must bid next. And so on until the last bid is made. Notice how this process effectively separates creation from claiming. The parties may be given a day to determine whether or not to improve their bid.

The process requires an ordinal listing of A/B contracts and of A/C contracts. It helps if these contracts are also efficient. A also must designate the intertwining of levels. The listing or explanation of the A/B contracts need never be shown to C and the A/C contracts need not be shown to B. Of course, C should determine how high it is willing to bid for A by keeping in mind its negotiations with D. Party A might also wish to secure its reservation value by imposing a minimum starting bid for B and/or C. The process could be expedited if instead of the English open, ascending, outcry auction the B and C bidders employed a closed sealed-bid auction in which the high bid wins at the second high price. This procedure would have to be conducted by an outside, impartial agent who would be informed before the bidding took place how A would preferentially link the two tables (each listing the extreme-efficiency contracts).

Core Concepts

In negotiations with three or more parties, if you decide not to come to an agreement with all of your adversaries, you might still forge an agreement with a subset of the other parties. Even if there are two other negotiating parties—say, B and C—you might consider what you could do with B alone, or with C alone, or with both. You must also contemplate what B and C could do without you. Whom do you approach first? What enticements should you offer to B and C to keep them from defecting? If B and C form without you, how can you upset the coalition?

Von Neumann and Morgenstern (1944) investigated the complexities involved under the heading of "n-person games in characteristic function form," also known as *cooperative* game theory. They consider the case where all subsets can form coalitions; where each coalition has a combined monetary worth; where money is transferable; and where all is common knowledge. This chapter has reviewed that theory and shown how vicious behavior can be as the players jockey for a favorable coalition with favorable returns. The chapter then examined normative behavior and developed the *core of the game,* which seeks a monetary assignment to each of the players where, for any coalition A, the members are assigned values that sum to at least the monetary value of A—or else they would object to the assignment and form a binding coalition. The trouble is that there may be many monetary assignments in the core; or none; or even the case where there is exactly one but where this unique assignment has weak, behavioral, predictive value.

The chapter then reviewed some arbitration schemes (like the Shapley Value and some we offered as contenders) for resolving games in characteristic game form.

We then moved more toward reality, assuming a lack of a transferable monetary unit and a lack of common knowledge—specifically, some data about the payoffs to coalition A were known to A alone and at a cost. We suggested some modest prescriptive/descriptive advice to one of the players. We examined the special case where our client, A, has a choice between cooperating with B or C or both and is engaged in intertwining negotiations. Admittedly much was left unanalyzed; this is a fertile field for further analysis.

24

Voting

When people disagree but must act collectively, they often resort to various voting mechanisms to resolve their conflict. There is a vast literature on voting procedures, most of which presents variations of an original masterpiece written by Kenneth Arrow (1951). Our purpose in this chapter is to initiate readers who are not familiar with this literature to some of the intricacies of the problem.

To tie this discussion to the subject matter of this book, imagine an intricate negotiation among several parties that results in the identification of several plausible final contracts. Each party rank orders these contracts, and the task is to ascertain the group ranking of the contracts associated with each profile of individual rankings.

Majority Rule

Let's begin with a hypothetical case study.

Wyzard, Inc.

Messrs. Wysocki, Yarosh, and Zullo, joint owners of Wyzard, Inc., have to decide whether to start construction of a new Wyzard factory on a site in the town of Cohasset. They all agree that it is imperative for them to start construction of the factory in the next year, but there is some debate about where the new factory should be located.

It had long been anticipated by the joint owners of Wyzard that a new factory would have to be constructed, and three years ago they purchased a plot of land in the town of Allston as a site for the factory. Just two months after purchasing the Allston property, their realtor, Mr. Pumper, told them about another property that was available in the town of Brockton; he offered them the opportunity to swap the Allston property for the Brockton property plus a commission of $5,000. This swapping deal was viewed very favorably by Wysocki and Yarosh, but unfavorably by Zullo.

As early as 1974, when Wysocki, Yarosh, and Zullo started their joint venture, they had anticipated they would have differences of opinion, and they agreed at that time to resolve disagreements by majority rule. They have great respect for one another and have never resorted to strategic voting; each issue is considered separately and voted on, and no logrolling has ever taken place. They also agreed from the outset that if one of them was outvoted by the others, he would go along with the majority, even if he felt strongly about the issue. Since Zullo was on the losing side of the debate over the Brockton and Allston sites, he gracefully accepted the decision to pay Pumper a $5,000 commission and the three partners agreed to switch to Brockton. But Zullo did some investigating of his own, and with Pumper's help he discovered in the town of Cohasset another site, also owned by Pumper, which he thought was far superior to the Brockton site. Yarosh agreed with Zullo, but Wysocki thought otherwise. Subsequently, Wyzard signed papers with Pumper swapping the Brockton site for the Cohasset site—plus another $5,000 commission to Pumper.

Now, a year later, the three partners meet to discuss the timing for the construction of their new factory. Wysocki is uncomfortable. "I'm unhappy about our situation," he declares. "I still feel that after all our wheeling and dealing we would have been better off with the Allston site."

"What did you say?" demands Zullo. "I always wanted Allston! So why are we going to build in Cohasset?"

"Now wait a minute, you guys," interrupts Yarosh. "Cohasset was our agreed-upon choice. We agreed by majority vote that Brockton was better than Allston and that Cohasset was better than Brockton, and we've already paid Pumper $10,000 in commissions."

"I know that," retorts Zullo, "but I agree with Wysocki that Allston is better than Cohasset."

"Look," says Yarosh in a pained manner, "I trusted you two to vote honestly, and here you are scheming against me. Would you really pay Pumper another $5,000 so that we could go back to Allston? That's the silliest thing I ever heard of! What caused you to change your minds?"

"I don't know what you're complaining about, Yarosh. Wysocki and I aren't engaged in any conspiracy. I haven't changed my mind and I'm being perfectly honest. Do you want me to lie to you?" counters Zullo.

"Maybe I'm to blame," says Wysocki, "because we seem to be in a ludicrous situation. I really would prefer Allston to Cohasset—but my favorite is still Brockton."

Zullo bangs on the table and says heatedly, "I formally propose that we vote on asking Pumper to give us back our original Allston site. Let's not argue. We long ago agreed on a democratic procedure for resolving conflicts: by good old-fashioned majority vote. So let's get on with it."

This illustration fuses two ideas: (1) majority rule results in intransitive group preferences if the profile of individual rankings exhibits a cyclical preference pattern; and (2) a decision agent that insists on intransitive paired preferences can become a money pump.

The preference rankings for alternatives A, B, and C by individuals W, Y, and Z are shown in Table 24.1. Using majority rule, A yields to B, which yields to C, which yields to A, and so on in a circular pattern. Wysocki, Yarosh, and Zullo are not strategically misrepresenting their votes; in the vernacular of political science, they are not voting "insincerely." The anomaly arises because of the voting mechanism: majority rule.

Let's change the setting. Suppose that three legislative committee members are about to recommend bill A. One of the legislators would rather amend A so that it becomes bill C, but he knows that C will not supplant A by majority rule. Instead, he can first suggest modified bill B, which will beat A, and then he can introduce bill C, which he thinks can beat B. The legislator honestly prefers B to A, so he is voting sincerely; but he is playing strategic games. Is this done in legislatures? Yes, unfortunately. The trouble is that majority rule is so vulnerable to manipulation.

A single individual can also exhibit intransitivities. There are lots of examples where a person might say that he or she prefers C to B, B to A, and A to C. Some of these people might change their minds once this intransitivity is pointed out to them. Others insist, however, in holding firm: "If I'm intransitive, so be it—this is how I feel." An adamant individual might even rationalize his or her preferences: "I am interested in (W)ater accessibility, the availability of a suitable (Y)ard, and in proper (Z)oning. B is better than A on the W and Y qualities; C is better than B on the Y and Z qualities; and A is better than C on the W and Z qualities. I think all qualities are equally important. So, you see, I'm not mixed up after all."

Once preferences have been established, the idea of the money pump be-

Table 24.1   A preference profile that results in an intransitive ordering by majority rule

		Individual	
Preference	W	Y	Z
First choice	B	C	A
Second choice	A	B	C
Third choice	C	A	B

comes applicable (Savage 1954). How much are you willing to pay to go from A to B? From B to C? From C to A? From A to B? And so on.

We're being pretty harsh on majority rule. We're purposely leaving aside all its positive aspects, such as simplicity, impartiality, and understandability. All we want to point out here is the long-known result that sometimes majority rule can generate intransitivities in paired comparisons: B over A, C over B, and A over C, and so on. Symbolically, we have

$$A \rightarrow B \rightarrow C \rightarrow A \rightarrow B \rightarrow \ldots$$

Let's look at some alternatives to majority rule—alternatives that also will exhibit anomalies.

Independence of Irrelevant Alternatives

Wysocki, Yarosh, and Zullo are still upset at their abortive attempt to find a suitable site for their new factory. Their choice problem has become even more complex because their real estate agent, Mr. Pumper, has discovered two additional sites in the towns of Dedham and Essex to add to the existing potential sites of Allston, Brockton, and Cohasset.

Wysocki's daughter Pamela, an MBA student, counsels her father and his partners: "You got into trouble last month because you used majority rule to compare pairs of alternatives. Why don't each of you just rank the five alternatives from best to worst, giving 5 points to the best, 4 points to the second best, and so on? Then all you have to do is total up the points and see which site wins."

That's what the partners do. This time they're very careful about their rankings. They take into account not only the physical environments and surrounding amenities, but also the tax structures in the different towns. Their individual rankings are as shown in Table 24.2; the totals are shown in the far right column.

"Well," Wysocki says gleefully, "I guess we're going to build in Allston."

Just then Pumper rushes into the meeting and breathlessly and apologetically announces, "I hope you fellows didn't decide on Essex, because I just found out that the property is not zoned for light industry."

"No matter," explains Yarosh. "Essex was not competitive."

Zullo, feeling miserable about the loss of his preferred site, Brockton, plaintively asks Pamela, "If we knock Essex out of the competition, how badly does Brockton do then?"

"Well," says Pamela, "let's see . . . Oh no!"

To everyone's surprise, it turns out that when the remaining four sites are re-

Table 24.2   Individual rankings of five alternative sites

| | Individual ranking (5 = best) | | | Total points (maximum = best) |
	Wysocki	Yarosh	Zullo	
Allston	5	5	2	12
Essex	4	4	1	9
Brockton	3	3	5	11
Cohasset	2	1	4	7
Dedham	1	2	3	6

ranked, Brockton emerges as the highest-ranked choice. With Essex out of the competition, the points range from 4 for the best to 1 for the worst. Allston gets 9 points; Brockton 10 points; Cohasset 6 points; and Dedham 5 points. So using Pamela's weighting scheme, Allston is best among the full range of competitors; but Allston falls behind Brockton if Essex is removed from the list of contenders.

This anomaly was observed long ago and is quite familiar to theorists. It's worth repeating here, though, because we're talking about mechanisms for resolving conflict, and many people don't realize that it's impossible to devise a foolproof scheme.

Insincere Voting

Wysocki and Yarosh are still wondering how they ever got into the mess they're in. They both prefer Allston over Brockton, but Pamela's scheme, which seems unassailably fair, dictates that Brockton is the winner once Essex is knocked out. Wysocki feels a bit defensive about Pamela's plan.

"How did Zullo ever rank Dedham ahead of Allston?" Yarosh asks incredulously.

"Maybe crafty Zullo voted strategically," muses Wysocki.

"I've a great idea," exults Yarosh. "Let's tell Zullo that on reflection we absolutely agree with him that Dedham is better than we originally thought. He can't complain about that. Let's change our rankings and move Dedham right up behind Allston. Then Pamela's scheme will favor Allston."

"That's a good suggestion. But should we be doing this—acting not quite honestly?"

"Well, Zullo started it!"

It's hard enough to come up with voting rules that are impervious to insincere voting by a single individual. When coalitions of voters coordinate their misrepresentations, it presents even tougher challenges to designers of voting schemes.

Arrow's Impossibility Theorem

The Problem

Assume the following: There is a group of N individuals—where N is at least three—who wish to give a group (ordinal) ranking of a set of alternatives (at least three of them) based on a *profile* of their N individual (ordinal) rankings of these alternatives. The mapping from the profile of the N individual orderings to the group ordering will be referred to (by us) as an *arbitration rule, R*. We, following Arrow, will shortly impose some reasonable requirements on this arbitration rule, which, alas, will turn out to be impossible to fulfill. No arbitration rule can simultaneously satisfy all these desiderata.

We shall assume the ordinal rankings—whether for an individual or the group—will allow ties (for example, alternative x is preferred to alternatives y and z and these latter two are tied).

For three alternatives, each voting member has a choice of thirteen transitive preference orderings as shown in Table 24.3.

A *profile* of orderings consists of one ordering of alternatives by each voting member. With three alternatives and three members, there are $13 \times 13 \times 13$ or 2,197 different profiles.

A decision rule (social welfare function) associates a group ordering—one of the thirteen possible orderings—to each of the 2,197 different profiles.

Conditions of the Arbitration Rule, R

Condition 1.
- *There are at least three alternatives.*
- *The arbitration rule R is defined for all possible profiles.*
- *There are at least two individuals.*

Condition 2: Positive Association.
- *If, in any individual ordering, an alternative, x, is pushed up—all else remaining fixed—then x should not be pushed down in the group ordering.*

Table 24.3 Preference orderings for three alternatives, A, B, and C

1	2	3	4	5	6	7	8	9	10	11	12	13
ABC	AB	AC	BC	A	B	C	A	A	B	B	C	C
	C	B	A	BC	AC	AB	B	C	A	C	A	B
							C	B	C	A	B	A

Note: Ties are shown on the same line.

<u>Condition 3: Independence of Irrelevant Alternatives.</u>
- *If, in two different profiles, all individual paired comparisons between two alternatives, x and y, are identical, then the associated group ordering of x and y should be the same.* [Stated alternatively: The group ordering of any two alternatives, *x* and *y*, should depend only on the individual orderings of x and y.]

<u>Condition 4: Citizen's Sovereignty.</u>
- *For each pair of alternatives, x and y, there is some profile of individual orderings such that the group prefers x to y (that is, individual votes count!).*

<u>Condition 5: Nondictatorship.</u>
- *There is no individual with the property that whenever he or she prefers x to y, the group also prefers x to y.*

Arrow's Impossibility Theorem states that conditions 1, 2, 3, 4, and 5 are inconsistent. That is, there does *not* exist any arbitration rule that possesses the properties demanded by these conditions. Stated alternately, if an arbitration rule satisfies conditions 1, 2, and 3, then it is either imposed or dictatorial.

The most problematical condition is the independence of irrelevant alternatives. The following example, taken from Goodman and Markowitz (1951), further illustrates this point. A host has two dinner guests to whom he is willing to serve either coffee or tea but not both. Instead of asking which each prefers, coffee or tea, this subtle host gets them to rank a whole class of drinks. Two possible profiles of responses are shown in Table 24.4.

For profile 1, the host deems it fair to serve coffee, and for profile 2, to serve tea. He reasons that the other drinks are not irrelevant, and their introduction permits him to appraise the relative intensities of preferences for coffee versus tea.

Yet, is such reasoning really plausible? One can argue that this procedure introduces a notion of interpersonal comparison of utility, and, if it is desirable to do that, then this is surely a naive way of doing it. Indeed, if a statement such as "1 prefers coffee to tea more than 2 prefers tea to coffee (in profile 1)" has any meaning at all, then an alternative rationalization of profile 1, as shown in Figure 24.1, can be given to show that the host should have served tea.

Strengths of preferences.   One way to get around the Arrow paradox is to change the problem by requiring additional information from the individual members of the group. This is the approach we have taken in this book. Instead of asking individuals for their ordinal preferences, we ask the individuals to provide their *strengths of preferences*. We introduce *cardinal* scoring of individual preferences, and in this domain the Arrow paradox disappears. So imagine our

Table 24.4   Preferences for drinks

	1's preferences	2's preferences
Profile 1	Coffee	Tea
	Postum	Coffee
	Milk	Postum
	Lemonade	Milk
	Hot chocolate	Lemonade
	Coca-Cola	Coca-Cola
	Tea	Hot chocolate
Profile 2	Postum	Tea
	Milk	Postum
	Coffee	Coca-Cola
	Tea	Lemonade
	Lemonade	Hot chocolate
	Hot chocolate	Milk
	Coca-Cola	Coffee

consternation when in the case on NASA's Voyager mission at the end of this chapter the space scientists thought that getting ordinal data about preferences for trajectory paths was just fine, but they had reservations about giving more detailed cardinal preferences. There just aren't any free lunches.

In two-party negotiations an arbiter might reasonably elicit cardinal preferences from the contending parties. But as the number of disputants increases, as is the case in many community disputes, it is difficult to get the parties to register their strengths of preferences; we are stuck with ordinal data that abstract away strong feelings or intensity of feelings.

In a classic treatment of the problem, Robert Dahl (1956, p. 90) writes:

> What if the minority prefers its alternative much more passionately than the majority prefers a contrary alternative? Does the majority principle still make sense?
>
> This is the problem of intensity. And, as one can readily see, intensity is almost a modern psychological version of natural rights. For, much as Madison believed that government should be constructed so as to prevent majorities from invading the natural rights of minorities, so a modern Madison might argue that government should be designed to inhibit a relatively apathetic majority from cramming its policy down the throats of a relatively intense minority.

Strategic Voting

Strategic ordering of bills in a legislature.   Political scientists may not feel too disturbed by the possible intransitivities arising from the use of simple majority

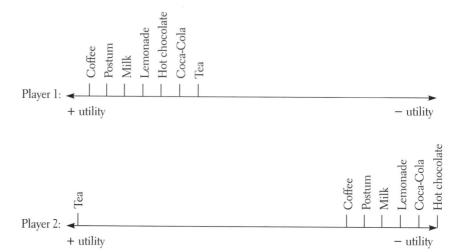

Figure 24.1.   Preference profiles for beverages.

rule, since in most legislative bodies individuals are asked to select a single alternative, not to order them by preference. Often a single alternative to an existing law is suggested, and when there are several alternatives, they are put forward in succession, each being pitted against the current status quo. Some may feel, therefore, that intransitivities cannot arise under present practice. This is a naive view, however, as we shall now show. Let x be the existing law, and let y and z be possible replacements. Suppose the legislative body divides into three equal groups, which, if called upon, would register the profile:

Group 1	Group 2	Group 3
x	y	z
y	z	x
z	x	y

Suppose y is first pitted against x and then z is pitted against the winner. This ultimately leads to z, since, by majority rule, y loses to x and x to z. But suppose z is first pitted against x and then y is pitted against the winner. This ultimately leads to y, since x loses to z and z to y. Therefore, as is well known to politicians, the final outcome can depend upon the order of presentation of the bills.

If a defeated bill is reintroduced and then wins, the interpretation often made is that some people have changed their minds. This may be so, but it need not be. The interpretation usually made ignores the different status quos in the two cases. Thus, given the usual application of the simple majority rule in legislative bodies, observers may be quite unaware of intransitivities even when they do exist.

Insincere voting revisited.   Arrow points out that it can benefit an individual legislator to misrepresent his true feelings in legislatures that vote on successive motions by simple majority rule. His example can be examined here. Consider the following profile:

Individual 1's ordering: x, y, z
Individual 2's ordering: y, x, z
Individual 3's ordering: z, y, x

Suppose that the motions come up in the order y, z, x. If all individuals voted according to their orderings, y would be chosen over z and then over x. However, individual 1 could vote for z the first time, ensuring its victory; then, in the choice between z and x, x would win if individuals 2 and 3 voted according to their orderings, so that individual 1 would have a definite incentive to misrepresent.

The 1980 presidential election provides an example in which there was extensive strategic, insincere voting. Some people preferred Anderson over Reagan over Carter, while others preferred Anderson over Carter over Reagan. Some of these voted for Anderson, but others voted for Reagan or Carter rather than for their favorite. The voting mechanism invited this insincere voting.

Randomization.   Randomization can be used to encourage sincere voting, but it isn't practical. Let A run against B. Suppose that a candidate will be selected by a random device where the probability that A will win is equal to the proportion of votes that A gets. So if A gets 60 percent of the vote, his chances of being selected are .60. (We're not advocating this scheme—just explaining it.) But now if you favor B and think that you are in a distinct minority, you still have a motivation to vote for your preferred candidate. If this scheme were used with Anderson, Carter, and Reagan, the Anderson supporters would want to vote for their man. If Anderson got 12 percent of the vote, he could be elected with probability .12. Of course, if he were lucky, then lots of people might be very unhappy. The system wouldn't work, but still it would generate sincere voting. The message is that sincerity in voting is a desirable but not a sufficient desideratum.

Strength of preference and logrolling.   We can argue that in some measure intensity of preference does receive expression in actual practice through "logrolling." A senator who feels strongly about bill q and indifferent about bills r and s will trade his votes on r and s for the desired votes of senators indifferent about q. Thus, his strong preference for bill q is recorded, and it may be passed even though according to the true tastes of the senators it should have been defeated. Is this good? That depends upon the bill q—at times we grumble "Shameful!" and at others chuckle "Beautiful strategy!"

In legislatures in the United States, strengths of preference are not directly registered. If 51 percent of legislators are mildly for A and 49 percent are adamantly opposed, then A wins. This is a deficiency in the system, so legislators will try to work around the system by trading or logrolling their votes. Some observers think that logrolling distorts the system; others believe that it makes an intolerable system more palatable. Some want to recognize that logrolling occurs and think it should be institutionalized so that legislators can fully register the nuances of their preferences—they want to establish a pseudo-market in vote trading, with tradeoffs openly posted. There are schemes that encourage honest revelations. To repeat: this is important, but not the only desideratum.

Filibustering, and its threatened use, is another method by which an intense minority can exert pressure on the majority. A significant difference between filibustering and logrolling, however, is that an intense minority can defeat even an intense majority with filibustering but not with logrolling.

Implications for many-party negotiations. From a pragmatic point of view, when dealing with large numbers of people who deserve to have a voice in the outcome, we are stuck with variants of majority rule, and unfortunately, this invites strategic, game-theoretic behavior. It also engenders problems in designing processes for multiparty negotiations, as we shall illustrate in Chapter 25 on multiparty interventions.

When many people disagree in the course of trying to make a collective decision, and when there is no institutional mechanism for resolving their conflicts of interest, the contending parties can try to negotiate an outcome directly. They can also try to negotiate the adoption of a mechanism (for example, a voting scheme, an auction or competitive bidding procedure, or a pricing system) that might facilitate the resolution of the conflict, or at least structure the ensuing negotiations. The analytical challenge is to design such a mechanism that is fair, equitable, and efficient and that will encourage honest revelations by individuals and groups.

We have seen that intervention in negotiation includes not only facilitation, mediation, and arbitration but also rules manipulation. Much of what we have discussed here can be broadly classified under the heading of rules manipulation for conflict resolution.

A Real Case: The Voyager Mission to Outer Planets

In late 1980 the front pages of newspapers were excitedly reporting new discoveries about the planet Saturn. Information was being transmitted to earth by a space probe whose trajectory had been selected by an intricate arbitration procedure. J. S. Dyer and R. F. Miles (1976) give a fascinating account of the way in which collective choice theory was used to select trajectories for the Mariner Jupiter/Saturn probes; much of what follows is based upon their account.

The collective choice problem. In September 1973 the National Aeronautics and Space Administration (NASA) announced plans for two exploratory space-craft to be launched in August and September 1977. Their trajectories would take them past Jupiter in 1979, and close to Saturn in late 1980 or early 1981. The Jet Propulsion Laboratory (JPL), which was responsible for managing that part of the space program for NASA, attached great importance to the selection of the trajectories, because the trajectory characteristics would significantly affect the scientific investigations.

NASA chose some eighty scientists, divided by specialization into ten scientific teams, to help select an appropriate pair of trajectories. Each of these teams had its own special scientific interest (radio science, infrared radiation, magnetic fields, plasma particles, and so on), and each team had its own preferences for differing pairs of trajectories. The JPL plan was to have each team articulate its own preferences for trajectory pairs and then to let the Science Steering Group (SSG) choose a compromise pair. The SSG membership comprised one leader from each of the ten teams.

Of the thousands of possible trajectory pairs, the JPL engineers, after some iterative, informal discussions with the scientific teams, reduced the competition to thirty-two contending pairs.

Let's imagine the problem confronting Dyer and Miles, who had to recommend a scheme for combining the preference rankings for each of the ten scientific teams to get a preference ranking for the entire group, knowing what they do know about the anomalies pointed out by the Arrow Impossibility Theorem.

Collective choice procedures and the independence of irrelevant alternatives (IIA). Should the arbitration scheme for combining the profiles of the team preferences to arrive at a group preference ordering satisfy the IIA principle? Let's examine what this means in this case.

To apply the principle here, suppose that trajectory pair 17 is deemed the best overall by the SSG; then the steering group is informed by the JPL engineers that trajectory pair 28 is no longer possible. Is it conceivable that the nonavailability of 28 could cause the SSG to shift from pair 17 (which is still available) to some other trajectory pair? In this context, the principle of independent alternatives seems compelling: if 17 is best overall, it should remain best after 28 is deleted—unless, of course, the reason that 28 has been removed has implications for the desirability of 17.

Yes, Dyer and Miles concluded, the IIA principle seems compelling in this context. What should be done?

One way out of the dilemma, which was followed, was to require more detailed preference inputs from the scientific teams. Instead of requiring them merely to give an ordinal ranking of the thirty-two trajectories, they were asked to provide their strengths of preferences for these trajectories as well. Each team was asked to provide a utility scale for the thirty-two trajectories. For norming purposes, they also added a thirty-third alternative, which they dubbed the "Atlantic Ocean alternative," the trajectory that would provide no information at all. The teams presumably ranked the Atlantic Ocean Special at the bottom of the heap and this was assigned a utility of 0 by each team. Let

$U(i,j)$ = the utility given by team i for trajectory alternative j,
for $i = 1, \ldots, 10$, and $j = 0, 1, \ldots, 32$ (where $j = 0$ refers to
the Atlantic Ocean Special), and $U(i,0) = 0$ for all i.

The Nash solution, which satisfies the IIA principle, chooses j to maximize (the value of the group score for j)

$$U^*(j) = \text{product of the } U(i,j)\text{'s over } i$$

$$= U(1,j) \times U(2,j) \times \ldots \times U(10,j),$$

and we now see the importance of each team's assigning 0 to the Atlantic Ocean Special. This sets up the joint reservation value to measure excesses for each team.

The Nash solution satisfies the principle of the independence of irrelevant alternatives, and in addition, it treats each team on a par. If the teams were to be randomly labeled with noninformative letters, and if each team's array of utility values for the trajectory pairs were listed, then would it be appropriate for the SSG to know the identities of the different teams? What do you think? We think that it would. After all, some scientific purposes might be more important than others. If so, then the Nash solution, which by definition treats all teams symmetrically, abstracts away too much. John Harsanyi (1956) gave an intuitively appealing rationalization for group-scoring each trajectory pair by taking a weighted average of the ten scores for a particular trajectory pair, and then choosing the alternative that maximized this group score (that is, the weighted average). The weights, of course, would have to be supplied by the arbitrator—or, in this case, by the SSG. They would somehow reflect the relative importance of the different scientific teams.

Different collective choice rules ranked the different trajectory pairs differently. The most commonly accepted rules (Nash and variations of Harsanyi, with some simple interteam weightings) rated three particular trajectory pairs among the top three—but with differences in the rankings of these three. The SSG examined the formal evaluations and selected one of these three top alternatives; however, it did not use any formal procedure to make this final choice. The two individual trajectories of the winning pair were labeled JSI and JSG, where J stood for Jupiter, and S for Saturn, and I and G for two of Jupiter's satellites, Io and Ganymede, which were to be encountered on the corresponding trajectories.

Since the teams were somewhat skeptical about the use of utilities, the SSG improvised a compromise that paid some attention to the utility submissions but did not fully subscribe to some formal mechanism. Thus the procedure followed was not fully in conformity with what Dyer and Miles had in mind.

Part of the difficulty came from normalizing the preference inputs so that each team scored the Atlantic Ocean Special at 0 and their best of the 32 real trajectories at 100. Doing so entails no loss of generality, but the 32 real alternatives get squished up near the top of the 100-point scale. The teams became aware that the higher value they gave to the least desirable of the 32 real alternatives, the more they would demean the Atlantic Ocean Special. Some teams admitted assigning the utility of their least desirable alternative strategically—meaning giving it a higher value than it deserved—because they suspected that the other teams would be similarly inclined. That's reality!

Postscript.   Dyer and Miles record that the scientists felt overwhelmingly that the process used was fair and that *ordinal* rankings of the alternatives helped in understanding and in communicating. But the teams thought that the cardinal utility information was a superfluous addition. This presents a terrible problem for methodologists. Aggregation mechanisms based on ordinal data can give rise to anomalies we might want to avoid, and the requirement for cardinal input information is designed to overcome these anomalies. A preferred way out of Arrow's Impossibility Theorem is to require each individual voter to give *cardinal* rather than *ordinal* rankings of alternatives. Roughly going from ordinal to cardinal values forces each individual voter to furnish not only preferences but strengths of preferences.

In explaining the meaning of utilities, Dyer and Miles had to introduce the notion of probabilistic choice—for example, comparing middle alternative B with a .50 chance at A and a .5 chance at C—and this might have seemed an irrelevancy. In hindsight, we think we would have cut corners and taken the rankings of the thirty-three trajectory alternatives, equally spaced them on a scale from 0 to 100 (say), called this the "default" assignment, and then invited each scientific team to rearrange the spacings between trajectories on the line to reflect its strengths of preferences, whatever that means. (See the discussion in Chapter 2 on desirability versus utility scaling.)

Core Concepts

When people disagree but must act collectively, they often resort to various voting mechanisms to resolve their conflict. Imagine a group of negotiators who identify a set of possible final contracts to choose among. They record their individual rankings, and now someone has to decide on a group ranking. Kenneth Arrow showed the anomalies that could arise from some obvious and commonly used amalgamation procedures like majority rule or a rule that sums the ordinal ranking scores. Indeed, Arrow has shown that some reasonable desiderata for any such amalgamation procedure going from a profile of individual preference orderings to a group preference ordering are collectively self-contradictory. There is no satisfactory solution for the conventional voting problem.

Since there are no ideal solutions for the problem as posed, astute players (for example, legislators) can play intricate games—like exploiting the ordering of bills to be voted on; or by influencing an outcome by registering insincere preferences along the way.

One inadequacy of standard voting systems is that a small majority of almost indifferent players can overpower a passionate, large minority. One way to cope with this lack of sensitivity in the system for acknowledging strengths of preferences is by the use of logrolling or filibustering. Another way to avoid the anomalies of the voting game is to change the game by requiring the players to record their strengths of preferences initially with the use of cardinal utility assignments and by the subsequent use of the Nash arbitration procedure that maximizes the product of utilities. This was tried, somewhat unsuccessfully, in the choice of trajectory paths for the exploration of Saturn.

25

Pluralistic Parties

Complex bilateral negotiations often involve three agreements: one across the table and one on each side of the table. Many egregious errors occur because of what takes place—or doesn't take place—on either side. Ideally, external (across the table) and internal (on each side of the table) negotiations should be synchronized and coordinated to maximize joint gains in the external negotiation. But decision makers frequently ignore these dependencies.

More often than not the protagonists on one side of the table have different responsibilities, tradeoffs, and visions. They may have different predictions of the future and have different risk tolerances. For the most part, internal negotiations are conducted more in the FOTE manner than the external ones. But not always. Sometimes internal negotiations may be very strategic and divisive while the culture of the external negotiations may be far more collegial.

A completed negotiation across the table may leave winners and losers on each side. Presumably, the fact that an external deal is made means that the net outcome is positive for each side (that is, it beats their collective BATNA). The winners gain more than the losers lose. To get agreement on one side of the table, the winners may have to creatively compensate the losers. This is the topic we now consider.

Throughout this book we have made the simplifying assumption that the players on each side of the table are unitary entities or monolithic. That's the pattern of the top panel in Figure 25.1. In contrast, we could consider the case in the middle panel, where negotiator A* is the agent for principals A1, A2, and A3, say, and A2 in turn is the agent for two principals A21 and A22. Life can get more complicated. Principal A21 should be working through her agent A2, and she in turn is one of the principals for A*, who is supposed to be negotiating with B; but A21, not trusting A2 or A*, has opened a secret back channel with B1. It's easy to get more complicated but we'll keep things fairly simple. We want to present the situation as simply as we can while still capturing the essence of what it means to synchronize internal and external negotiations.

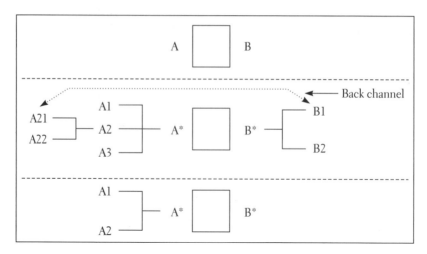

Figure 25.1.   Pluralistic parties.

Case Study: An Agent with Two Principals

The case we'll examine has the structure shown in Box 25.1. The external negotiations are between entities AAA and BBB. To be a bit specific, let these entities
be corporations. In the external negotiations, AAA's negotiator, A*, acts as the
agent for his two principals, A1 and A2. To reach an agreement, A* needs the
consensus of both A1 and A2. On the other side, B* acts for BBB—this party is
monolithic. Our concern is with advising party AAA.

Keeping with the spirit of introducing just enough complexity, we assume
that there are just two issues to be resolved: price (P) measured in dollars and
time measured in days (D). Both will be treated as continuous variables. Everyone on the AAA side wants lower values of P and D, but as you might have
guessed, A1 and A2 have different tradeoffs for P and D. A1 and A2 must instruct
A* how to negotiate with B*, who wants higher values of P and D.

Internal Differences

If A*, who will be negotiating for AAA, doesn't come to an agreement with B*,
then AAA will use another supplier for their needs at a price of $80 and a deliv-
ery date of 80 days.

A1 values a saving of $4 in P as equivalent to reduction of one day (that is,
$4P = 1D$). On the other hand, A2 feels differently about the tradeoffs: a $1 sav-
ing is worth 4 days to him (that is, $1P = 4D$). Look at Figure 25.2. Days are on
the horizontal axis, price on the vertical, and the reservation values for A1 and
A2 are displayed. Point X, for example, is acceptable to A1 but not to A2; point Y
is acceptable to both but barely to A2; point Z is acceptable to A2 but not to A1.

Box 25.1   Synchronization Problem

Parties
    AAA   and   BBB
Principals
    For AAA: A1 and A2   BBB is monolithic
Negotiation agents
    For AAA: A*   For BBB: B*
Issues to be resolved
    Price (P) and Days (D)
Main themes
    A1 and A2 have different tradeoffs for P and D
    A1 and A2 have to instruct A* how to negotiate with B*

In order to equalize the gains to A1 and A2, the price, P, must be equal to the days, D. In DP coordinates, the line of equality of payoffs is shown with some illustrative points (70, 70), (60, 60) and so on.

Internal payoffs (by formula).   The first step in examining the internal payoffs of the negotiation is to find the reservation values for A1 and A2. We do this by formula. Expressed algebraically, the payoff to A1 is

$$100 - P/4 - D.$$

Letting $(P, D) = (80, 80)$ in this expression, we see that A1 must get a positive return (that is, at least 0) with this payoff—this is A1's reservation value.
    A similar story holds for A2, whose payoff is

$$100 - P - D/4.$$

This also gives A2 a reservation value of 0.
    In the background material for the case it turns out that A1 and A2 are engaged in handling a totally independent budgeting problem in their firm, and they can trade value between them. This means that if there is an imbalance between A1 and A2, it's possible to use the budget to transfer funds between them.

Empirical Behavior in the Laboratory

In simulated negotiation exercises conducted in a classroom or laboratory setting, the AAA team (comprising agent A* and principals A1 and A2) observe

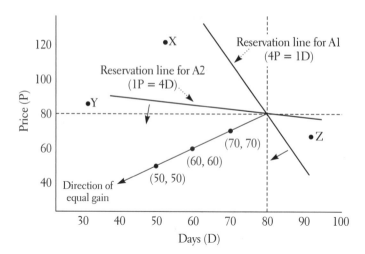

Figure 25.2.   Internal differences: Tradeoffs between days and price for A1 and A2.

that, for internal parity, price (P) and days (D) must be kept close together. The principals instruct their agent A* to achieve low values of a common value for P and D. Agent A* starts with a reservation value of (80, 80) for P and D and tries to push this down to (70, 70), to (60, 60), to (50, 50), and so on. It turns out that the reservation value for B* is (60, 60) or higher, and a lot of effort goes into a distributive bargaining exercise looking for a compromise between 60 and 80, A* wanting 60 and B* wanting 80.

Some subjects not only share the pie but enlarge it a bit. It turns out that B* has a tradeoff for P and D, where 2D's equal 1P, so that if A* yields an increase of one unit of P, it can be offset by a reduction of 2 units of D. The trouble is that if P and D are not in balance, then principals A1 and A2 will be differentially affected. The idea of tolerating an imbalance in the payoffs to A1 and A2 that could be corrected by a later side payment between A1 and A2 occurs to a few negotiating groups in the laboratory, but even then the idea is not fully exploited. So most negotiating groups leave potential gains on the table—the pie is not fully enlarged.

How Negotiations Should Have Been Conducted

How then should the parties handle a situation like this? The answer lies in agreeing to adjust the distribution internally, separately from the external negotiation. This frees the agent to concentrate on maximizing the total payoff to the side.

Internal compensation.   Suppose that A* has a choice that yields A2 a payoff of 10 and A1 a payoff of 40—that's point Q in Figure 25.3—or a choice of 15 to

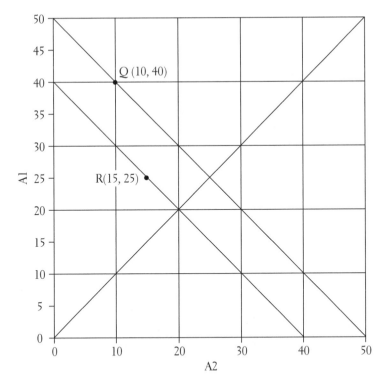

Figure 25.3. Internal compensation between A1 and A2.

A2 and 25 to A1—that's point B. Is it clear which one A* should choose without showing favoritism to one of his principals? The choice is abundantly clear as long as the proviso that A1 and A2 can trade in money through the independent budget is included. By choosing point Q, monetary exchanges allow A1 and A2 to move along the line through Q. Similarly, they can move along the line through R if R is selected.

With the possibility of a side deal, the mission of A* should be to maximize the sum of the payoffs to his principals:

$$\left(100 - \frac{P}{4} - D\right) + \left(100 - P - \frac{D}{4}\right) = 200 - \frac{5}{4}(P+D),$$

and this means he should try to minimize P + D and not worry about the balance he achieves between them. This is a profound conclusion that is most often violated in practice. Namely, the instructions to an agent should reflect the possibility of making internal side payments on one side of the table. When students engage in this simulation, the agent A* does not get proper instructions, with the result that most final agreements reached in the external negotiation are far from efficient—joint gains are left on the table.

A simpler mandate for the agent. In conducting his external negotiations, Agent A* should be given the mission statement: "Try to minimize P + D, keeping in mind that the sum should not exceed 80 + 80, or 160." For logistical reasons, A* is also informed that D cannot fall below 30 days.

In sharing information about tradeoffs, in a POTE-like manner, A* and B* inform each other that given any (P, D) point, each has a constant substitution rate. A* is willing to substitute 1P for 1D whereas B* is willing to give up 2D's for 1P. And of course this difference can yield joint gains: a reduction of 3D's for an increase of 2P's will make both sides happier.

The exchange of a little more information leads A* and B* to the conclusion that the criterion of efficiency suggests that D should be pushed down to its limit of 30 and P maneuvered for equity concerns between A* and B*. Note that A1 is happy with small values of D, but A2 is going to need a lot of internal compensation.

Use of SOLVER. Let's look at a solution of this internal/external simulation exercise if the parties use an analytical external helper operating under conditions of FOTE. It's not necessary to use the full power of SOLVER in this case, because the external marching orders for agent A* are so simple: just minimize P + D subject to the reservation condition that this sum should not exceed 80 + 80. But let's solve it anyway and then we'll show how this generalizes (using SOLVER) to instances when the agent cannot be given such simple instructions.

The problem is as shown in Table 25.1. The control variables are P, D, and a transfer payment of X from A1 to A2, located in cells E2:E4. We start off with an illustrative assignment for P, D, and X of 100, 50, and 10 respectively. B* imposes a constraint: D must be at least 30. A* adds the constraint that X must be between −25 and +25. The value to A1 (VA1), the value to A2 (VA2), and the value to B* (VB*) are as shown in E7:E9 using formulas shown in the boxed entry under column G. You should observe that for B* an increase in 2 days is equivalent to an increase of P = 1. This will mean, now knowing what you do, that A* and B* should push down the number of days, and the efficiency criterion suggests that the final contract should be fulfilled in 30 days. But SOLVER doesn't know this as yet, and we won't tell that to SOLVER but have it figure it out.

Next, using FOTE, the reservation values are shown in E11:E13. We now define for SOLVER what we call the excess values—i.e., the differences between payoffs and reservation values. The excess values appear in F16:F18. The illustrative contract is not feasible since the excess value to A2 is EVA2 = −2.50.

Now we're ready for SOLVER to tackle the problem. We tell SOLVER, "The variables at your control are P, D, X; P must be non-negative; D must be at least 30; X must be between −25 and +25." This defines the set of legitimate contracts—the zone of possible agreement. In other words, we program the

Table 25.1   Use of SOLVER

A	B	C	D	E	F	G
1 Control variables						
2	Price		P =	100		
3	Days		D =	50	D>30	
4	Transfer		X =	10	X<25, X>−25	
5						
6 Payoffs						
7	Value to A1		VA1 =	15	100 − P/4 − D −X	
8	Value to A2		VA2 =	−2.5	100 − P − D/4 + X	
9	Value to B*		VB* =	35	D/2 + P − 90	
10 Reservation values						
11	Res. val. to A1		RVA1 =	0		
12	Res. val. to A2		RVA2 =	0		
13	Res. val. to B*		RVB* =	0		
14						
15 Excess values						
16	Excess val. to A1			EVA1 =	15	(VA1 − RVA1)
17	Excess val. to A2			EVA2 =	−2.5	(VA2 − RVA2)
18	Excess val. to B*			EVB* =	35	(VB* − RVB*)
19				EVA2 − EVA1 =	−17.5	
20				EVB* − EVA1 =	20	

spreadsheet so that it can find the payoffs and excesses and as part of *feasibility* we require the excesses to be non-negative.

We now need to address the question of equity: what should be maximized? We impose the condition—you may prefer some other requirement—that the excesses of the three parties be equal, and we ask SOLVER to maximize the common value—or what amounts to the same thing, to maximize EVA1. (See Table 25.2.) SOLVER reports in no time flat that P = 96, D = 30, X = 25 and the common payoffs are 41.2. When we ask SOLVER to do the same thing with X = 0—when there is no side payment within party AAA—then the common value of the payoffs is 33.6. These solution values are shown in Table 25.2. Of course, SOLVER stands ready to solve more complicated problems.

Fair Division with Three Parties

After first analyzing the fair-division problem among three parties, AAA, BBB, CCC, we shift our attention to a case of relevance for this chapter: namely, we consider the variation of this problem when AAA and BBB are a team jointly negotiating with CCC. In this case we shall also assume that transfer payments can be made internally between AAA and BBB. In other words we consider the case where AAA and BBB constitute a pluralistic party.

Table 25.2   Solutions generated by SOLVER

Problem #1: Select P,D,X	Solution
Subject to the constraints	P = 96
P >=0	D = 30
D<=30	X = 25
X >= −25, X <= 25	Values
EVA1=EVA2=EVB*	A1 A2 B*
To maximize EVA1	41.2 41.2 41.2

Problem #2: Select P,D,X	Solution
Subject to the constraints	P = 69
P >=0	D = 69
D<=30	X = 0
X=0	Values
EVA1=EVA2=EVB*	A1 A2 B*
To maximize EVA1	33.6 33.6 33.6

Three Parties: The Standard Problem and Its Analysis

In this problem, AAA, BBB, and CCC are given ten items to share "equally."
Imagine that each scores the items by sprinkling 100 points over the items, just
as we did earlier for Janet and Marty. The data of the problem are exhibited in
the left-hand box (A5:D14) in Table 25.3. In E5:G14 we exhibit an illustrative or
representative allocation (or contract). To be legitimate, the sum of the entries
has to sum to one in each row. For example, the entry in H5 is the sum of entries
E5, F5, and G5. In this allocation AAA is given items 2, 3, 4, and 8; BBB gets 6,
7, 9 and 10; CCC gets 1 and 5. By using the sumproduct command, the spreadsheet knows how to get the payoff values for the three parties: AAA gets 57.50;
BBB gets 67; and CCC gets 38. For these values we list the sum, the minimum,
and the product in preparation for SOLVER to make the calculations.

In using SOLVER the variables to be controlled are in the boxed region
E5:G14 containing 0's and 1's. For legitimacy these entries must be non-negative and sum to 1.0 for each row. Pure, as distinct from mixed, allocations add
the requirement that entries in the allocation box must be integers. We can now
ask SOLVER to get to work and find the legitimate allocation that optimizes a
specified objective.

Comparison of Solutions

We exhibit in Table 25.4 five responses that SOLVER produced when asked to
maximize: the sum, using mixed (m) values; the minimum, using pure (p) and
(m) values; and the product using (p) and (m) values. CCC's return can vary

Table 25.3   A fair-division problem with three parties

	A	B	C	D	E	F	G	H
1								
2								
3						Allocations		
4	Item	AAA	BBB	CCC	AAA	BBB	CCC	Sum
5	1	18	2	20	0	0	1	1
6	2	16	4	13	1	0	0	1
7	3	22	2	15	1	0	0	1
8	4	16.5	1	15.5	1	0	0	1
9	5	17	4	18	0	0	1	1
10	6	2	30	10	0	1	0	1
11	7	4	10	4	0	1	0	1
12	8	3	20	3	1	0	0	1
13	9	1	14	1	0	1	0	1
14	10	0.5	13	0.5	0	1	0	1
15	Total	100	100	100				
16			Values					
17		AAA	BBB	CCC	Sum	Min	Prod	
18		57.5	67	38	162.5	38	146,395	

Table 25.4   Comparison of solutions

	Values						
Max	AAA	BBB	CCC		Sum	Min	Prod
sum(m)	54.5	87	38		179.5	38	180,177
min(p)	54.5	57	48		159.5	48	149,112
min(m)	52.3	52.3	52.3		156.9	52.3	143,056
prod(p)	54.5	87	38		179.5	38	180,177
prod(m)	47.5	87	44.6		179.1	44.6	184,209

from 38 to 52.3. We're inclined to use the min(m) calculation, but ex ante, not knowing the outcome of the comparison table, we can make a good case for the product(m).

So far this is fairly routine analysis that we partly covered earlier in the book. The punch line for this analytical abstraction, however, is yet to come.

Fair Division with Possible Side Payments

We now assume that BBB can make a transfer payment to AAA. How does this change the picture? Intuitively you should see that it makes good sense now to

Table 25.5   Fair division with three parties and transfer payments

	A	B	C	D	E	F	G	H
1								
2								
3			Values			Allocations		
4	Item	AAA	BBB	CCC	AAA	BBB	CCC	Sum
5	1	18	2	20	0	0	1	1
6	2	16	4	13	1	0	0	1
7	3	22	2	15	1	0	0	1
8	4	16.5	1	15.5	.23	0	.77	1
9	5	17	4	18	0	0	1	1
10	6	2	30	10	0	1	0	1
11	7	4	10	4	0	1	0	1
12	8	3	20	3	1	0	0	1
13	9	1	14	1	0	1	0	1
14	10	0.5	13	0.5	0	1	0	1
15	Total	100	100	100				
16			Values		**Transfer payment from BBB to AAA: X = 7**			
17		AAA	BBB	CCC	Sum	Min	Prod	
18		51.795	60	49.935	161.73	49.935	155,183	

give most, if not all, of items 6 to 10 to BBB and have BBB transfer points to AAA. Let's see how this works with SOLVER (see Table 25.5). The free variables in SOLVER's control are the allocations as before—we still require legitimacy—plus the use of a transfer payment of X from BBB to AAA.

If payments can flow only one way, we must impose the additional legitimacy requirement that X be non-negative. In the display we called on SOLVER to get as much as possible for AAA subject to the feasibility constraints that BBB get at least 60 and that CCC get at least 50. We just chose those artificial reservation values as being reasonable. AAA ended up with 68.73 (the table with this solution is not shown). Now for some further exploration.

In Table 25.4, without the possibility of a side payment, it was possible by using a mixed contract for all three parties to achieve 52.30. In Table 25.5, with transfers available, when SOLVER was asked to maximize the minimum in mixed contracts, it yielded 59.06 to all parties (again, not shown). Next we imposed the condition that CCC get 52.30 and sought the maximum common value for AAA and BBB. Under these constraints they were able to achieve 64.30 each (not shown). We could go on to explore other possible solutions. But what is fair?

If I (HR) were BBB, I think I could rightly complain to AAA and to CCC that I should get the most because I value the items so differently from the way they do. I am the one who creates the synergy that allows each of us to soar over the hurdle of 100/3, or 33.3. This moves the group toward the Nash solution that gives BBB a total of 87. What do you think is fair?

So now we arrive at the punch line in this example. Joint gains can be achieved if the external negotiations across the table between AAA and BBB on one side, and with CCC on the other side, can be synchronized with an internal negotiation on the AAA/BBB side of the table. That's a result well worth thinking about.

Generalizing from the fair-division problem.   Imagine a template that examines the issues and resolutions as they pertain to parties AAA, BBB, and CCC. But now also assume that AAA and BBB can enter into side deals—deals that may involve creative methods of compensation between AAA and BBB. This complexity can be accommodated—at least in principle—by adding new issues to the template, together with their accompanying resolutions, that affect only AAA and BBB and for which CCC has no discernible interest. Now the problem, with this extended template, can be analyzed very much like the way it was done with the fair-division problem in Table 25.4, keeping in mind that the transfer payment X from AAA to BBB is just such an added issue.

We now switch gears and go from an ultra simple problem of the use of internal compensation to a complex real case.

The Panama Canal Negotiations

Negotiations between the Panamanian government and the United States over the status of the Panama Canal took place in the mid-1970s. There were many issues under discussion, and the parties to the negotiation were not monolithic: external negotiations had to be coordinated with internal negotiations. In this section we describe techniques used by Ellsworth Bunker to deal with internal constituents who posed a threat to an external agreement.

Prelude to Negotiations

In 1959 mobs of Panamanians invaded the Panama Canal Zone, then under U.S. sovereignty, and demanded a revision of the existing treaty governing the status of the Panama Canal. Throughout the 1960s and early 1970s a series of fruitless bilateral negotiations took place against a background of vociferous opposition to the American presence in Panama. In 1973 the United Nations Security Council proposed a resolution guaranteeing the "full respect for Panama's effective sovereignty over all of its territory." The U.S. ambassador vetoed the proposal, but widespread international opposition to the U.S. position forced the secretary of state, Henry Kissinger, to take action. He appointed a new negotiator—the highly respected ambassador-at-large Ellsworth Bunker. His task was to reconcile the emotionally charged demands of the Panamanians with the interests of various U.S. parties: the Department of State, Department of Defense, Joint Chiefs of Staff, Congress, and nongovernmental interest groups, to name but a few.[1]

To do this, Bunker had not only to come to an agreement with the Panamanians, but to bring antagonistic forces within the United States to some grudging compromise. The Department of Defense, clearly, would have to be allowed a role in the negotiations: the hard-liners, in return for relaxing their rigid adherence to the status quo, would have to be given a voice in formulating the U.S. negotiating position. The hard-liners consisted of two groups: the "Zonians" and the "Southern Command." To facilitate Pentagon participation, Bunker set up a "Support Group" in the Department of State to help prepare possible U.S. positions. The Support Group included representatives from the Department of State (Panama Desk, Legal Section, and so on) and members from a Department of Defense ad hoc group responsible for developing and coordinating Department of Defense positions on the Panama issue. Formed by Secretary of Defense Melvin Laird and called the Panama Canal Negotiations Working Group (PCNWG), this ad hoc group included representatives from the Office of the Secretary of the Army (which in turn represented the interests of the Zonians), Security Affairs, and the Office of the Joint Chiefs of Staff (representing the hard-line interests of the Southern Command). For a long time, the PCNWG was chaired by the deputy undersecretary of the Army, who was sympathetic to the Zonians.

Members of Congress were aware of a general opposition in the country to any treaty that would entail significant U.S. concessions. The American public had indicated in several polls that it regarded the Canal as a symbol of American ingenuity, a piece of peculiarly American property that should by no means be given up to Panama. By the time of Bunker's appointment, a number of resolutions had been proposed in both the House and the Senate opposing the negotiation of a treaty that would dispose of U.S. sovereign rights in the Canal Zone. Senator S. I. Hayakawa summed it up prettily when he proclaimed, "We stole it fair and square."

Besides the concerns of the general public, Bunker had to consider the commercial and parochial interests of a variety of groups with powerful lobbies on Capitol Hill. The American Institute of Merchant Shipping, for example, was very apprehensive about the possibility of the Panamanians gaining control over the pricing structure of Canal services, which in the long run would mean higher toll rates for the use of the Canal and might eventually make U.S. intercoastal trade through the Canal unprofitable. Another group, the Canal

---

1. The descriptions of the setting of the problem and its denouement are drawn from "Panama Canal Treaty Negotiations," a case study prepared by Mark G. McDonogh under the joint supervision of Douglas Johnston and Howard Raiffa.

Zone Central Union, represented the interests of U.S. employees of the Panama Canal Company. Any new treaty that enlarged the Panamanian role in the administration of the Canal Zone would lead to gradual displacement of U.S. employees by Panamanian nationals and the elimination of special commissary privileges and retirement benefits. To gain an understanding of the problems facing these interest groups, Bunker and other government officials participated with them in a number of seminars run by independent think tanks such as the Brookings Institution.

Initial Negotiations and Internal Preparations

In his first meeting with Panama's foreign minister, Juan Antonio Tack, on November 26, 1973, Bunker received the impression that an agreement was possible. Before negotiating specific points, however, he felt that the two sides should agree on some general principles to guide their exploration of specific alternatives. In a round of negotiations that took place January 1–6, 1974, the two sides agreed on a United States–Panama Joint Statement of Principles, which Kissinger and Tack signed in Panama on February 7. (Commentators considered Kissinger's presence to be of "symbolic importance" to Panama, because it suggested an equality between the negotiating parties.)

After these initial rounds of negotiation, by the end of June 1974 the United States and Panama had managed to agree on a definition of major issues relating to the intended joint statement. The issues were: (1) *duration:* the length of time before a new treaty would expire and all rights would revert to Panama; (2) *jurisdiction:* the number of years before the United States would give up certain jurisdictional rights in the Canal Zone, such as those of criminal jurisdiction and police authority—rights not directly linked to Canal operation; (3) *defense role of Panama:* the degree to which Panama would assume responsibility for Canal defense; (4) *land and water:* the percentage of the Canal Zone that was to be turned over to Panama when a new treaty was ratified; (5) *expansion rights:* the deadline for a U.S. decision on whether to expand the Canal by adding a third set of locks or a new sea-level canal; (6) *expansion routes:* possible routes that could be used by the United States in the event it decided to build a new sea-level canal; (7) *use rights:* the jurisdictional rights required by the United States for the efficient operation of the Canal; (8) *compensation:* the amount of money the United States would pay to Panama for the right to operate and defend the Canal; (9) *U.S. defense rights:* the resources (facilities, personnel, and so forth) that the United States would be permitted to retain to defend the Canal, and the extent to which it would be allowed to guarantee the neutrality of the Canal; (10) *U.S. military rights:* the degree to which the United States could retain military rights not directly related to local defense of the Canal.

Bunker subsequently prepared for another round of negotiations with the Panamanians. What might have gone through his mind at that time? The issues had been clearly designated and grouped into ten categories. For each category, Bunker had some idea of the bargaining ranges involved. For example, on the issue of compensation, Panama might have been seeking an annual fee of about $75 million, while the United States was considering $30 million. Bunker might reasonably have wanted to get a more precise feel for tradeoffs between issues: how much should the U.S. side be willing to give up on Issue X for a given incremental change on Issue Y?

But who was the "U.S. side"? Was it Bunker, the Department of State, the Department of Defense, or the Department of Commerce? The tradeoffs of a military man with a mission are likely to be very different from the tradeoffs of a representative of the Department of State: one cannot expect that high-level officials from different branches of government will attach the same value to certain issues. Bunker attempted to devise a comprehensive value function (implying tradeoffs) for the U.S. position that he could use in external bargaining, but he could not reach an internal consensus. Worse, guardians of special interests all want to establish reservation values on those issues of primary concern to themselves. If Bunker had formally asked each constituent representative for a reservation value on each issue separately, we can guess what might have happened. The guardian of Issue X would stake out a bargaining position and exaggerate his needs; so would the guardian of Issue Y, exaggerating her needs. If the former exaggerated and the latter didn't, then when they were both compelled by higher-ups to relax their demands, the guardian of Issue X would end up better off. But the guardian of Issue Y, anticipating this, would likewise play the internal negotiating game. It's in the nature of the situation that if a compromise has to be settled externally, some internal faction will be disappointed with the result.

It may not always be desirable for a collective U.S. team to agree to a proposed treaty: a reservation value should be established for the overall contract, but not necessarily for each issue. Bunker would have been severely hobbled if he had had reservation values for each of the issues separately—especially if the set of all reservation values would have yielded a composite contract that was completely unacceptable to the Panamanians and was merely wishful thinking on the U.S. side. The secretary of defense might have been unwilling to trade military preparedness for, say, a gain in commerce. He might, however, have been exasperated with individual members of the Joint Chiefs of Staff wanting to put separate reservation values on the needs of the Navy, Air Force, Army, or Marines.

At about this time, the U.S. negotiators enlisted the aid of a consulting firm, Decisions and Designs Incorporated (DDI), to help them formulate a negotiating strategy. The DDI analysts interviewed members of the U.S. negotiating team, including Ambassador Bunker, and on the basis of the responses con-cocted a point scoring system (a value function) for the U.S. side. This was done, with slight modifications, very much as we did in Chapter 12. For ease of analy-sis, an additive scoring system over the issues was used without any real checks to see if there were interactions that would render the additive form inappropriate. The analysts should ideally have checked for preferential independence before blithely using an additive scoring system, and they probably would have discov-ered some dependence between issues; but it is likely that the additive form pro-vided good, convenient approximation.

There is no evidence that Bunker cleared the resulting additive scoring sys-tem with the Support Group, the PCNWG, or congressional committees. The scoring system reflected the tradeoffs that Bunker's personal negotiating team deemed appropriate, with all viewpoints and pressure informally incorporated. A consensus, if attempted, would not have been achieved, but Bunker and his team wanted a means of articulating some of their tradeoffs because they anti-cipated a need for such knowledge in the external negotiation process. Besides assessing component value functions over each of the issues and assigning im-portance weights for the U.S. side, Bunker's team also recorded their own per-ceptions of the Panamanian positions. The consulting analysts, using the addi-tive scoring systems, generated the efficient frontier of possible treaties and constructed a dozen or so efficient treaties—that is, treaties whose joint evalua-tions fell on the efficient frontier.

Course of the Negotiations

In an effort to keep the negotiations on track without disruption, the Panama-nian and U.S. negotiators decided to concentrate initially on those issues that would be easier to resolve, and negotiate the harder ones later. For the round of negotiations scheduled for November 1974, the U.S negotiators had prepared a package that they believed would go far toward meeting Panamanian demands on issues of comparatively minor significance to the United States. The package included the return of some jurisdictional rights to Panama within a period of less than five years after the treaty went into effect, and also included terms to in-crease Panamanian participation in the administration and defense of the Canal. In return, the U.S. negotiators expected to get a Status of Forces Agreement (an administrative agreement governing the conditions under which a foreign mili-tary force is subject to, or exempt from, the laws of the country in which it is sta-tioned), and the unilateral right of the United States to be guarantor of the secu-rity of the Canal when the treaty expired.

In the session of November 6, Bunker encountered strong Panamanian re-sistance to the package. To avoid risking a break-off in the negotiations, and to demonstrate goodwill that would be necessary for later Panamanian concessions, he decided to concede some issues to Panama without insisting on a quid pro quo. Thus, that same day, Bunker and Tack initialed three "threshold agreements" on jurisdiction and on Panamanian participation in the defense and operation of the Canal.

Internal Conflict

Following the initialing of the threshold agreements, Bunker requested presidential guidance to proceed with the negotiations. This guidance was expected to emerge from a series of National Security Council (NSC) meetings, which were to serve as a forum for the presentation of the positions of the Department of State, the Department of Defense, the Central Intelligence Agency, and so on. As it turned out, however, these NSC meetings brought into the open the Department of Defense's strong resentment about Bunker's concessions and its dissatisfaction with its negotiating role in general.

The first NSC meeting, in April 1975, was acrimonious, and revealed major differences between the State Department and Defense Department positions. The Pentagon representatives argued that the U.S. negotiators, just to keep the talk going, were conceding too much too soon without receiving anything in return: although Bunker's team had initialed a draft Status of Forces Agreement with Foreign Minister Tack on March 15, Deputy Secretary of Defense William Clements felt that the Panamanian concession on this issue was minimal compared with those (on jurisdiction, canal operation, and canal defense) which the United States had made. Also, the Pentagon officials complained that Bunker, in making the concession on canal defense, had acted independently and had overridden the final U.S. negotiating positions agreed upon by the two departments.

At issue was the clause of the threshold agreement that stated that the United States and Panama would "commit themselves to guarantee the permanent and effective neutrality of the interoceanic canal . . . and . . . make efforts to have this neutrality recognized and guaranteed by all nations." Before the negotiations started, the Department of Defense had agreed to a package that would give the United States the *unilateral* right to guarantee the permanent neutrality of the Canal. In addition to emphasizing the security risks involved in the multinational agreement, Clements maintained that without the unilateral right, the treaty would become a political issue in the presidential primaries of 1976. Clements also suggested that the Defense Department's representation on the mid-level State Department Support Group was inadequate for the protection of its interests and that some other arrangement would have to be made.

Bunker's team and the State Department, on the other hand, contended that the Pentagon's complaints derived from an unwillingness to accept the negotiating parameters set forth in the 1974 Kissinger-Tack principles, which stated the intention of "increasing the Panamanian participation in the defense of the Canal." U.S. military leaders regarded internal civil disturbances and acts of sabotage as a much more credible threat to the Canal than any attack by a foreign power. They were reluctant to place much faith in the reaction of Panamanian forces to sabotage or attack by extremist Panamanian nationals. The State Department argued in response that any treaty agreement that met the basic Panamanian nationalist concerns would defuse the motivation for sabotage operations (Duker 1978, p. 14), and also maintained that the Defense Department was being needlessly unyielding on the issue of land and water by insisting that almost all of the lands and waters in the Canal Zone were needed to operate and defend the Canal.

To gain support for their position, Pentagon officials leaked the substance of the intragovernmental conflict to the press, and this stimulated congressional opposition to the negotiations. By June 1975 Senator Strom Thurmond "had already gathered 37 Senators on the 1975 model of his resolution to block a new treaty and had personally warned Kissinger not to send up a treaty" (Rosenfeld 1975, p. 7). In the House of Representatives, Congressman M. G. Snyder offered an amendment to the State Department's appropriation bill, which provided that none of the appropriated funds would be used for negotiating "the surrender or relinquishment of any U.S. rights in the Panama Canal Zone." Although neither piece of legislation passed, they indicated the opposition that any future agreement would face in the absence of Defense Department support.

Looking back over the negotiations thus far, one can see that they were conducted in stages. When a treaty cannot be resolved it is nevertheless important, for international political reasons, to avoid risking a complete break-off and to demonstrate goodwill; thus, representatives of the two sides may agree on face-saving partial agreements. This was done with the Tack-Kissinger agreements in the early part of 1974 and with the three threshold agreements at the end of that year. One of the difficulties in settling the easier issues first is that there remain fewer opportunities for logrolling with the residue of tougher issues. Critics often protest that too much is given away in these interim agreements, but what these agreements buy has linkage value in foreign policy: sometimes a government desperately needs some peace and quiet so that its leaders can concentrate on more important problems. Of course, the other side might be aware of this need and might exploit it.

After signing the three threshold agreements, Bunker and his team faced new internal problems. The remaining issues were packaged and the bargaining ranges on the unresolved issues were shifted somewhat. The tradeoffs, too, shifted, and Bunker's team went through the exercise of reassessing component value functions and importance weights for the two sides. Once again this exercise was used to prepare for the next round of negotiations, but once again the results were not used in any formal way during those negotiations, and apparently no formal analysis was done on reservation values.

Political Leaders as Agents for Multiple Principals

These Panama Canal talks present an interesting view of the way in which internal conflicts are continually mediated throughout the negotiation process. Let's look more broadly at the pressures that are brought to bear on the external negotiator. Often he cannot get a clear set of internally generated instructions suitable for external use and consequently must feel his way along, buffeted by external and internal pressures. Occasionally, in an internal deadlock, someone has to back down. How does this happen?

Suppose that in the course of some international treaty negotiations the Joint Chiefs of Staff dig in their heels and absolutely refuse to make further concessions. The external negotiator, an ambassador, has no power to push them further and must enlist the aid of higher-ups (in Bunker's case, these would have been President Ford and later President Carter). The president can try to cajole the Joint Chiefs to yield a bit, but as guardians of a mission they sincerely believe that any further concessions would be detrimental to the security of the country. The president, with wider perspectives to balance, thinks otherwise. He can try to convince the Joint Chiefs, but he cannot comfortably fire or threaten to fire his top staff; they'll withdraw from the government and lend their support to the opposition party. So the president's power, too, is limited. But he knows that, although the military firmly believe in the value of their demands, perhaps an extra aircraft carrier or two, or maybe some additional Army funding might counteract the perception of weakness in the proposed treaty. In other word, the mediation of internal conflicts can be resolved by linkages to other problems.

These sorts of linkages are made frequently, and can be useful and effective strategies: they are the very art of compromise. Of course, if a president is weak and "buys" the acquiescence of his staff with outlandish side payments, then he might encourage a contest among potential recipients to see who can get the most. Such payments are only appropriate within reason. If one argues that each problem should be resolved unto itself, that logrolling between issues is reprehensible, then one seriously curtails potential zones of agreement. It is far better to negotiate acceptable deals through linkages than to resolve conflicts one by one through sheer exercise of power. The president of a country, the chief executive officer of a state-owned enterprise, the head of a firm, and the president of a university all frequently act as mediators in internal conflicts — "mediators with clout" whose power comes from their ability to link together negotiation problems.

Let's switch gears once again; this time from reality to stylized abstraction as we review core concepts.

Core Concepts

Complex bilateral negotiations often involve three agreements: one across the table and one on either side. What occurs—or does not occur—on each side of the table can lead to serious mistakes on the part of the negotiators. Ideally, synchronization and coordination of the external (across the table) and internal (on each side of the table) negotiations will achieve the greatest joint gains in the external negotiation. But these dependencies are frequently disregarded by those making the decisions.

This chapter has examined the simplest case to make the above observations more concrete. This occurs with a bilateral negotiation involving two separate issues (money and time) and where one side has two principals working with an agent. In laboratory negotiations most groups of players do not adequately synchronize internal and external negotiations and end up in inefficient outcomes. The principals' advice to the agent entering into external negotiations frequently does not adequately reflect the internal possibilities of side transfer payments from one principal to the other.

Once again we have illustrated how SOLVER can readily find efficient and equitable outcomes if the parties adopt FOTE behavior—possibly using a neutral joint analyst.

We have thus returned to the fair-division problem but now allow for the possibility of transfer payments within a coalition of players. The availability of this internal flexibility can be exploited in determining an efficient external agreement. More generally, the possibility of external agreements between a subset of negotiators can often be internalized by adding new issues to the template that affect only those parties.

Negotiations between the Panamanian government and the United States over the status of the Panama Canal took place in the mid-1970s. There were many issues under discussion and the parties to the negotiation were not monolithic: external negotiations had to be coordinated with internal negotiations. Any external agreement results in gains and losses to the internal constituents and the losers can form a coalition to block any external agreement. This calls for an internal negotiation where the winners can compensate the losers to break the power of the blocking coalition. The chapter has described how Ambassador Bunker and others dealt with some internal constituents (mostly the Defense Department) who posed a threat to an external agreement.

Multiparty Interventions

In Part IV we examined mediation, arbitration, and facilitation in two-party disputes. Here we consider the same forms of intervention but in multiparty contexts. The move from two parties to n parties makes a big difference. The preceding chapters have underscored the dramatic increase in the structural, cognitive, substantive, and procedural complexity when we move from two party to multiparty negotiations. More complexity puts a premium on formal analysis that can usefully simplify and meaningfully synthesize critical data for making better-informed decisions. This applies equally to negotiators and to intervenors.

Here's an example of the complexity involved. Imagine AAA and BBB began a negotiation, only to find that they had to invite CCC and DDD to take part as well if a satisfactory outcome was to be achieved. Assume the parties have not been able to reach a compromise, and have jointly decided to invite an external helper to assist them. The challenge for the external helper is to assist the parties to reach a Pareto-efficient, jointly feasible, and subjectively fair compromise solution. This is no easy task.

Roles of an Intervenor in a Multiparty Dispute

The roles of an external helper in two-party negotiations, described in Chapter 17, are readily transferable to the multiparty forum. In many-party negotiations the helper can do all that is done in the two-party setting and more. It is often harder for many parties to agree, and therefore the need for a helper grows as the number of negotiating parties increases. In Chapter 21 we examined in what ways groups perform poorly (losses from coordination, conceptualization, and motivation) and suggested remedial actions that could be taken by the group. These could be listed here as activities of the many-party intervenor as well.

In two-party interventions we remarked that the intervenor must negotiate with the parties the roles he or she might play. The same is true for many-party negotiations. But now these negotiations may be more complex, divisive, political, and legalistic (for example, there may be rules for open meetings). In summary, they are harder.

We examined earlier some of the interests of the intervenor in two-party negotiations; these interests also occur in the multiparty case. Even though the intervenor may have an interest in the substantive outcomes of the negotiations, he or she may be able to perform in a professionally neutral manner.

In multiparty negotiations, one of the parties may wear two hats: (1) as an interested player; and (2) as a chair of meetings, where the chair assumes many of the roles of the facilitator/mediator. This role is often assumed by one of the lesser (or smaller) players. Uruguay or Chile could, for example, play this role in meetings of the Latin American countries; so might Malta in the Mediterranean community or Luxembourg in the European. In two-party negotiations, the external helper can often help each side prepare for negotiations. The same applies with more than two parties, but now a new complication (or opportunity) arises: subsets of the parties may wish to prepare and brainstorm together with the help of a facilitator/mediator. There may be natural coalitions of parties around some subset of issues. Coalitions can, of course, complicate the tactics and dynamics of claiming; but negotiators, working in coalitions, can also brainstorm together about jointly beneficial, creative options.

We examine two case studies illustrating some of the above observations. Our goal will be to illustrate how an intervenor can be useful, and to show some of the techniques that skilled intervenors use.

A Hypothetical Community Dispute

The town of Pleasantville is getting a little less pleasant to live in for its 20,000 inhabitants. George Ericson, owner of Ericson Farms, recently died, and Ericson Estates, which holds title to the land, plans to sell the property for development. Pleasantville has expanded over the years, and the farm now lies within the town's limits. The property is zoned for single-family units with a 15,000-square-foot minimum. Ericson Estates has hired a developer who would like to build a shopping mall and office buildings on the property, but needs the town's permission for a zoning bylaw change. The enticement for Pleasantville is that the developer's plans include a gift of land to the town for a much needed town cemetery and space for town recreational facilities in a mini-park. In addition, Pleasantville is in dire need of further income for school construction and for the repair of its sewerage system. If the town accepts the Estates proposal, the hope is that this would increase the tax base considerably. If the town refuses, then Ericson Estates says that it will erect one hundred single-family units on the property. That project would also bring new revenue to the town, but in addition it would significantly increase the school population and might require the building of a new school. The opponents of the plan argue that the shopping mall and office development will change the atmosphere of the town, increase the need for services, and hopelessly snarl traffic; at the same time, it's not clear what the increase in the tax base will be. There are many uncertainties, and people are quickly taking sides with only sketchy information. All generally agree, however, that both the present commercial development and residential development plans are subject to negotiation.

Pleasantville is governed by two hundred elected Town Meeting members and a board of five elected selectmen. Any plans to change the zoning bylaws requires a two-thirds approval of the Town Meeting.

Introduction of a Facilitator/Mediator

Selectman John Davis, who teaches law at Boston University, has suggested that Pleasantville consider bringing in a colleague of his, Robert Allen, who could act as an adviser to the town in its negotiations with the Ericson Estates. Allen teaches a course at BU on the mediation of community disputes.

After some encouragement from the other selectmen, Davis approached Allen, who was cool to the idea of giving partisan advice to the town but enthusiastic about the possibility of acting as a facilitator/mediator hired by both sides. Davis was a bit confused about what this would entail, but nevertheless he agreed to ask Helen Ericson, the principal negotiator for the Ericson Estates, to come to a breakfast meeting with Allen and himself. She was skeptical about the idea, because obviously Allen was an old acquaintance of Davis's and she thought that this would stack the deck against her interests. Nevertheless, knowing that she could always back out, she agreed to the meeting.

The interests of the facilitator/mediator.   The more Allen talked, the more intrigued Ericson and Davis became. Allen explained that he would like to foster a collaborative relationship between the town and the Estates in order to seek joint gains for both. He indicated that his interests were:

- to help both sides generate joint gains
- to help train four of his graduate students in the art and science of facilitation/mediation
- to write a case study for research and pedagogical purposes
- to be, and appear to be, neutral
- to enhance his professional reputation
- to provide a service to the town

Allen also indicated that he himself would work pro bono but that his graduate students would have to be modestly compensated at a cap of $10,000 for all. Allen insisted that he would want the town and the Estates to split this expense. He estimated that he personally might spend up to thirty hours on the project.

Allen agreed to moderate any open meetings with the public and, if requested, to speak at Town Meeting, provided, of course, that the Planning Committee and Ericson Estates thought this would be helpful. Allen reminded them on departing that either side could always back out of any agreement with him.

Both Davis and Ericson left the breakfast meeting feeling that they personally would like to work with Allen, but they had deep misgivings about what they thought was his wildly naive enthusiasm and idealism. They still felt the dispute to be much more zero-summish than Allen.

A week later Allen got an amber light: both sides decided to use his talents, but would do this adaptively. Allen's role would be restricted at first and then expanded, perhaps, as agreed upon along the way by both the town and the Estates. Four graduate students cleared enough of their calendars to help their academic supervisor in this joint learning experience.

Joint meeting to negotiate the role of the F/M team. On September 20, Allen met with representative members of the town and with the Ericson Estates Executive Committee to negotiate what roles his facilitation/mediation team could play. He reviewed for them his personal interests in working with them, and indicated that although he would try to remain impeccably neutral, he would play an active role in making suggestions that might help create a larger "pie." Both sides nodded their hesitant agreement. Then Allen suggested that each side prepare separately, spending most of their time articulating their interests (objectives), and for this activity he volunteered one of his assistants to help each of them, if they desired.

Preparation

Protocol to help each side prepare for negotiations. Allen had suggested that either side could request facilitation help from his team in preparing for negotiations, but a representative of the other side would then have to be invited to such meetings. "Why would we ever prepare with the other side present to learn our secrets?" Davis queried, with the nodding approval of Ericson. "Because a lot is to be gained by truthfully sharing interests," explained Allen. "What is good for one side is not necessarily bad for the other side. Either side would always be free to go into executive session in complete privacy. I think that you are going to find that there is an opportunity here that can be exploited, and that most of the joint gains will be achieved while preparing for negotiations and truthfully sharing these reflections." The town accepted Allen's offer but the Estates demurred at this time.

Establishment of the Ericson Farms Planning Committee. The town first met in closed session (that is, with neither a facilitator nor an observer from the Estates) and decided to appoint an Ericson Farms Planning Committee, comprising ten members and headed by Sam Whalen, a management consultant. The team was balanced on backgrounds: two lawyers, a real-estate person, a member of the town's planning board, a teacher, two abutter homeowners, an expert on traffic, and two business persons. There was deliberate attention paid to balance on gender and race. The committee was appointed by the five selectmen. The committee in turn selected three of its members as an Executive Committee that would meet more often (in closed session) and would actually carry out the negotiations.

The committee meets with the F/M team. In an open session, with a representative of the Ericson Farms Planning Committee present, Greg and Alice, two members of the facilitation/mediation team, led the discussion about interests. Greg acted as facilitator, and Alice as scribe. She used a flip-chart system, with charts displayed on the walls and held up with tape. They agreed to keep in mind two broad options: a *commercial* development with land set-asides, and a *residential* development using the standard 15,000-square-foot minimum. Both options had to be investigated.

Interests of the town. Greg suggested that the group enter into the spirit of brainstorming—half-baked ideas were welcome—and encouraged participation from all. Alice closely monitored the discussion—which at first was slow in developing until the members got into the spirit—and recorded the interests expressed on flip charts. Group members concentrated on the articulation of the town's interests, and they came up with the following list:

- Financial concerns—up-front outlays and tax base
- How much will the town net?
- Effect on property values and distributional effects
- Use of town's services and resources
  - Schools
  - Fire, police
  - Sewerage
- Traffic
  - Total
  - Rush hours
  - Safety
- Town ambience

- Equity of burden (for example, abutters)
- Open-land usage (how much)
  - For cemetery
  - For recreational use (for example, tennis, track, soccer field)
  - For senior center
  - For arts center
- Low-income housing (for town employees)
- Housing for elderly
- External funding
  - From the state (for selected development)
  - From the federal government
  - From foundations and private (local) philanthropy

For each of these interests, Greg asked the standard Why, What, and How questions. The How question generated some innovative options.

Tasks to be performed and formation of subcommittees. The participants agreed that in order to balance the tradeoff between development and open space for town use, it was important to understand what kind of development and what kind of open-land use was being proposed. Group members decided an inventory of what was needed in the town was crucial. Here was an opportunity to establish some synergies between separate town problems. Greg also suggested that their problem was not entirely new; his team could help investigate what other towns had done in similar situations.

After deliberating about the tasks to be done and keeping their interests in mind, committee members voted favorably on three action items:

1. Creation of sub-committees centered around some of these interests
2. Enlargement of the committee to share the burden of subcommittee involvement and to provide needed expertise
3. Establishment of liaisons with other town committees (for example, the School Committee, the council on the elderly, even a committee of youth) to find out what they might want

Greg tried to get the Planning Committee to think more proactively in terms of opportunities to exploit rather than problems to solve. Each of these newly created subcommittees was urged to appoint an internal facilitator, a scribe (not only to record subcommittee discussions and decisions but to submit reports in writing back to the parent committee), and a timekeeper to pay special attention to the deadlines that had been set. Some members asked about the availability of Greg and Alice for subcommittee functions. "It's O.K. with us if the demand for our services is not too high elsewhere, but the Estates would also have to be invited as observers."

The committee then went into executive session, and Greg, Alice, and the Estates' observer were excused. In closed session the Executive Committee talked about negotiation strategy and what power moves they could make if the Estates played hardball.

The Planning Committee develops options for negotiations.   Over the next month the members of the Executive Committee honed their negotiation strategy. They developed four competing plans for commercial development and two for residential development. Some of the plans were quite fascinating. When the youth were canvassed they registered a desire for places to hang out—places where they could talk over a coke and dance outdoors in the summer. Some of the suggestions involved creation of townhouses for the elderly in a small retirement community that would not take up much space and not contribute to rush-hour traffic. The seniors preferred shuttle board courts rather than more tennis courts, and they thought an outdoor space for an aerobics class would be great as well. Also discussed was the possibility of a day-care center for toddlers, a senior center, a youth center, and an arts center, either separately or in combination. It was clear that underground parking was a must.

Ericson Estates declines the F/M's help in preparation.   Ericson Estates did not make use of the facilitation/mediation offer in their planning meetings, but they did send representatives to observe the open planning meetings of the town. Helen Ericson, who had lived in the town all her life and who planned to continue to live there, became especially interested in the town's planning process. Although she did not make any suggestions, she had strong opinions about what should be done. Her fiduciary responsibility, however, was to the Estates.

The Negotiated Agreement and the F/M Role

Allen moderated the early stages of formal negotiations, helping with the proposed negotiation process and helping to organize the issues to be resolved, but he assumed an observer status when the negotiators confronted hard tradeoffs. Allen said he was there to create a larger pie but not to get involved in the nitty gritty of dividing the pie. Of course, he could have helped in their claiming game, but that was not the agreement he had made with them in establishing his original mandate.

The negotiated agreement between the Estates and the town was a combination of these one-sided, predigested plans the town brought to the table. They agreed on a commercial plan with land set-asides for town usage; on the specifics of the commercial development; on traffic patterns *within* the development; on the acreage of the mall; on restricting the usage of signs; on the amount of space for offices; on the number and size of townhouses for the elderly and on units for moderate-income housing (giving town employees preferential status). But what they did not do jointly in these negotiations was to specify what exactly the town would do with roughly one-third of the land allocated for its exclusive use; nor were there any plans made for external traffic control or for any compensation for disadvantaged abutters. Those matters were left for the town to decide.

The negotiated contract also specified the intent of the Estates if the Town Meeting failed to ratify the agreement. It involved the construction of single-family homes allowed by the present zoning requirements.

Public Hearings and the Ratification Process

The facilitator/mediator, Robert Allen, conducted public meetings, required by town law, on the merits of the negotiated, but as yet not ratified (partial) contract. The Planning Committee took the lead in presenting the negotiated agreement. As expected, there was a group of citizens, not all abutters, who were incensed about the process and its conclusion. However, most of the people who liked the conclusion also liked the process. Allen had to allow the dissenters time to vent their anger, and he tried to articulate the concerns of the negatively inclined. Their collective argument went as follows: it's a tough tradeoff to decide how much free land the town should get for giving the Estates a variance for the proposed commercial development; and a lot depends on what specific uses the town will make of this land. Therefore, because the town's specific intent was unknown, they argued that Town Meeting members should reject the agreement.

It became clear that the vote would fall short of the two-thirds required for ratification. Many townspeople believed that the Estates had not bargained in good faith. For example, the rumor was spread that if the Estates did not prevail with the commercial development, they would design the development (including playgrounds and a nursery) to attract large families with school-age children in order to overburden the schools or necessitate the building of a new school.

Competing plans for the town land and a voting process.   The town decided to be more specific about its plans for the one-third of the Farm land it would have control over, and in open meetings all sorts of suggestions were offered. The Town Committee convened once again, with Allen acting as moderator, and developed four rather different plans about what to do with the open land; what to do about traffic mitigation and control; and what to do about compensating the disadvantaged abutters in the form of real-estate tax relief or help with moving expenses. We label these plans, respectively, as R for tilting toward recreation; S for tilting toward the seniors; Y for tilting toward the youth; and C for tilting toward conservancy and open space. Each of these plans had mixtures of varying degrees of the four orientations (R, S, Y, and C). The idea was to have the Town Meeting members vote on these alternatives. These votes would not require a two-thirds majority, but still, given what we have learned about the weaknesses of majority rule, it's important to know what voting procedure should be followed. In what order should the votes be taken? A suggestion was made and defeated that each Town Meeting member be given 100 points to allocate simultaneously among the four alternatives, with the alternative that accumulated the highest number of points declared the winner. It was felt by most that this procedure involved too much strategizing and would encourage insincere voting. Instead, the following was adopted: each Town Meeting member would be given two ballots to be cast for two of the four proposals, R, S, Y, and C, with the option of using only one ballot. Then there would be a runoff ballot (majority rule) for the top two contenders. Of course we know from Chapter 24 that the final result could depend on the voting procedure adopted.

The Ericson charitable offer.   Before the votes were taken, Helen Ericson made a magnanimous offer. She said that she would make a personal gift in honor of her father of $2 million to the town for a George Ericson Performing Arts Center, to be located on the open portion of land. This required a modification of the existing plans. But more important, it squelched some of the more vitriolic anti-Ericson feeling. A committee was formed to actively seek other philanthropic giving for the development of the open land. This effort was quite successful, but not during the time frame of these negotiations.

The denouement.   Votes were taken. There was a runoff between modified plans R and S, and in a majority vote R prevailed. The runoff votes only required a simple majority. Amendments were added to give the more seriously affected abutters some graduated tax relief up to a maximum of 20 percent. The contribution of the facilitation/mediation team was gratefully acknowledged by the town, not so much for their efforts during the negotiations but for their help in structuring and conducting the planning process of the town.

Energy Negotiations in the European Union

We can now move from a fictitious case to a nonfictitious one. This case begins with a narrative summarizing the course of the negotiations and the main substantive issues on the table. This is followed by the formal analysis of these negotiations, including a full application of the FOTE methodology, from structuring the negotiation template, to preference analysis involving real people, to linear and nonlinear programming allowing comparison of different feasible, efficient, and fair solutions to the complex multiparty negotiation.

Background to European Energy Liberalization

In 1985 the European Community made a historic decision to create a single European regulatory framework allowing free trade in good and services. In many sectors, such as transport, telecommunications, and financial services, there were compelling economic reasons for the majority of European governments and industrial lobbyists to support market liberalization and increased competition. Winners far outweighed losers. But some sectors of the European economy were not ready for a strong dose of market-led reform. In the energy sector, and particularly in the natural gas market, there was vociferous opposition to proposals for the introduction of competition into monopolistic energy markets.

The key decision on liberalization had to be made by the Energy Council, composed of energy ministers from the fifteen member states of the European Union and the European commissioner responsible for energy. According to the decision-making rules of this Council of Ministers, the decision to proceed could be made by a qualified majority of weighted votes—unanimity was not required. However, the strategic importance of energy supplies and the negative consequences of forcing a government (or governments) to implement an agreement it had previously rejected meant that in political terms something close to unanimity was required.

Obstacles to Agreement: The European Commission's 1988 Policy Paper

In a 1988 white paper, the European Commission's Energy Directorate pinpointed the obstacles to negotiating a deal on liberalization of Europe's gas markets.

1. *Variations in production capacity.* Only two countries, the United Kingdom and the Netherlands, produced large enough quantities of natural gas to be self-sufficient. And only the Netherlands exported gas to other European Union countries. Germany and Italy produced over a quarter of their natural gas needs but imported the rest.

2. *Dependence on a small number of import sources.* Apart from the countries just mentioned, other European countries imported their natural gas. A lack of diversification exposed these countries to strategic security of supply risks—they imported primarily from Russia, Algeria, Libya, the Netherlands, and Norway. Greece, Spain, and Finland had only one main source of supply.

3. *Diversity in consumption patterns.* In the United Kingdom, France, and the Netherlands, small consumers accounted for the lion's share of annual gas usage. However, in other states, like Greece, Ireland, and Spain, virtually all gas was consumed by large industrial concerns, with Germany and Italy falling somewhere in between.

4. *Different levels of market maturity.* In the Netherlands, the United Kingdom, Germany, Belgium, France, and Italy, gas penetration of households was high; far-reaching transmission and distribution networks had been constructed. Elsewhere, in countries like Greece, Spain, Ireland, and Portugal—and when they joined the European Union in 1995, Finland and Sweden—distribution was limited to the capital city or to a select group of large industrial customers.

5. *Contrasts in market structure.* The majority of European states had public monopolies in natural gas—the entire transmission system was owned and operated by a government-controlled national gas company with a monopoly over gas imports and commercialization. Only Germany had competition between commercial gas companies on a regional basis, but even there Ruhrgas dominated. In 1986 the UK's Conservative government had privatized the leviathan British Gas Corporation, which began a drive toward competition in the domestic gas market. This became a model for liberalizing the European market.

These multiple, deep, cross-cutting conflicts of interest involving sixteen parties made it exceptionally difficult for the European Commission to draft a broadly acceptable proposal. After getting nowhere for several years, Europe's energy ministers decided in 1993 to put gas liberalization negotiations on the back burner—pending an agreement on an internal market in electricity. When agreement was reached on an electricity directive in June 1996, the six-month Irish presidency of the Energy Council restarted the gas negotiations.

The History of Negotiations

The Irish presidency of the Council: July–December 1996. Irish officials consulted widely, redrafted the European Commission's proposal, and brought together the conceptual elements of a possible deal. According to the Irish document, the Council of Ministers had to grapple with nine issues: (1) What should be the minimum percentage of each national market open to competition immediately following the implementation of the gas directive? (2) What should be the minimum percentage of each national market open to competition ten years after the gas directive takes effect? (3) What exemptions from market liberalization should be incorporated into the gas directive with regard to long-term contracts where the customer guarantees to buy a minimum volume of gas, known as take-or-pay contracts? (4) Should exemptions from market liberalization be granted to geographically limited regions with undeveloped or immature gas markets? (5) Should offshore natural gas pipelines be treated differently from onshore gas pipelines in the directive? (6) Should the directive allow governments to impose public service obligations—guaranteeing gas supply and service at a reasonable cost, independent of geographic location—on natural gas companies? (7) Should the directive offer exemptions from market liberalization for natural gas distribution? (8) Should the directive grant exemptions from market liberalization to "immature" emerging national gas markets and to national networks unconnected to the Continental network? (9) Should natural gas production activities be unbundled from the integrated accounts of natural gas undertakings?

The Netherlands presidency of the Council: January–June 1997. Finding a compromise in the gas negotiations fell to the Dutch government in its January-to-June presidency of the Council. The ace in the presidency's pocket was that in a pinch, it could invoke the rule of qualified majority voting—only 67 out of a total 82 weighted votes would be required to get the deal done. Deal-making expediency, however, had to be weighed against the political fallout of forcing one member state, or more, to swallow an unpalatable compromise. Presidency negotiators were expected to aim to achieve consensus—or at least virtual consensus—not build minimal winning coalitions.

In its pre-presidency publicity drive, the Dutch government trumpeted the priority it would give to reaching an agreement on the internal market in natural gas. This dossier, the Dutch claimed, would be one of the key deals by which their efforts at European leadership could be judged. The omens were mixed. Emmet Stagg, the Irish energy minister and outgoing mediator for the talks, confidently stated that there was sufficient political determination to achieve an agreement in the near future. But one of his officials was more ambivalent. "I've been known to bet on horses," he said, "but I wouldn't bet on reaching an agreement on this by the end of the Dutch Presidency." What should the Dutch do?

As disinterested mediators they were expected to craft a compromise, carefully balancing the interests of fifteen member states, satisfying the European Commission, and guarding against the temptation to advance their own interests. After a preparatory meeting in March 1997, the presidency team placed a new proposal for agreement on the agenda for a May 1997 ministerial meeting. However, the Dutch had not resolved the tension between their own very significant interests in gas production and the dramatically different substantive interests of the large gas-importing countries. Instead of managing the negotiations from their Brussels embassy, they retained control in the Dutch Economics Ministry in The Hague. Unsurprisingly, the new proposal met with a frosty reception from European ministers representing antiliberalization and gas-importing member states. Given the partisan slant to this single negotiating text, and the Dutch team's unwillingness to amend it, meaningful negotiation under Dutch mediation was impossible.

The Luxembourg presidency of the Council: July–December 1997.   The next team to try its hand at the gas directive game belonged to Luxembourg. Would it fare any better? Or worse, perhaps? Like the Dutch, the Luxembourgers gave the negotiation top priority. Luckily their minister for energy, Robert Goebbels, had a great deal of experience in Brussels negotiations. Their lead diplomat on the issue, Jean-Marc Hoscheit, deputy ambassador to the European Union, was an old hand at complex negotiations as well.

The first thing these savvy Luxembourg negotiators did was to schedule a special meeting of energy ministers in October 1997. They plied them with fine food and wine in a luxurious Luxembourg resort. Negotiations during the ministerial lunch went well on a number of issues, including the thorny question of exemptions for take-or-pay contracts. Some member states were beginning to show more flexibility on allowing the Commission to play a decision-making role. However, amid the clatter of cutlery and swilling of wine, official factotums and multilingual interpreters found it hard to capture the exact details of informal deal making. In the afternoon, the meeting restarted in a more formal setting. Apparent agreements over lunch unraveled at the negotiating table as beneficial misunderstandings were brought to light by more sober discussion. Nevertheless, agreement on a number of outstanding issues was within reach that day. Then the French industry minister spoke.

In apologetic tones, he explained that his country's newly elected Socialist government, led by Lionel Jospin, could not at present accept the compromise proposal. This, he maintained, was only partly related to the substance of the deal. It was primarily a question of timing. First, for political reasons the Socialists could not immediately sign up to an agreement the conservative government had so recently opposed. The gas directive would affect many of the Socialists' loyal public sector voters—liberalizing the gas market was poor thanks for their votes. Second, Jospin explained that public sector trade unions were due to hold elections at the end of November. He couldn't afford to stir up a hornet's nest now, because radical antimarket Communists might advance their political cause within the union movement. These were the kind of arguments that the other political fixers found persuasive. The French were let off the hook—for the moment. But what should the Luxembourg mediators do?

Rather than reach a partial agreement, leaving two or three politically sensitive "crunch" issues to be negotiated in December, Goebbels concluded the informal meeting without resolving a single issue. He purposefully kept several issues on the table for the final session, underlined the impossibility of reaching a compromise without more flexibility from all parties, and braced himself to face the media and explain why the negotiations had "failed." (See Box 26.1.)

Box 26.1   External Helpers in Multiparty Negotiations

Ideally how would you prepare, as an external helper, for a multiparty negotiation? How would you structure the problem? Would your preparations focus on specific issues and parties? Or would you try to gather a comprehensive file of information? How would you summarize the information? What would be the most useful format for the lead negotiator operating under time pressure? If consensus isn't necessary, would you analyze possible blocking coalitions? Might you draft a shortlist of parties that could be pressured into agreement? Would it be useful to line up some side payments? Should you be aware of bilateral disputes that might flare up and derail the multiparty talks? How would you predict which parties will be the strongest opponents of your compromise text? Would you solve the issues sequentially or keep several up in the air to be resolved contingently and simultaneously?

Before the Luxembourgers made their final attempt at reaching a consensus, they initiated an intensive round of information gathering. This information gathering was guided by a matrix, or qualitative negotiation template, they had created. In the left-hand column was a list of the sixteen parties, with their weighted votes noted in parentheses, and across the top were the issues still to be negotiated. Officials in the Energy Ministry assiduously sought details from their foreign contacts so they could specify the "position" and "interest" of each party on every issue. These pieces of valuable data were added to the template until a complete picture of the upcoming negotiation had been compiled. Armed with this synthesis, Goebbels himself called his ministerial counterparts. He tried to get a feel for their tradeoffs and asked what their bottom line was on key issues. He then probed their no doubt exaggerated claims of what was minimally acceptable. The energy officials updated the template. They used color coding to identify "hot" issues and penciled in "reservation values."

During the December 8, 1997, Energy Council meeting in Brussels, the concise summary of the parties' preferences, tradeoffs, and reservation values gave Goebbels the upper hand. This quick reference matrix provided him with the flexibility to balance tradeoffs, juggle concessions, and draft new compromises while taking into account voting weights, priorities, and issue-based reservations. None of the other ministers had invested equivalent time and energy in preparation. Goebbels had successfully anticipated a disagreement between France and Germany over the right of German natural gas companies to construct and operate gas distribution facilities in France. After consulting the Commission—who had notional veto power over the entire text—the Luxembourgers put the French and German ministers in different office suites, shuttled between them, and iteratively demanded concessions. Since the dispute was not one of principle, but derived from the market for gas distribution in the Alsace region, resolution of this side deal was not significant for the other parties.

With the Franco-German split healed, the Luxembourg team felt it had crafted a proposal that would garner a near consensus. Confident that sufficient parties would be satisfied, Goebbels reconvened the formal session. He had deliberately kept multiple issues unresolved until this moment so he could give something to everyone. (He knew full well that each minister required a morsel of the shared success to sell to the braying press pack after the negotiation ended.) Would the parties find the compromise proposal acceptable? All but one of the European Energy ministers supported the final proposal. To be sure, some of them grumbled that the French had done rather too well—although they didn't say it in so many words. And the Netherlands minister inserted a general reservation—the diplomatic equivalent of sour grapes. But the Luxembourgers had successfully pushed the parties together and achieved a virtually consensual agreement. A diplomatic triumph for their six-month presidency.

Applying FOTE Analysis to the Natural Gas Negotiations

We hope that the foregoing case narrative has given you a flavor of the real difficulties of reaching an agreement between sixteen well-prepared, highly motivated, coherent parties who have a long track record of cooperative decision making in a shared institution. This case shows how hard it is for many collaboratively minded parties to find a mutually satisfactory agreement—imagine the difficulties of intractable conflict. The following section applies the FOTE methodology to the natural gas negotiation and illustrates how formal analysis can deliver tangible advantages to intervenors in multiparty negotiations.

Structural Analysis

Let's look at the analytics of this sixteen-party negotiation. First, we need to create a qualitative negotiation template from the issues and resolutions identified by the Irish presidency and incorporated by the Luxembourgers into their matrix. All issues and resolutions need to be clearly defined to ensure accurate preference elicitation in the next phase of analysis. In addition, because we intend to transform this qualitative template into a quantitative model, we need to ensure from the start that the qualitative template has additive properties. This means structuring and restructuring the template until preferential dependencies between issues have been eradicated. It also means that resolutions must be mutually exclusive. If the negotiation template is badly structured, you have little hope of producing meaningful quantitative analysis. The qualitative "template" is shown in Table 26.1.

Table 26.1 Negotiation template for the natural gas negotiations

1. Initial extent of market liberalization
   What should be the minimum percentage of each national market open to competition immediately following the implementation of the gas directive? (0%, 15%, 18%, 20%, 23%, 28%, 30%)

2. Final extent of market liberalization
   What should be the minimum percentage of each national market open to competition ten years after the gas directive takes effect? (25%, 30%, 33%, 45%, 50%)

3. Exemptions for long-term take-or-pay contracts
   What exemptions from market liberalization should be incorporated into the gas directive with regard to long-term take-or-pay contracts?
   - No exemptions for take-or-pay contracts.
   - Exemptions for take-or-pay contracts existing before the entry into force of the directive, to be decided on by Member States.
   - Exemptions for take-or-pay contracts existing before the entry into force of the directive and exemptions for future take-or-pay contracts, both to be decided on by Member States.
   - Exemptions for existing or future take-or-pay contracts to be decided on by Member States with European Commission oversight.

4. Exemptions for emerging regions
   Should exemptions from market liberalization be granted to geographically limited regions with undeveloped or immature gas markets?
   - No exemptions should be accorded to "emerging regions."
   - A list of "emerging regions" benefiting from exemptions should be attached to the gas directive in an appendix.
   - Emerging regions should be granted exemptions from market liberalization by European Commission decisions based on strict criteria defined in the directive.

5. Treatment of offshore gas pipelines
   Should the directive treat offshore natural gas pipelines differently from onshore gas pipelines?
   - Entirely exclude offshore pipelines from the directive.
   - Include offshore pipelines in the directive under a specially defined technical regime.
   - Include offshore pipelines in the directive under the same technical regime as onshore pipelines.

6. Public service obligations
   Should the directive allow governments to impose public service provisions on natural gas companies that oblige them to guarantee gas supply and service at a reasonable cost independent of geographic location?
   - Do not allow governments to impose public service obligations on natural gas companies.
   - Include a provision in the directive allowing governments to impose public service obligations on natural gas companies.

7. Exemptions for natural gas distribution
   Should the directive offer exemptions from market liberalization for natural gas distribution?
   - Natural gas distribution should not be exempt from the directive's market liberalization proposals.

Table 26.1   (continued)

_____

- Member States may refuse to grant authorization to construct or operate distribution facilities if the authorization would conflict with public service obligations or damage the general economic interest of the country.

8. Exemptions for emerging markets and unconnected networks
   Should the directive grant exemptions from market liberalization to "immature" emerging gas markets and to national networks unconnected to the Continental network?
   - Emerging markets and unconnected networks should not be granted exemptions from market liberalization.
   - Emerging markets should be granted optional exemptions limited in time from the market liberalization provisions of the directive.

9. Unbundling and transparency of accounts
   Should natural gas production activities be unbundled from the integrated accounts of natural gas undertakings?
   - Natural gas production activities should be unbundled from integrated accounts.
   - Natural gas production, as such, should not be unbundled from integrated accounts.

_____

These nine issues in the table have between two and seven resolution levels and collectively represent the major political issues on the table in the gas directive negotiations. The full set of contracts is calculated by multiplying the resolutions on each issue ($7 \times 5 \times 4 \times 3 \times 3 \times 2 \times 2 \times 2 \times 2$) = 20,160 contracts. According to this model of the gas directive negotiation, 20,160 different contracts could be agreed, each one having different consequences.

Preference Analysis

What needs to be done now is to apply the FOTE methods set out in Part III. Collecting preference data from sixteen negotiating parties may seem burdensome, but the payoffs should be huge. We used face-to-face interviews and a preference analysis software package to accurately and systematically collect preference data. Strengths of preference for resolutions were collected, then importance weights for issues. When answering tradeoff questions, interviewees were frequently asked to take the range of resolutions into account. Some preference data had to be discarded because the interviewee didn't have sufficient knowledge of the issues or was unable to fully comprehend how the methodology worked.

We don't show the raw data because it is confidential, but in Figure 26.1 we show a comparison of the issue weights for each party. The data can be interpreted as follows. Each section of a column represents the importance weight of an issue; the longer the section of the column, the more important it was for the negotiator to achieve his or her most preferred result on that issue. Can you

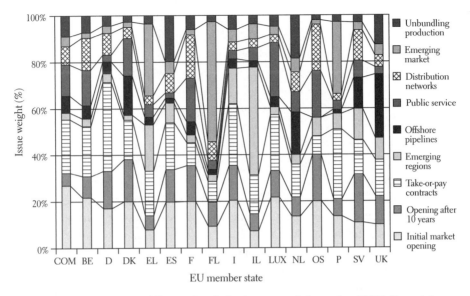

Figure 26.1. Issue weights differ significantly for sixteen negotiating parties. (COM–Commission, BE–Belgium, D–Germany, DK–Denmark, EL–Greece, ES–Spain, F–France, FL–Finland, I–Italy, IL–Ireland, LUX–Luxembourg, NL–Netherlands, OS–Austria, P–Portugal, SV–Sweden, UK–United Kingdom.)

think of some reasons why having a graphical summary of issue tradeoffs for every party would be useful for an intervenor (or a negotiator) in a complex multiparty context who has to make decisions on the fly?

Measuring reservation values.   Having structured the negotiation template and elicited preference data, our next step in the analysis is to measure each party's reservation value—the minimum amount each party would accept on each issue. The overall reservation value of each party was assessed by asking a negotiator what the minimal acceptable resolution was on each issue—then probing the answer. This produces a list of resolution levels that define a preference value on each issue. Summing these values produces an overall reservation value. We should note that this is not necessarily equivalent to the BATNA.

Optimization analysis: Efficiency, feasibility, equity.   Let's recap the key concepts for assessing outcomes. An efficient outcome cannot be improved for one party without making another party worse off. A feasible outcome is evaluated by each and every party to be equal to or better than its reservation value. An equitable outcome enshrines a fair distribution between the parties of the excess value created by their agreement. An intervenor's mission is to assist the parties to achieve an efficient, feasible, and equitable outcome through a mutually acceptable process. In a multiparty context the combination of many parties, many issues, many resolutions, and a multitude of incompatible interests makes it extraordinarily difficult for an intervenor to fulfill his or her mission. But help is at hand. The FOTE methodology increases the capability of an intervenor to achieve better outcomes for all parties.

Why did the Dutch proposal made in May 1997 wreck their credibility as disinterested mediators? From the narrative we know that it was rejected by a large number of ministers from member states. Could they have anticipated this rejection? Figure 26.2 shows the preferences of the parties for the Dutch proposal. It is immediately apparent that the "compromise proposal" had very little prospect of being accepted by the parties because it was not "feasible" for five of them—including heavy-hitters like France. In essence, even a quickly executed FOTE analysis would have saved the Netherlands mediating team the many days its members spent crafting their proposal and the political embarrassment when the proposal was rejected.

With the Dutch example we have demonstrated how FOTE can assist an intervenor to predict how multiple parties will value a proposal for agreement. Now assume the baton has been passed on to the Luxembourg presidency and we are advising their team. How can we use FOTE prescriptively to develop an option for agreement that is efficient, feasible, and equitable? The answer lies in solution concepts from axiomatic game theory. The axiomatic approach considers the set of all possible outcomes (in this case the 20,160 contracts) and selects one particular outcome based on a priori considerations defined as axioms. The axioms are basic (unprovable) assertions, selected because they are intuitively reasonable or fair. Different sets of axioms have been proposed, and we enter into the realm of philosophy in asserting that one is better than another. In other words, no solution dominates. We have advised the Luxembourg team to use the Nash solution (Nash 1950) and the maximin solution (Raiffa 1996). We treat the Nash and maximin concepts as indicators of efficient, feasible, and fair outcomes and intend to use them to generate deep insights into how we can best resolve the negotiation. Of course it would be naive to think we can simply translate quantitatively derived "optimal" contracts into concrete texts for negotiation.

After running the analysis on our raw preference data in a spreadsheet (requiring some linear and nonlinear programming), we can produce the information in Table 26.2. Each column is dedicated to one of the negotiating parties (from COM to UK). The word "Pure" indicates that the arbitration procedure is strictly limited to select one, and only one, resolution on each issue. The word "Mixed" indicates that the arbitration procedure may include mixed resolutions on each issue—but the weighting still has to sum to 1 for each issue. Running through the five outcomes, we therefore have the EU Directive (as agreed under the Luxembourg presidency), the Nash solution, and the maximin solution, all with pure contracts, then the Nash and the maximin solutions with mixed contracts. In mathematical terms, pure contracts are a special case of mixed contracts. For each outcome, "value" is the preference value of the contract with a

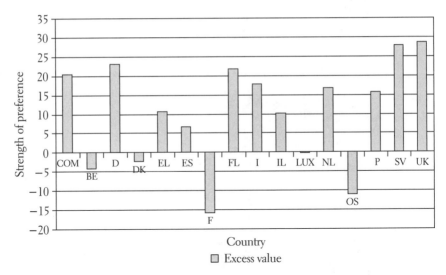

Figure 26.2. Dutch presidency proposal fails the feasibility test for five parties. (For country abbreviations, see Figure 26.1.)

minimum of 0 and a maximum of 100; "excess" is the excess value above the reservation value; and "POP" is the proportion of the potential. Take some time to compare the numbers in the table—there is a great deal of analysis behind this.

How can the analysis be used to assist intervenors in the negotiation? Let's start with solutions based on pure contracts. Examine Figure 26.3, which compares the EU Directive as agreed by the European energy ministers with the maximin solution with pure contracts against the benchmark of the proportion of the potential achieved.

Consider first the POP scores for the EU Directive—the actual agreement. The Netherlands (NL) has a negative POP score—the final outcome was below its reservation value. Some other parties have exceedingly high POP scores, above 80 percent (see Denmark, DK; Greece, EL). These negotiators must have been ecstatic when the directive was agreed to. With all other parties above at least 20 percent of their POP score, the figure demonstrates how competent the Luxembourg proposal was. Of course, the problem with this outcome is that, in our terminology, it is not "feasible." We might also consider that it is inequitable.

Let's turn our attention to the maximin solution with pure contracts. The most important fact is that a feasible outcome was possible—the POP score for the Netherlands is now well above the 20 percent threshold. However, this solution significantly redistributes value across the other parties when compared with the EU Directive. Look at the decrease for Denmark and the increase for Ireland. For this reason we need to emphasize that axiomatic solutions should be used as indicators of the possible or as a framework to provide intervenors with deeper understanding of where compromises can be reached. You might like to use the data in Table 26.2 to make further comparisons of the solution concepts

Table 26.2 Comparison of solutions: Values, excesses, and POP scores

Optimization	COM	BE	D	DK	EL	ES	F	FL	I	IL	LUX	NL	OS	P	SV	UK
Pure																
EU Directive (1998)																
Value	34.5	40.2	31.6	46	59.8	37.1	46.2	58.8	39.4	44.3	37.7	35.4	53.3	57.1	38.4	40.2
Excess	7.9	12.9	7.5	15.8	12.1	8.8	3.8	9.9	14.5	12.0	4.8	−1.0	19.0	17.2	14.6	8.3
POP	36%	60%	30%	91%	90%	34%	24%	48%	65%	63%	26%	−8.0%	77%	66%	52%	33%
Nash solution																
Value	34.9	40.7	39.2	33.1	58.3	37.0	47.8	63.9	35.2	41.6	39.4	38.3	49.3	58.7	46.3	46.8
Excess	8.4	13.5	15.1	2.8	10.6	8.7	5.4	15.0	10.3	9.3	6.6	1.9	15.0	18.8	22.5	15.0
POP	38%	62%	60%	16%	79%	33%	34%	74%	46%	49%	36%	16%	61%	72%	81%	59%
Maximin solution																
Value	40.3	37.6	34.1	37.8	54.9	38.2	45.9	62.3	37.9	48.7	37.0	39.1	49.3	53.2	34.1	42.0
Excess	13.8	10.4	10.0	7.5	7.1	9.9	3.5	13.3	13.0	16.4	4.1	2.7	15.0	13.3	10.3	10.1
POP	62%	48%	39%	43%	53%	38%	22%	65%	58%	86%	23%	22%	61%	51%	37%	40%
Mixed																
Nash solution																
Value	40.9	37.5	41.1	33.5	56.9	38.3	46.4	66.1	40.6	41.0	36.7	40.9	45.3	57.2	45.1	48.2
Excess	14.4	10.3	17.0	3.3	9.1	10.0	4.0	17.2	15.7	8.7	3.9	4.5	11.0	17.3	21.2	16.4
POP	65%	47%	67%	19%	68%	38%	25%	84%	70%	46%	21%	37%	45%	66%	76%	65%
Maximin solution																
Value	39.4	38.6	39.2	34.6	57.5	37.9	46.8	65.0	40.1	41.7	37.4	39.5	46.6	57.3	44.2	46.6
Excess	12.8	11.4	15.1	4.3	9.8	9.6	4.4	16.1	15.2	9.4	4.6	3.1	12.3	17.4	20.4	14.8
POP	58%	53%	59%	25%	72%	37%	27%	79%	68%	49%	25%	25%	50%	67%	73%	58%

Note: COM–Commission, BE–Belgium, D–Germany, DK–Denmark, EL–Greece, ES–Spain, F–France, FL–Finland, I–Italy, IL–Ireland, LUX–Luxembourg, NL–Netherlands, OS–Austria, P–Portugal, SV–Sweden, UK–United Kingdom.

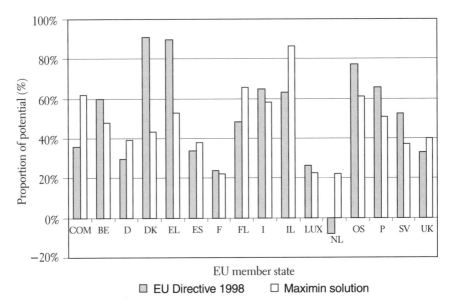

Figure 26.3. Nash and maximin solutions offer insight into feasible, efficient, and fair agreements. (For country abbreviations, see Figure 26.1.)

and the EU Directive. Do you find excess values or POP scores to be a more compelling benchmark for assessing outcomes?

The next item to consider is how the intervenor can benefit from the additional insight offered by mixed contracts. Naturally, one expects mixed contracts to enable the parties to collectively achieve better outcomes (although some parties are likely to be better off under the constraint of pure contracts). More important, if no agreement under pure contracts is jointly feasible, relaxing this constraint without changing the issues may allow the parties to reach agreement. An intervenor can use mixed contracts to identify issues with a high conflict of interest and interpolate compromise resolutions. In this case the mixed contract is comparable to a loosely worded phrase in a treaty which is deliberately left open to multiple interpretations, or a contingent agreement to be triggered by specific events which the parties subjectively assess to have different likelihoods of occurring. In short, mixed contracts increase the ability of the intervenor to explore a range of options for agreement.

Core Concepts

This chapter has introduced few new concepts. Rather, it has shown how the arsenal of techniques introduced throughout the preceding chapters can be used to handle big, messy negotiations.

Our first example was a fictional town zoning dispute. It demonstrated how one might handle issues surrounding complex parties—that is, a side (or sides) that breaks down into a disorderly mess of constituents with different interests and different skill levels. One problem is how to aggregate their interests enough to allow for meaningful negotiations. Another is how to get approval when there are far too many actors involved to get anything done with them all at the table.

Our second problem was a real one—the case of energy market liberalization in the European Union. Here each party is relatively organized and able to reach decisions—the problem is that there are fifteen clearly distinct, separate parties. In practice, structured information gathering and diplomatic dealmaking made it work. We showed how some FOTE template analysis might have made things easier, and perhaps more efficient.

The sixteen parties sitting in the European Union's Energy Council could not agree on how to liberalize Europe's natural gas markets. One after another, Council presidents failed to achieve agreement through facilitation during their six months in the chair. But Luxembourg's minister of economics, Robert Goebbels, achieved success during the Luxembourg presidency by adopting the role of a neutral joint analyst (NJA) in contrast to the facilitative roles in previous negotiations.

In a FOTE-like manner the Council members agreed on a template for negotiation and then Goebbels shuttled back and forth, helping each party to qualitatively evaluate the template. Armed with this confidential information, Goebbels estimated for each national delegation its issue-specific reservation levels. Then, very much in the spirit of the analysis in Chapter 22 on consensus, he led each country to share with him its qualitative within-and-between issue evaluations of the possible resolutions in the template and some information about its reservation levels. He then developed a matrix showing each member's preferences for resolutions and their intensities, not unlike the decision aids used in Chapter 22, and used this to make a series of contract proposals leading to an acceptable agreement.

One of us (DM) followed up on this work (Metcalfe 2001). He purified the template, interviewed negotiators from each delegation, quantified the members' evaluations, quantified their reservation values, used linear programming to find efficient and equitable contracts (using the Nash and maximizing the minimum POP criteria), and he compared these solutions with the contract ingeniously developed by Minister Goebbels.

27

Social Dilemmas

How much are you willing to sacrifice to help others? How should you trade off a tangible hurt to yourself for the benefit of society? These are the themes of this chapter. They hearken back to the social dilemma game introduced in Chapter 4, but now we are dealing with not two players but many players.

The essence of the social dilemma is that each player has a dominant strategy, but when each acts rationally by choosing this strategy, the entire group suffers. Each player has a choice: do what makes sense selfishly, or make a personal sacrifice for the good of the whole. If they all make a sacrifice, then each will do better than if they had all acted selfishly.

The Multiparty Social Dilemma Game

The game to be shortly described is an *n*-person generalization of the two-person social dilemma game (commonly known as the prisoner's dilemma game). Recall that in the two-person version each party could cooperate (C) or defect (D) where D dominates C, but the payoff from {D, D} is worse for each than the payoff from {C, C}: the pair of strategies {C, C} is Pareto optimal but not in equilibrium; the pair {D, D} is in equilibrium but not Pareto optimal. Individual rationality leads to inferior outcomes for the group; hence the dilemma. We now generalize to many players.

The Basic Game: No Preplay Communication, Private Choice

Imagine that you are one of many (N) identical players. A convenient number N to keep in mind is N = 100. You and each of the others have two choices: C for cooperate and D for defect. (When I, HR, play the game in a classroom setting and I want to let the drama unfold, I label the choices X and Y.) The rules of the Basic Game are:

- The game is to be played just once.
- No preplay communication is allowed among the players.
- Anonymity. The choices of the players are known to the instructor but not to one another. The instructor informs the players of the total number of players who choose C, but particular identities remain confidential.
- The payoffs (in dollars) are
    to a D-player: [Number of C's].
    to a C-Player: [Number of C's] minus .4N.
- All the above is common knowledge.

Further remarks about the rules: The payoff units can be cents, dimes, or dollars, and if the instructor wants the players to think big, the units can be in make-believe billions of dollars. To be specific, let the units be in dollars and let the players be told that actual money will be exchanged. How would you play?

Illustration. N = 100. Number of C's is 30. Each of the 70 D-players gets $30; and each of the 30 C-players loses $10.

If all played C, each would have gained $60. C's need at least 40 of them to make money. The parameter .4 that appears in the payoff to a C-player can be manipulated by the instructor to illustrate different pressures.

The C's generate capital for all to share. If a D were to shift to C, he (or she) would provide $1 more to each of the 100 players but at a sacrifice of $40 to himself (or herself).

Empirical results. For the basic game, about 15 percent of MBA student-subjects choose the noble strategy C. But maybe half of these are not really noble but confused.

Preplay Communication, Private Choice

Now let the players be able to communicate with each other before making their choice. A leader or preacher comes to the fore. In one class he actually jumped up on a table and made an appeal to the group: "We have to stick together; if all of us choose to cooperate, then we all shall gain; the D's are free riders and socially irresponsible." "Yea, yea," the group responded.

Now once again the players are asked to respond privately. As you might expect, uplifting speeches help, and the number of C's might zoom up to 50 percent. This means that the C's each end up with $10 but the free-riding D's get $50. The C's are unhappy and some are furious.

Preplay Communication, Public Choice

Now allow preplay, edifying speech making but this time let each player see who chooses what. Now the free riders are visible. Results differ from class to class, but a typical response may be 75 percent C's. In one class I ran, some character made an impassioned plea for everyone to choose C and then he gleefully turned around and voted D. Some in the class were outraged at this "despicable" behavior. "Aw c'mon," he responded, "it's only a game." There were people in the class that twenty years later still would not speak to him.

Repeated Play

Now let's have repeated play of the basic game. In some classes I've run, the proportion of C's tumble down, with repeated play, to 0 percent. Some initial cooperators get discouraged and reluctantly choose D to protect themselves, and this negative feedback discourages other C-players. But this doesn't always happen. Occasionally some D's are impressed with the initial proportion of C's and feel guilty about exploiting them and switch to C. This dynamic has in some cases resulted in increases in the proportion of C's over time.

I admit that I want to show my classes that over time the proportion of C's might increase, and to accomplish this I might reduce that .4 parameter (in the payoffs to C's) to .25 and at the same time allow for preplay discussion and publicly observed choice.

One way to get more intense involvement is to ask each player to make allocations of the payoffs to internal social programs (for example, education or welfare). The purpose of this variation is to stress that there may be compelling internal reasons to be less cooperative in the external game.

I have not experimented in the classroom with the introduction of an external helper—in this case a facilitator who can preach the nobility of the cooperative choice. Now back to the real world—or at least *toward* the real world—to examine examples of social dilemmas.

Commons Problems

The Mathematics of Sheep Grazing

Rancher Brown uses the Commons land for the grazing of his sheep. So do other sheep ranchers. The value of each sheep grazing on the Commons goes down as the number of sheep goes up. To be specific, if there are X sheep using the Commons, let the *value* (in units of hundreds of dollars) *per sheep*, VPS, using the Commons be

$$VPS(X) = 1 - .001X; \tag{27.1}$$

so if X were 300, say, the value of each sheep would be

$$1 - .001 \times 300 \text{ or } .7, \text{ or } \$70.$$

We see that as soon as X goes over 1,000, the value goes negative. The *total value*, *TV*, of all sheep to all owners is

$$TV(X) = X(1 - .001X). \tag{27.2}$$

So if $X = 300$, the total value of the herd would be $300 \times \$70$ or $\$21,000$.

A monopolist's analysis.   Observing that $TV(X)$ is quadratic in X and that

$$TV(0) = TV(1,000) = 0,$$

the maximum total value of the entire herd occurs at the midpoint of the two roots, 0 and 1,000, or at $X^* = 500$ for a total value of $\$25,000$. See Figure 27.1.

If Rancher Brown owned all of the sheep on the Commons he would want to increase his sheep until they totaled 500. Going above that would be counterproductive.

The competitive analysis.   If Rancher Grant is grazing K sheep on the Commons and now Rancher Brown is contemplating adding sheep of his own, how many sheep should Rancher Brown introduce to maximize his own return? A little analysis shows[1] that Brown's best choice against Grant's herd of K sheep is

$$X^*(K) = .5 \times (1000 - K). \tag{27.3}$$

---

1. If Rancher Grant is grazing K sheep and Brown adds X more, then Brown's return from these K sheep is

$$X \times [1 - .001(K + X)], \quad {}_\circ$$

and this is maximized at $X^*(K)$, the midpoint of the two roots, giving

$$X^*(K) = .5 \times [0 + (1000 - K)] = .5 \times (1000 - K).$$

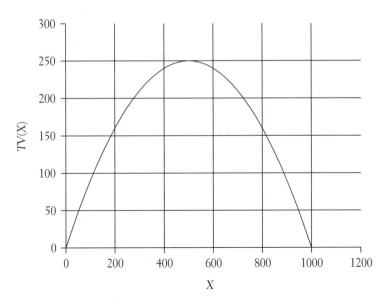

Figure 27.1. Total value of different numbers of sheep.

So, for example, if Grant were grazing 400 sheep on the Commons, Brown would want to add 300 of his own. We could say 300 is best against 400.

If Brown introduced these 300 sheep, then Grant, in response, would want to reduce his holdings from 400 to 350, and then Brown would want to go down to 325, and it can be shown that equilibrium is reached when each grazes 333.33 . . . sheep on the Commons. Neither would want to change if the other held fixed at 333.33 . . . But while {333.33, 333.33} is an equilibrium pair, it is not Pareto efficient. They maximize their joint holdings if together they graze 500 sheep. But if, for example, each grazed 250, for an optimum total of 500, there would be an incentive for each to cheat and add additional sheep of his own.[2]

Now with two ranchers it may be easy enough to enter into a binding agreement, but what happens when a third and then a fourth rancher wants to add his sheep to graze on the Commons?

Entry. The more ranchers that use the Commons, the harder it is to coordinate and the more incentive there is to cheat. The smaller the totality of sheep

2. It is not difficult to show with N ranchers that an equilibrium occurs when each rancher grazes $1000/(N + 1)$ sheep for a total of

$$1000N/(N + 1)$$

grazing sheep, far in excess of the 500 sheep the monopolist would want to graze.

grazing on the Commons, the more incentive there is for a new rancher to use the Commons. When a new rancher enters, he increases his own worth at the expense of the existing ranchers. The new entrant hurts larger holders the most, and this sets up a competition between the haves (that is, the rich ranchers) and the have-nots.

Population control.   Instead of looking at a patch of grazing land as the Commons, let's think big and view our planet as the Commons, replacing sheep by people. Each family must decide on the number of children it will add to the Commons. It may be in the family's best interest to have lots of children, even though this results in too many souls for our planet to sustain. At this writing our population has reached 6 billion people, with projections that it will rise to 8.5 billion in the next half century. If China had not instituted its policy of one child per family, the prospect would have been a projected rise to 10 billion rather than 8.5 billion seeking to share limited resources. Individual uncoordinated family rationality does not yield group efficiency.

But this example brings up an interesting question: how should we measure utility when the extra "grazers" added to the Commons are not merely affecting the payoffs of the others, but are enjoying payoffs as well? We don't count the satisfaction of the sheep themselves. But shouldn't we count that of the additional people? Would we prefer a planet with 1 billion people at, say, 80 percent happiness or one with 2 billion at 50 percent? We don't even know how to begin to answer that.

A variation of the Commons Problem is that of management of renewable resources, and the decimation of the whales is an illustrative case.

On Not Catching Whales

A simple abstraction.   Let's simplify by making believe once again. This time we posit the existence of only one kind of whale, the Blue Whale. It is coveted by several whaling nations, and it is in danger of extinction through over-hunting. Imagine, contrary to the facts, that we are in a deterministic world, and we know with great clarity the size of next year's whale population, given this year's and the size of the catch. If there were no catch, the population would peak out into an equilibrium value—at the point where there is just enough food for all the whales. If the population gets too small there is also a problem of regeneration, because of mating problems.

To give this discussion more mathematical specificity, imagine time, t, is in integral years, and at time t the whale population is w_t. Assume, without any whales being caught, that growth from year to year is governed by the equation

$$w_{t+1} = w_t + .000002 \; w_t \, (100,000 - w_t) - 1500.$$

The population is at an equilibrium (that is, $w_{t+1} = w_t$) when $w_t = 8,167$ or $w_t = 91,833$. In between 8,137 and 91,833 the population increases; below 8,133 the population decreases (it's hard to find a mate); and also above 91,833 the population also decreases (it's hard to find food). The growth from year to year is

$$w_{t+1} - w_t = .000002 \; w_t \, (100,000 - w_t) - 1500.$$

Now introduce reductions due to hunting. The population w_t at time t would remain fixed if the catch were equal to the natural growth at w_t. This sustainable catch would be

$$SC(w_t) = w_{t+1} - w_t$$

and $SC(w_t)$ is maximized when $w_t = 50,000$, with $SC(50,000) = 3,500$. So, in the abstraction, if the population rises to 50,000 then 3,500 whales could be caught forever more.

I (HR) have used the above abstraction to simulate in laboratory experiments the dynamics of catching whales, with student-subjects playing the roles of whaling nations. We start with a monopolist catching all whales; then go on to uncoordinated duopolists, then to many whaling nations with rogue entrants. The student-subjects are instructed to maximize the net present value of their catches, and they are given different discount rates so that some nations are more impatient than others. It's not a pretty story because typically the players, acting in their own interests, quickly decimate the population. This is not unlike what occurs in sheep grazing on the Commons.

Back to the real world. The charter of the International Convention for the Regulation of Whales (1946) set up the International Whaling Commission (IWC) with a mandate to (a) guard against overexploitation of whales and (b) avoid economic and nutritional distress and to make an orderly development of the whaling industry. The IWC had limited powers to make regulations but no power to enforce them. It served as the meeting ground for multiparty negotiations. See McHugh (1974).

As a contrast to the free-for-all, strategic dynamics of the whale industry, the Fur Seal Treaty (1911, revised in 1957) specified that furs are to be produced under single management and the proceeds split among the signatory nations (United States, USSR, Japan, United Kingdom, and Canada).

A variant of the Commons Problem is the class of pollution problems that have a similar theme: individual rationality begets group irrationality.

The Case of Global Warming

The problem of global warming was identified in the early 1970s as one of the world's most serious environmental problems. Two preliminary observations. First, the accumulation of CO_2 and other greenhouse gases threatens to change the climate of the globe. From a global perspective gases mix in the upper atmosphere so that it matters little to the United States whether a unit amount of the offending gases is discharged in the United States or elsewhere (for example, China). Second, there are many uncertainties, and their resolutions will unfold slowly over time.

Scientific evidence has been mounting that the problem is real, but still countries like the United States cannot garner the political will to reduce their production of carbon dioxide and other greenhouse gases. We do not do X—like controlling the mix of cars on our highways, using coal more effectively, switching to natural gas more quickly, reducing our profligate energy consumption, and so on—because the economic and political costs to us in doing X is more than the certainty equivalent of the very uncertain benefits. But this cost-benefit calculus does not factor in (1) the benefits of our doing X to other countries (we have some obligation to help others); (2) the fact that our doing X might influence another country in doing Y, which will reduce the buildup of the gases and thereby benefit us; (3) the fact that our decision to do X might affect the atmosphere of other collaborations we have with other countries (and this is worth something in the calculus); and (4) the importance of allowing us to preach: we're doing our share so why don't you do likewise? The United States cannot preach to others when it's so delinquent itself.

Gradually over time the developing countries (for example, China) will be producing an alarming amount of greenhouse gases as they develop their copious coal deposits. Early decisions made by China (about the use of natural gas and the technology for cleaning coal) will affect the longer-term buildup of greenhouse gases, and it is in our interest to influence those decisions through collaborative negotiations. Our ability to do so may be related to our doing X ourselves.

The countries of the world cannot get their collective act together—there are just too many free riders. But we would argue that the world should not despair. We must learn not to let the existence of free riders inhibit a Noble Coalition of those countries that want to act responsibly from taking action of its own. The Noble Coalition should continually put pressure on the free riders to join or to gain partial membership in its club of do-gooders. Just as in the case of the multiparty, repeated social dilemma games, it's important for the Cooperators not to dissolve simply because of the existence of the exploitative Noncooperators.

An aside. Besides the commons problem (overpopulation, pollution, degradation of the environment), there is also a class of similar problems in business competition: cut-throat pricing, competitive advertising, and so on. The generic advice for those competing firms is to collude, to plan together, to collaborate. In these cases, of course, what may be bad for the competitive firm may be good for the consumer.

The LULU Problem

LULUs, NIMBY, and BYBYTIM

A state has a problem: what should it do with the hazardous wastes being generated by its highly profitable and prestigious industry? Presently this waste is being transported to neighboring states at huge costs, but this is getting tougher and tougher to do. The governor has appointed a prestigious scientific committee to suggest potential sites for this locally undesirable land use (LULU) facility. The potential sites fall in the townships of Aspen, Baileyville, Camille, Donneybrook, and Eagleston. A LULU in any one of these sites would solve the problem for the entire state. But as the committee expects, each locality will holler "Not in my backyard" (NIMBY) or "Better your backyard than in mine" (BYBYTIM).

One suggestion: Resolve the problem through the use of an auction mechanism.

The descending auction. The state has decided to use its general tax revenues to hold an open, descending, outcry auction in which each locality is invited to bid for the amount it would want as compensation to locate the facility in its region. Suppose Baileyville opens the bidding at $25 million. Donneybrook counters with a bid of $15 million, and Aspen bids $13 million. Baileyville returns with a bid of $12 million, and the bidding stops. Sold to Baileyville for a compensation of $12 million. Successive bids might be required to be at most one week apart.

Instead of the towns declaring amounts they would accept with successive decreasing declarations, the process could be reversed with an auctioneer: "Do I have any takers at $5 million? No? Well, let's try $7 million? . . . How about $11 million? Still no. How about $12 million?" Now let's imagine that Baileyville would accept $12 million, but it holds off until the auctioneer's declaration reaches $14 million.

Creative compensation. Easier said than done. Let's look at Baileyville. Baileyville is not monolithic. How should it decide its reservation value—that minimum value it would be willing to accept to permit the LULU to be built?

Baileyville formed a committee to determine its reservation value and bidding strategy. The committee reviewed the preliminary specs for the LULU and ascertained that 100 of the township's 5,000 homes would be directly affected. After reviewing the needs of the town, the committee members decided that they would need at least $12 million to accept the LULU. They would allocate $4 million to compensate the minority of irate citizens who would feel the impact directly. The $4 million would be given in the form of $40,000 property tax relief to one hundred families. Ninety of these one hundred are willing to go along, but the other ten are holding out for more. One family says no amount of bribery money will change their minds. They are offered a $25,000 grant for relocation. The property value of their home, which they purchased two years ago for $110,000, would actually go up because of the $40,000 proposed tax relief. A new owner would be free of property taxes for years to come. Baileyville's remaining $8 million would be used for sewers, roads, and lighting ($3 million); for increases in the salaries of municipal employees ($2 million); for new recreational facilities ($2 million); and to buy land for a cemetery ($1 million).

Through the bidding process Baileyville was awarded the LULU for $14 million, $2 million above its reservation value. The allocation of that extra $2 million went mostly for renovations of the township's much neglected schools.

Asymmetries in the Five Sites

Once again let's suppose that the governor's committee identifies these five possible sites, but now suppose some are better than others; some are better located; some would require more initial work. These differences can be accommodated in the bidding process by giving differential bonus points to the five competing towns.

Towns	Bonus points
Aspen	0
Baileyville	2
Camille	1
Donneybrook	5
Eagleston	2

Suppose, for example, Donneybrook makes an offer of $16 million. Since Donneybrook has 3 bonus points more than Baileyville, for Baileyville to be competitive, it would have to announce an acceptance value of $13 million. Camille would have to announce at $12 million to be tied with Donneybrook.

These bonus points would be common knowledge before the bidding commences.

But there is a problem that has to do with environmental justice: not all LULUs should be located in poorer, more disadvantaged places, like Baileyville. How about locating the next LULU in one of the richer towns, like Camille? Baileyville already has more than its share of LULUs and Camille has none. We believe that Baileyville is in a strong position to demand compensation—and we mean lots of compensation from the richer Camilles—to address the myriad problems it faces. It will not be unconscionable if the next LULU ends up in Baileyville, but it will be unconscionable if Baileyville is not highly compensated—and it would not be unreasonable for the amount of compensation to be outlandishly high at first, to establish a principle. One problem is the source of these compensatory funds. They should come from the broad base of all those taxpayers who will profit from the LULU. The compensation might be so high that another poor town like Eagelston would be jealous, and Eagleston might enter into competition with Baileyville for the location of the next LULU. So the price of compensation might come down and be more realistic.

The state of Nevada should be open minded about putting nuclear wastes at the Yucca Mountain site, provided it is amply compensated. The state is worried about the health of its citizenry. In that case, let the compensatory package explicitly deal with that concern. More and better hospitals, clinics, and medical schools. How about safety? Be sure to add into the compensatory package more police and fire protection, better lighting, better housing for the poor, and on and on. Where is this money going to come from? States like Massachusetts will profit from a resolution to the problem of finding a location for spent nuclear fuel, so they should pay their fair share in compensating the losers. The managerial task is to establish a mechanism to collect contributions from the diffuse group of winners to make the deal more attractive to the potential losers. If the compensatory package mounts high enough, who knows? The state of New Mexico may cry, "Why not us?"

Personal Ethics and the Social Dilemma

The Misplaced Mattress

Imagine this scenario: One miserably hot afternoon, you are driving from Boston to Cape Cod. Traffic is crawling along a stretch of two-lane, one-way highway that usually presents no problem. Finally after an interminable time you discover the trouble: a mattress is in the middle of the road; cars have to squeeze to one side to pass it. After you clear the problem, should you stop your car and move the mattress? There will be at least some inconvenience to you and possibly a bit of danger involved if you stop. "Why should I be the fall guy? Let someone else be the good guy. Anyway, I'm already late, because of the delay, for my appointment."

This scenario, concocted by Thomas Schelling, depicts another aspect of the social trap problem. What would you do? Would you stop and move the mattress?

It would be interesting to see how various cultural groups would react to this problem. A Mormon student we know claimed that if the mattress were on a highway in Utah it would be moved very quickly. It is part of the Mormon culture to try to do good for others, to take responsibility for the common cause. Would Japanese behave differently than Chinese or East Indians or Israelis or Kenyans or Swedes? How much are we individually willing to sacrifice for the good of others when we get no immediate tangible reward other than the self-satisfaction of doing good?

Personal Charities

Some give a lot to charity; others, in similar financial circumstances, give little. Foundations and personal philanthropy account for much of the support of medical, educational, and art institutions as well as for support of the needy. As a society we should be proud of the amount of individual charitable giving. But the process is flawed, because there are many free riders who prefer to keep their money and let others support these causes, and because some causes do not adequately attract the support of the givers. One alternative, of course, is to depend less on charitable giving and more on governmental interventions for welfare and public institutions. The government would drain away some of the money from the givers and the nongivers (the free riders) and allocate it as it sees fit. The argument against the government's assuming the supporting role is that it involves too much bureaucracy and the government is not close enough to the needs of the people; better to administer welfare through the churches than through the government. It's not one or the other, of course, but mixtures of the two, and the U.S. government tilts more to the voluntary side than many of the European countries. Those governments tax more and spend more on welfare but perhaps not as much on medical research. The U.S. government provides generous tax exemptions for charitable giving, and this influences behavior.

Some Personal Reflections

Occasionally I find myself in the position exemplified by the *n*-person social trap: I can act to improve my immediate well-being a good deal, but only at the expense of hurting a lot of others a little bit and where the totality of damage to others is greater than my personal gains. For example, I drive my car and shun public transportation. As an analyst, I think about these issues. I'm not a purist, and my tradeoffs depend on the amount I'm being called upon to sacrifice and the totality of harm to others and how this collective harm is distributed.

I believe that most utilitarian calculations in situational ethics are too narrowly conceived. In a loose sense, all of us are engaged in a grandiose, many-person, social dilemma game, where each of us has to decide how much we should sacrifice to benefit others. The vast majority of us would like to participate in a more cooperative society, and all of us may have to make some sacrifice in the short run for that long-run goal. We have to calculate, at least informally, the dynamic linkages between our actions now and the later actions of others. If we are more ethical, it makes it easier for others to be more ethical. And as was the case in the multiperson social dilemma game, we should not become excessively distraught if there are a few cynical souls who will tangibly profit from our combined beneficent acts.

If you act to help others, you may hurt yourself in the short run; but if your act is visible to others, you may profit from it in the long run because of the cyclical reciprocities it may engender. In that sense, your noble-appearing action may be in your long-run selfish interest. Yet much traditional moral teaching holds that doing good should not be advertised.

I think we should not demean visible acts of kindness, even though in part they may be self-serving, because those actions may make it easier for others to act similarly, and the dynamics will reinforce behavior that is in the common interest. An action that represents a moderate sacrifice in the short run may represent only a very modest sacrifice in the long run, when dynamic linkages are properly calculated. Many people, including myself, are also willing to make small (long-run) sacrifices for the good of others, all things considered. The visibility of beneficent acts thus plays a dual role: it reduces the tangible penalties to the actor, and it spurs others to act similarly; these two facets then interact cyclically.

A student of mine who heard this sermonizing said, "All this sounds a bit cynical to me. I don't want to teach my children to make personal sacrifices with lots of visibility to entice others to do good. Goodness should be its own reward." And when I responded, "But why not exploit the contagion factor of doing good?" he quipped, "So when the box goes around in the movie theatre for muscular dystrophy, you want me to say in a loud voice to my date: 'Should I put in five or ten dollars, sweetie?'"

I should say that I also believe that empathizing with others should be reflected in your own utility calculations: a sacrifice in long-range tangible effects to yourself, if it is compensated by ample gains for others, could be tallied as a positive contribution to your cognitive utilitarian calculations. I guess that's just an analytical way of saying that doing good is its own reward. But doing good *and* getting others to do some seems even better.

Core Concepts

How much are you willing to sacrifice to help others? How should you trade off a tangible hurt to yourself for the benefit of society? This chapter has returned to the social dilemma game introduced in Chapter 4, but this time the focus was on not two parties but many.

We have analyzed a simple, highly stylized social dilemma game with many players from a behavioral game theory perspective. Each of N players must choose either to cooperate (C) or to defect (D); the C's create wealth (at some expense to themselves) and the D's free-ride. Individual rationality leads to social irrationality, just as in the two-party version. Variations were considered: whether preplay communication is permitted (if so, some players could act as leaders and preach cooperation); whether the players know who is choosing what before or after choices are made; whether the game is repeated (allowing the possibility of temporal collusion).

Several abstract examples, with a bit of real-world context, were also examined: the Commons problem (overgrazing of sheep on a common plot, overproduction of people to share planet earth); the (mis)management of renewable resources (the overharvesting of whales, and depletion of forests); the pollution problem (as in the accumulation of greenhouse gases and their effect on climate); the location of locally undesirable land uses (LULUs) and the question of environmental justice when LULUs are concentrated in poorer neighborhoods; the decision to sacrifice oneself for the benefit of many (as in the mattress problem or the reporting of a crime—will a hero please come forth?); and the support of charities (how much are you willing to give for the good of society?).

References

Allais, Maurice. 1953. "Le comportement de l'homme rationnel devant le risque: Critique des postulats et axioms de l'école Américaine." *Econometrica* 21: 503–546.

Arrow, Kenneth. 1951. *Social Choice and Individual Values.* Cowles Commission Monograph 12. New York: John Wiley.

Arrow, Kenneth, Robert H. Mnookin, Lee Ross, Amos Tversky, and Robert Wilson, eds. 1995. *Barriers to Conflict Resolution.* New York: Norton.

Axelrod, Robert. 1967. "Conflict of Interest: An Axiomatic Approach." *Journal of Conflict Resolution* 11 (January): 87–99.

—— 1970. *Conflict of Interest: A Theory of Divergent Goals with Applications to Politics.* Chicago: Markham.

—— 1997. *The Complexity of Cooperation.* Princeton: Princeton University Press.

Baird, Bruce F. 1989. *Managerial Decisions under Uncertainty.* New York: Wiley.

Barclay, Scott, and Cameron R. Peterson. 1976. "Multi-Attribute Utility Models for Negotiations." Technical report 76-1 (May). McLean, Va.: Decisions and Designs.

Bartos, Otomar J. 1974. *Process and Outcome of Negotiations.* New York: Columbia University Press.

—— 1977. "Simple Model of Negotiation: A Sociological Point of View." *Journal of Conflict Resolution* 21, no. 4: 565–579.

Bazerman, Max H. 2002. *Judgment in Managerial Decision Making.* 5th ed. New York: Wiley.

Bazerman, Max H., and Margaret A. Neale. 1992. *Negotiating Rationally.* New York: Free Press.

Bell, David E., and Howard Raiffa. 1980. "Marginal Value and Intrinsic Risk Aversion." In Bell, Raiffa, and Tversky (1988).

Bell, David E., Howard Raiffa, and Amos Tversky, eds. 1988. *Decision Making: Descriptive, Normative, and Prescriptive Interactions.* Cambridge: Cambridge University Press.

Bok, Sissela. 1978. *Lying: Moral Choice in Public and Private Life.* New York: Vintage.

Bove, Alexander A., Jr. 1979. Article in *Boston Globe*, July 16.

Brams, Steven J., and Alan D. Taylor. 1996. *Fair Division.* Cambridge: Cambridge University Press.

—— 1999. *The Win/Win Solution.* New York: Norton.

Brandenberger, Adam M., and Barry J. Nalebuff. 1996. *Co-opetition.* Cambridge, Mass.: Harvard Business School Press.

Burrows, James C. 1979. "The Net Value of Manganese Nodules to U.S. Interests, with Special Reference to Market Effects and National Security." In *Deepsea Mining.* Cambridge, Mass.: MIT Press.

Callières, François de. 1716. *On the Manner of Negotiating with Princes.* Translated by A. F. Whyte. Boston: Houghton Mifflin, 1919; originally published Paris: Michel Brunet.

Center for Strategic and International Studies. 1978. *Where We Agree.* First Report of the U.S. National Coal Policy Project. Washington, D.C.: Georgetown University.

Chatterjee, Kalyan. 1978. "A One-Stage Distributive Bargaining Game." Working paper 78-13 (May). Graduate School of Business Administration, Harvard University, Cambridge, Mass.

—— 1979. "Interactive Decision Problems with Differential Information." Ph.D. dissertation, Harvard University.

Chatterjee, Kalyan, and William Samuelson. 1981. "Simple Economics of Bargaining." Distribution paper. Boston University.

Chelius, James R., and James B. Dworkin. 1980. "An Economic Analysis of Final-Offer Arbitration." *Journal of Conflict Resolution* 24, no. 2 (June): 293–310.

Contini, B. 1967. "The Value of Time in Bargaining Negotiations: Part I, A Dynamic Model of Bargaining." Working paper 207. Center for Research in Management Science, University of California, Berkeley.

Cormick, Gerald W., and Jane McCarthy. 1974. *Environmental Mediation: A First Dispute.* Seattle: Office of Environmental Mediation, University of Washington.

Cormick, Gerald W., and Leota Patton. 1977. *Environmental Mediation: Defining the Process through Experience.* Seattle: Office of Environmental Mediation, University of Washington.

Corsi, Jerome. 1981. "Terrorism as a Desperate Game." *Journal of Conflict Resolution* 25, no. l: 47–85.

Cross, J. G. 1965. "A Theory of the Bargaining Process." *American Economic Review* 55: 66–94.

—— 1966. "A Theory of the Bargaining Process: Reply." *American Economic Review* 56: 530–533.

—— 1969. *The Economics of Bargaining.* New York: Basic Books.

Dahl, Robert. 1956. *A Preface to Democratic Theory.* Chicago: University of Chicago Press.

Deutsch, Morton. 1977. *The Resolution of Conflict: Constructive and Destructive Processes.* New Haven, Conn.: Yale University Press.

Diehl, M., and W. Stroebe. 1987. "Productivity Loss in Brainstorming Groups: Towards the Solution of a Riddle." *Journal of Personality and Social Psychology* 61: 392–403.

Drucker, Peter F. 1981. "What Is Business Ethics?" *The Public Interest* 63 (Spring): 18–36.

Druckman, Daniel. 1977. *Negotiations: Social-Psychological Perspectives.* Beverly Hills, Calif.: Sage Publications.

Duker, Robert P. 1978. "The Panama Canal Treaties: An Honorable Solution?" National War College, Washington, D.C.

Dunlop, John, and James J. Healy. 1953. *Collective Bargaining: Principles and Cases.* Homewood, Ill.: Richard D. Irwin.

Dyer, J. S., and R. F. Miles, Jr. 1976. "An Actual Application of Collective Choice Theory to the Selection of Trajectories for the Mariner Jupiter/Saturn 1977 Project." *Operations Research* 24: 220–224.

Edwards, Harry T., and James J. White. 1977. *The Lawyer as a Negotiator.* St. Paul, Minn.: West Publishing Co.

Ellsberg, David. 1961. "Risk, Ambiguity, and the Savage Axioms." *Quarterly Journal of Economics* 75: 644–661.

Environmental Mediation: An Effective Alternative? 1978. Report of a conference held in Reston, Va., January 11–13. Palo Alto, Calif.: RESOLVE, Center for Environmental Conflict Resolution.

Farber, H. S. 1978. "Bargaining Theory, Wage Outcomes, and the Occurrence of Strikes." *American Economic Review* 68: 262–271.

Feuille, Peter. 1975. *Final-Offer Arbitration: Concepts, Developments, Techniques.* Chicago: International Personnel Management Association.

Fisher, Roger. 1978. *International Mediation: A Working Guide.* New York: International Peace Academy.

Fisher, Roger, Elizabeth Kopelman, and Andrea Kupfer Schneider. 1994. *Beyond Machiavelli.* Cambridge, Mass.: Harvard University Press.

Fisher, Roger, and Alan Sharp, with John Richardson. 1998. *Getting It DONE: How to Lead When You Are Not in Charge.* New York: Harper Collins.

Fisher, Roger, and William Ury. 1979. "Principled Negotiation: A Working Guide." Harvard Law School, Cambridge, Mass.

—— 1981. *Getting to Yes: Negotiating Agreement without Giving In.* Boston: Houghton Mifflin.

Fisher, Roger, William Ury, and Bruce Patton. 1991. *Getting to Yes.* 2nd ed. New York: Penguin.

Fried, Charles. 1978. *Right and Wrong.* Cambridge, Mass.: Harvard University Press.

Friedman, James W. 1977. *Oligopoly and the Theory of Games.* Amsterdam: North-Holland.

Goffman, Erving. 1972. *Strategic Interaction.* New York: Ballantine.

Goodman, L. A., and Harry Markowitz. 1951. *Social Welfare Functions Based on Rankings.* Cowles Commission Discussion Paper, Economics, no. 2017.

Groves, Theodore, and John Ledyard. 1977. "Optimal Allocation of Public Goods: A Solution to the Free Rider Problem." *Econometrica* 45: 783–809.

Gulliver, P. H. 1979. *Disputes and Negotiations.* New York: Academic Press.

Hackman, J. R., and R. E. Walton. 1986. "Leading Groups in Organizations." In *Designing Effective Work Groups,* edited by P. Goodman. San Francisco: Jossey-Bass.

Hammond, John, Ralph Keeney, and Howard Raiffa. 1999. *Smart Choices.* Cambridge, Mass.: Harvard Business School Press.

Harsanyi, John C. 1955. "Cardinal Welfare, Individualistic Ethics, and Interpersonal Comparisons of Utility." *Journal of Political Economy* 63: 309–321.

—— 1956. "Approaches to the Bargaining Problem before and after the Theory of Games: A Critical Discussion of Zeuthen's, Hicks', and Nash's Theories." *Econometrica* 24: 144–157.

—— 1965. "Bargaining and Conflict Situations in the Light of a New Approach to Game Theory." *American Economic Review* 55: 447–457.

—— 1977. *Rational Behavior and Bargaining Equilibrium in Games and Social Situations.* Cambridge: Cambridge University Press.

Howard, Ronald A., and James E. Matheson, eds. 1983. *The Principles and Applications of Decision Analysis.* 2 vols. Strategic Decisions Group, Palo Alto, Calif.

Jenkins, B. M. 1974. "Terrorism and Kidnapping." Paper series P-5255. RAND Corporation, Santa Monica, Calif.

Kahneman, D., and A. Tversky. 1979. "Prospect Theory: An Analysis of Decision under Risk." *Econometrica* 47: 263–290.

Kalai, E. 1977. "Proportional Solutions to Bargaining Situations: Interpersonal Utility Comparisons." *Econometrica* 45: 1623–30.

Kalai, E., and M. Smorodinsky. 1975. "Other Solutions to Nash's Bargaining Problem." *Econometrica* 43: 510–518.

Karni, E., and A. Schwartz. 1977. "Search Theory: The Case of Search with Uncertain Recall." *Journal of Economic Theory* 16: 38–52.

—— 1978. "Two Theorems on Optimal Stopping with Backward Solicitation." *Journal of Applied Probability* 14: 869–875.

Katz, Ronald. 1979. "Financial Arrangements for Seabed Mining Companies: An NIEO Case Study." *Journal of World Trade Law* 13: 218.

Keeney, Ralph L. 1992. *Value-Focused Thinking.* Cambridge, Mass.: Harvard University Press.

Keeney, Ralph, and Howard Raiffa. 1976. *Decisions with Multiple Objectives: Preferences and Value Tradeoffs.* New York: John Wiley.

Kochan, Thomas A., and Todd Jick. 1978. "The Public Sector Mediation Process: A Theory and Empirical Examination." *Journal of Conflict Resolution* 22, no. 2 (June): 209–238.

Kolbert, Elizabeth. 1999. In "Talk of the Town." *New Yorker.* April.

Lantané, B., K. Williams, and S. Harkins. 1979. "Many Hands Make Light the Work: The Causes and Consequences of Social Loafing." *Journal of Personality and Social Psychology* 37: 822–832.

Lax, David, and James K. Sebenius. 1981. "Insecure Contracts and Resource Development." *Public Policy* 29, no. 4: 417–436.

—— 1986. *The Manager as Negotiator.* Free Press.

Livne, Zvi. 1979. "The Role of Time in Negotiations." Ph.D. dissertation. Massachusetts Institute of Technology, Cambridge, Mass.

Lorange, Peter. 1973. "Anatomy of a Complex Merger: Case Study and Analysis." *Journal of Business and Finance* 5.

Luce, R. Duncan, and Howard Raiffa. 1957. *Games and Decisions.* New York: John Wiley. Reprinted by Dover.

Mayer, Frederick W. 1988. "Bargains within Bargains: Domestic Politics and International Negotiation." Ph.D. dissertation. Kennedy School of Government, Harvard University.

McCullough, David. 1977. *The Path between the Seas*. New York: Simon and Schuster.

McHugh, J. L. 1974. "The Role and History of the International Whaling Commission." Chapter 13 in *The Whale Problem*, edited by William E. Schevill. Cambridge, Mass.: Harvard University Press.

Metcalfe, David. 2001. "Multiparty Negotiation Analysis." Ph.D. dissertation. Cambridge University.

Mills, Daniel Quinn. 1975. *Labor-Management Relations*. New York: McGraw-Hill Book Company.

Mnookin, Robert H., and Lewis Kornhauser. 1979. "Bargaining in the Shadow of the Law: The Case of Divorce." *Yale Law Journal* 88: 950–997.

Mnookin, Robert H., Scott R. Peppet, and Andrew S. Tulumello. 2000. *Beyond Winning: Negotiating to Create Value in Deals and Disputes*. Cambridge, Mass.: Harvard University Press.

Myerson, R. M. 1977. "Two-Person Bargaining Problems and Comparable Utility." *Econometrica* 45: 1631–37.

—— 1979. "Incentive Compatability and the Bargaining *Problem*." *Econometrica* 47: 61–74.

Nash, John F. 1950. "The Bargaining Problem." *Econometrica* 18: 155–162.

—— 1953. "Two-Person Cooperative Games." *Econometrica* 21: 129–140.

Nierenberg, Gerald. 1973. *Fundamentals of Negotiating*. New York: Hawthorne.

Nydegger, R., and G. Owen. 1975, "Two-Person Bargaining: An Experimental Test of the Nash Axioms." *International Journal of Game Theory* 3: 239–249.

O'Hare, Michael. 1977. " 'Not on My Block You Don't': Facility Siting and the Strategic Importance of Compensation." *Public Policy* 25: 407–458.

O'Hare, Michael, and Debra Sanderson. 1977. "Fair Compensation and the Boomtown Problem." *Urban Law Annual* 14: 101–133.

Osborn, A. F. 1957. *Applied Imagination*. New York: Scribner.

Owen, G. 1968. *Game Theory*. Philadelphia: W. B. Saunders.

Pratt, J., and R. Zeckhauser. 1979. "Expected Externality Payments: Incentives for Efficient Decentralization." Graduate School of Business Administration, Harvard University, Cambridge, Mass.

Prosnitz, Eric W. 1981. "Using Compensation for Siting Hazardous Waste Management Facilities and the Massachusetts Act." Cambridge, Mass.

Raiffa, Howard. 1951 and 1953. "Arbitration Schemes for Generalized Two-Person Games." Engineering Research Institute, University of Michigan report nos. M720–1, R30. Published in Annals of Mathematics Studies, Princeton University Press.

—— 1968. *Decision Analysis: Introductory Lectures on Choices under Uncertainty*. Reading, Mass.: Addison-Wesley.

—— 1973. "Analysis for Decision Making" (audiographics). Encyclopedia Britannica Educational Corporation.

—— 1977. "Interactive Decision Analysis" (audiographics). Unpublished.

—— 1981. "Decision Making in the State-Owned Enterprise." In *State-Owned Enterprise in the Western Economies*, edited by Raymond Vernon and Yair Aharoni. London: Croom Helm.

—— 1982. *The Art and Science of Negotiation*. Cambridge, Mass.: Harvard University Press.

—— 1996. *Lectures on Negotiation Analysis.* Cambridge, Mass.: Program on Negotiation at Harvard Law School.

Ramberg, Bennet. 1978. "Tactical Advantages of Opening Positioning Strategies: Lessons from the Seabed Arms Control Talks, 1967–1970." In *The Negotiation Process: Theories and Applications.* Beverly Hills, Calif.: Sage Publications.

Rapoport, Anatol. 1966. *Two-Person Game Theory: The Essential Ideas.* Ann Arbor: University of Michigan Press.

—— 1970. *N-Person Game Theory: Concepts and Applications.* Ann Arbor: University of Michigan Press.

Rosenfeld, Stephen S. 1975. "The Panama Negotiations—A Close Run Thing." *Foreign Affairs* 54, no. 1 (October): 5–6.

Ross, H. Laurence. 1970. *Settled out of Court: The Social Process of Insurance Claims Adjustment.* Chicago: Aldine.

Roth, Alvin E. 1977a. "Individual Rationality and Nash's Solution to the Bargaining Problem." *Mathematics of Operations Research* 2: 64–65.

—— 1977b. "Independence of Irrelevant Alternatives, and Solutions to Nash's Bargaining Problem." *Journal of Economic Theory* 16: 247–251.

—— 1978. "The Nash Solution and the Utility of Bargaining." *Econometrica* 46: 587–594.

—— 1979. *Axiomatic Models of Bargaining.* Berlin: Springer-Verlag.

Roth, Alvin E., and Marilda O. Sotomayor. 1990. *Two-sided Matching: A Study in Game-theoretic Modeling and Analysis.* Cambridge: Cambridge University Press. Paperback 1992.

Rubin, Jeffrey, and Bert Brown. 1975. *The Social Psychology of Bargaining and Negotiation.* New York: Academic Press.

Savage, Leonard J. 1954. *The Foundations of Statistics.* New York: John Wiley.

Sawyer, Jack, and Harold Guetzkow. 1965. "Bargaining and Negotiations in International Relations." In *International Behavior: A Social-Psychological Analysis.* New York: Holt, Rinehart and Winston.

Schelling, Thomas C. 1956. "An Essay on Bargaining." *American Economic Review* 46: 281–306.

—— 1960. *The Strategy of Conflict.* Cambridge, Mass.: Harvard University Press.

Sebenius, James K. 1980. "Anatomy of Agreement." Ph.D. dissertation. Harvard University, Cambridge, Mass.

Shaw, Marvin E. 1981. *Group Dynamics: The Psychology of Small Group Behavior.* 3d ed. New York: McGraw-Hill.

Shell, Richard G. 1999. *Bargaining for Advantage: Negotiation Strategies for Reasonable People.* New York: Viking.

Shubik, Martin. 1971. "The Dollar Auction Game: A Paradox in Non-Cooperative Behavior and Escalation." *Journal of Conflict Resolution* 15 (March): 109–111.

Simkin, William E. 1971. *Mediation and the Dynamics of Collective Bargaining.* Washington, D.C.: Bureau of National Affairs.

Slichter, Sumner H., James J. Healy, and E. Robert Livernash. 1975. *The Impact of Collective Bargaining on Management.* Washington, D.C.: Brookings Institution.

Smith, D., and L. Wells. 1975. *Negotiating Third World Mineral Agreements.* Cambridge, Mass.: Ballinger.

Stevens, C. M. 1963. *Strategy and Collective Bargaining Negotiation*. New York: McGraw-Hill.

Stone, Douglas, Bruce Patton, and Sheila Heen. 1999. *Difficult Conversations: How to Discuss What Matters Most*. New York: Penguin.

Sullivan, Timothy J. 1980. "Negotiation-Based Review Process for Facility Siting." Ph.D. dissertation. Harvard University, Cambridge, Mass.

Susskind, Lawrence, Sarah McKearnan, and Jennifer Thomas-Larmer, eds. 1999. *The Consensus Building Handbook*. Thousand Oaks, Calif.: Sage Publications.

Susskind, Lawrence, James R. Richardson, and Kathryn J. Hildebrand. 1978. *Resolving Environmental Disputes: Approaches to Intervention, Negotiation, and Conflict Resolution*. Cambridge, Mass.: Environmental Impact Assessment Project, Massachusetts Institute of Technology.

Thompson, Leigh. 1998. *The Mind and Heart of the Negotiator*. Upper Saddle River, N.J.: Prentice Hall.

Tversky, Amos, and Daniel Kahneman. 1974. "Judgement under Uncertainty: Heuristics and Biases." *Science* 185: 1124–31.

Ulvila, Jacob W. 1979. "Decisions with Multiple Objectives." Ph.D. dissertation, Harvard University.

Ulvila, Jacob W., and Warner M. Snider. 1980. "Negotiation of Tanker Standards: An Application of Multiattribute Value Theory." *Operations Research* 28 (January–February): 81–95.

U.S. Department of State. 1974. "U.S. and Panama Agree on Principles for Canal Negotiations." *Department of State Bulletin* 70 (February 25): 184–185.

U.S. House of Representatives. 1975. *Congressional Record*, 98th Congress, 2nd sess., October 7, pp. H9713–25, concerning the November 1974 Panama Canal negotiations.

Ury, William. 1991. *Getting Past No*. New York: Bantam Books.

Valacich, J. S., Dennis, A. R., and Connelly, T. 1994. "Idea Generation in Computer-Based Groups: A New Ending to an Old Story." *Organizational Behavior and Human Decision Processes*, 57.

Vickrey, William. 1961. "Counter Speculation, Auctions, and Competitive Sealed Tenders." *Journal of Finance* 16: 8–37.

Von Neumann, John, and Oskar Morgenstern. 1944. *Theory of Games and Economic Behavior*. New York: John Wiley.

Wall, James H., Jr. 1981. "Mediation." *Journal of Conflict Resolution* 25, no. 1: 157–180.

Walton, Richard E., and Robert B. McKersie. 1965. *A Behavioral Theory of Labor Negotiations*. New York: McGraw-Hill.

Watkins, Michael, and Susan Rosegrant. 2001. *Breakthrough International Negotiations*. San Francisco: Jossey-Bass.

Wessel, Milton R. 1976. *The Rule of Reason: A New Approach to Corporate Litigation*. Reading, Mass.: Addison-Wesley.

Wilson, Robert. 1968. "On the Theory of Syndicates." *Econometrica* 36: 119–132.

Young, Oran R. 1967. *The Intermediaries: Third Parties in International Crises*. Princeton: Princeton University Press.

———, ed. 1975. *Bargaining: Formal Theories of Negotiation*. Urbana, Ill.: University of Illinois Press.

Young, Peyton H., 1994. *Equity: In Theory and in Practice*. Princeton University Press.

Zajonc, Robert B. 1965. "Social Facilitation." *Science* 149: 269–274.

Zartman, I. William. 1975. "Negotiations: Theory and Reality." *Journal of International Affairs* 29 (Spring): 69–77.

—— 1976. *The Fifty Percent Solution*. Garden City, N.Y.: Doubleday.

——, ed. 1978. *The Negotiation Process: Theories and Applications*. Beverly Hills, Calif.: Sage Publications.

Zeckhauser, Richard J., Ralph L. Keeney, and James K. Sebenius, eds. 1996. *Wise Choices: Decisions, Games, and Negotiations*. Cambridge, Mass.: Harvard Business School Press.

Note on Sources

The material in Chapter 1 draws heavily on the Introduction and Overview Paper in Bell, Raiffa, and Tversky (1988), a collection of papers given at a conference held at Harvard Business School in 1983. The category of asymmetric prescriptive/descriptive is introduced and developed in Raiffa (1982) and (1996), respectively.

Chapter 2 draws heavily on the following three sources: Hammond, Keeney, and Raiffa (1998); Raiffa (1968); and Keeney and Raiffa (1976). For further discussion of decision analysis, see the works of Ronald Howard (especially Howard and Matheson 1983). Howard's contributions at Stanford have sustained the field, and his disciples have spread the word. Ralph Keeney also shares some of the kudos for bridging theory and practice.

Chapter 3 draws heavily on two sources: first, research done primarily by Robert Schlaifer and Howard Raiffa during the period 1960–1965 on the elicitation of subjective probabilities and utilities from experts for use in prescriptive decision making; and second, the considerable body of research by cognitive psychologists on behavioral decisions stemming from the seminal research of Amos Tversky (with the able collaboration of Daniel Kahneman). See Tversky and Kahneman (1974) and Kahneman and Tversky (1979). For a summary of the results of this school and a guide to the literature, see Bazerman (2002). These two strands of research have different motivations: Schlaifer and Raiffa were wedded to the advancement of prescriptive decision analysis and recognized the foibles of experts in providing judgmental inputs. They were concerned with finding the best ways for eliciting such information or compensating for such biases. At first behavioral psychologists, in an attempt to debunk the prescriptions of decision analysis, documented the fact that untrained individuals did not behave in ways that normative decision theory says that they should behave. I (HR) concur and therefore argue that this is all the more justification for developing prescriptive decision analysis to compensate for the anomalous behavior of individuals and for training people to behave somewhat "unnaturally." The eminently readable and informative books by Bazerman provide a nice balance between these two orientations.

Much of the material in Chapter 4 appears in modified form in Luce and Raiffa (1957). It actually appeared first in a report put out by the Engineering Research Institute of the University of Michigan (see Raiffa 1951). Much of it also appears in an unpublished audiographic course on interactive decision making (see Raiffa 1977).

The material in Chapter 5 has two sources: Raiffa (1982), chapters 1 and 2; and Raiffa (1996). The latter work introduces and develops a theory of negotiations for the polar ideal of FOTE analysis—Full, Open, Truthful Exchange.

The Elmtree House of Chapter 6 is a fictional composite of several realities. It appears in Raiffa (1982).

Chapter 7 is also taken from Raiffa (1982). Kalyan Chatterjee wrote about the double auction game in his thesis; I was his principal adviser. See Chatterjee (1978) and (1979). Later on he worked closely on this topic with another of my doctoral students, William Samuelson. See Chatterjee and Samuelson (1981).

The Sorensen Chevrolet case in Chapter 8 was written by John Hammond and further developed by myself. It appears in Raiffa (1982) and in Hammond, Keeney, and Raiffa (1999)—especially the dialogue between Debbie, Sam, and Jane.

Chapter 9 is also taken from Raiffa (1982) with some minor alterations. The strike game is my concoction and appears in Raiffa (1982), but the no-strike strike or the virtual strike is new.

I have lectured from time to time on auctions and bids and much of the material in Chapter 10 is an adaptation of my unpublished class notes.

The discussion in Chapter 11 of the three phases of negotiation and the central role of the template appears in Raiffa (1996) but is more fully developed here. The case study of AMPO vs. City appears in Raiffa (1982), and in an earlier version in Edwards and White (1977).

The material in Chapters 12 and 13 deepens the discussion in Raiffa (1982) and (1996). The fair-division problem and its analysis in Chapter 13, with and without the SOLVER program, are substantially more developed in this chapter than in Raiffa (1996).

Raiffa (1996) uses the example of Lisa vs. William from Lax and Sebenius (1986). A student taking a course from Professor Gordon Kaufman at the Sloan School of MIT developed the Nelson vs. Amstore version, whose analysis in Chapter 14 follows the prescriptions developed in Raiffa (1996). My thanks to Gordon Kaufman for the use of this material.

Much of the first half of Chapter 15 is an amplification of Raiffa (1982). The part on behavioral anomalies and cultural differences in negotiations is new.

Chapter 16 is a joint project with my two contributing authors, John Richardson and David Metcalfe. The chapter discusses how an analytically inclined protagonist might negotiate with different parties who are not so inclined.

Chapter 17, including the Camp David case study, reorganizes and amplifies much of what appears in Raiffa (1982).

The first part of Chapter 18 on evaluative interventions as well as the appendix on final-offer arbitration appears in Raiffa (1982). The part on nonevaluative interventions and the neutral joint analyst (NJA) is organized quite differently. I was assisted by Tom Weeks in helping with the division of Mrs. K's art collection.

Chapter 19 is a reworking of Raiffa (1982).

I'm indebted to Anthony Wanis St. John for his constructive critique of an earlier version of Chapter 20. Some of this chapter's material appears in a chapter I wrote for the edited volume by Arrow, Mnookin, Ross, Tversky, and Wilson (1995).

There is a vast literature on collaboration with many actors, the topic of Chapter 21. The problem is central to the topic of collaborative decision making: how to get groups to work productively together. I collaborated very closely with John Richardson on the first part of this chapter dealing with group anomalies, or on what needs to be fixed in the way groups behave.

Some of Chapter 22 appears in Raiffa (1982). The concrete but abstract case used in the first part of this chapter is a modification of a simulation developed by Larry Susskind and known to me as Deeport. The analytical structure of the case with the use of repeated formal votes came from suggestions from the Harvard Negotiation Roundtable, co-directed at the time by James Sebenius and myself.

Chapters 23 and 24 are reorganized Raiffa (1982).

The hypothetical, simple case study in the first part of Chapter 25 is based, in part, on the doctoral dissertation of Frederick Mayer (1988), supervised by Sebenius and myself. The use of SOLVER, a spreadsheet programming algorithm, is an added new frill. The case study about the Panama Canal negotiations is taken from Raiffa (1982). The fair-division problem, with possible side payments, is new.

The Ericson Farms case study in Chapter 26 takes its inspiration from some of the planning controversies in my home town of Belmont, Massachusetts. But it's all fiction. The case of energy negotiations in the European Union is adapted from the doctoral dissertation of David Metcalfe (2000).

Chapter 27 further develops Chapter 25 of Raiffa (1982).

Index